Sweet Science

Sweet Science

ROMANTIC MATERIALISM AND
THE NEW LOGICS OF LIFE

Amanda Jo Goldstein

The University of Chicago Press   CHICAGO & LONDON

The University of Chicago Press, Chicago 60637
The University of Chicago Press, Ltd., London
© 2017 by The University of Chicago
Published 2017
Printed in the United States of America

26 25 24 23 22 21 20 19 18 17   1 2 3 4 5

ISBN-13: 978-0-226-45844-1 (cloth)
ISBN-13: 978-0-226-48470-9 (paper)
ISBN-13: 978-0-226-45858-8 (e-book)
DOI: 10.7208/chicago/9780226458588.001.0001

The University of Chicago Press gratefully acknowledges the generous support of the Hull Memorial Publication Fund of Cornell University toward the publication of this book.

Library of Congress Cataloging-in-Publication Data

Names: Goldstein, Amanda Jo, author.
Title: Sweet science : romantic materialism and the new logics of life / Amanda Jo Goldstein.
Description: Chicago ; London : The University of Chicago Press, 2017. | Includes bibliographical references and index.
Identifiers: LCCN 2016051250 | ISBN 9780226458441 (cloth : alk. paper) | ISBN 9780226484709 (pbk : alk. paper) | ISBN 9780226458588 (e-book)
Subjects: LCSH: European literature—19th century—History and criticism. | Romanticism. | Materialism in literature. | Literature and science. | Blake, William, 1757–1827—Criticism and interpretation. | Goethe, Johann Wolfgang von, 1749–1832—Criticism and interpretation. | Shelley, Percy Bysshe, 1792–1822. Masque of anarchy. | Shelley, Percy Bysshe, 1792–1822. Triumph of life. | Lucretius Carus, Titus—Influence.
Classification: LCC PN751 .G65 2017 | DDC 809/.034—dc23 LC record available at https://lccn.loc.gov/2016051250

♾ This paper meets the requirements of ANSI/NISO Z39.48-1992 (Permanence of Paper).

For Tom

CONTENTS

Introduction: "Sweet Science"

Knowledge is not seeing, it is entering into contact, directly, with things; and besides, they come to us.

MICHEL SERRES, *The Birth of Physics in the Text of Lucretius*

a. Tingeing the Cup with Sweet

"Thy self-destroying beast formd Science shall be thy eternal lot"! So threatens William Blake's "Eternal Man" in an outburst meant to goad lethargic auditors to join the work of Apocalypse under way in *The Four Zoas* (c. 1796–1802).[1] Blake likes to level "Science" as a curse. He routinely deploys the three-headed empiricist Cerberus "Bacon. Locke & Newton" to demarcate and demonize three entangled strands of the contemporary scientific inheritance—experimentalist method, sensationalist psychology, and mechanist materialism—and to demonstrate that such "Science is Despair."[2] Empiricism, in Blake's books, trains subjects to incorporate present power relations at the level of their sense organs, and to use those same organs to verify those same relations as inalterably natural. So it is a surprise to find "Science"—not Imagination, Poetic Genius, Energy, Albion, or any of Blake's usual protagonists—presiding over the postapocalyptic "fresher morning" that dawns at the close of *The Four Zoas*. Over Blake's vision of a world renewed, "sweet Science reigns":

> The Sun arises from his dewy bed & the fresh airs
> Play in his smiling beams giving the seeds of life to grow
> And the fresh Earth beams forth ten thousand thousand springs of life
> Urthona is arisen in his strength no longer now
> Divided from Enitharmon no longer the Spectre Los
> Where is the Spectre of Prophecy where the delusive Phantom

Departed & Urthona rises from the ruinous walls
In all his ancient strength to form the golden armour of science
For intellectual War The war of swords departed now
The dark Religions are departed & sweet Science reigns
End of the Dream[3]

This study began as an attempt to grasp the science Blake called "sweet," the one to which this often fiercely antiscientific poet could imagine ceding place at the end of time.[4]

For that is what the passage seems to do: the prophet Los, Blake's usual poetic avatar, is nowhere to be found. Suddenly dismissing this force as a "delusive phantom," Blake rehearses some famous denunciations of poetic tropes in empiricist philosophical discourse: when Hobbes called them deceptive "*ignes fatui*," for instance, or Locke, "perfect cheats," within "all discourses that pretend to inform or instruct."[5] But in Blake's passage the poet Los is less "Departed" than hidden in plain sight, where the hide-and-seek chorus, "*Where* is the Spectre of Prophecy *where* the delusive Phantom," teases us to discover him. In fact, Los has been absorbed into the person of "Urthona," a writer, readers learned earlier, of "bitter words / Of Stern Philosophy."[6] That Blake now imbues this combined poet-philosopher with a strength he calls "ancient" is a clue that the phrase "sweet Science" in fact glosses a very old formula for the interanimation between poetry and natural knowledge. For the present book, Blake's phrase is a clue to a series of linked, Romantic-era experiments in the cooperation of scientific and poetic knowing—at a moment when natural science was turning, thrillingly, toward the problem of biological life.

"Sweet Science" is indeed ancient: some decades before Horace's lasting endorsement of the poet who mixes sweetness and usefulness (*qui miscuit utile dulci*), delighting (*delectando*) and instructing (*monendo*) at once, the Roman poet Lucretius minted the enduring topos of the "honeyed cup" to describe the way his epic poem, *De rerum natura* (c. 55 BCE), served up the teaching of the Greek materialist philosopher Epicurus. Like the "bitter" and "Stern Philosophy" of Blake's Urthona, Epicurean materialism, Lucretius wrote, would be too "bitter" (*amarum*) and "stern" (*tristior*) to swallow, were it not for the touch of poetic sweetness his *De rerum natura* provided.[7] This introduction begins to tell the story of Romantic "sweet Science" as a strategic redeployment of Lucretius's poetic materialism, unfolding the unrecognized presence and unfamiliar implications of that classical poetic physics within famously Romantic concerns: the problem of living form, the experience of history, and the increasingly strained relations between "the Poet" and "the Man of Science."[8]

As a matter of allusion, Blake's postapocalyptic fresher morning in *The Four Zoas* dawns rather exactly on the beginning of Lucretius's *De rerum natura*. There, Lucretius casts the philosopher Epicurus, just as Blake casts Urthona, as a notably *dis*arming intellectual warrior whose "lively power of mind" undoes the "savage works of war" (Blake's "war of swords") and the knots of "false religion" (Blake's "dark Religions").[9] Much as Urthona "rises from the ruinous walls" of the earth in Blake's epic, Epicurus rises "beyond the flaming walls of the world" to "traverse the immeasurable universe in thought and imagination" in Lucretius's. Yet each extraterrestrial excursion returns lessons of earthly finitude and pleasure: in *De rerum natura*, "the wonderworking earth puts forth sweet flowers" (*suavis daedala tellus/summittit flores*); in *The Four Zoas*, "the fresh Earth beams forth ten thousand thousand springs of life." In each episode, maternal and erotic power melts martial strength,[10] dispersing the awe-inspiring promise of an afterlife into the generative elemental motions of water, air, and light through *this world*—a world become sweeter in the abeyance of "dark Religions" that held it in contempt. This may begin to account for the extraordinary tranquility that the last three words, "sweet Science reigns," let fall upon Blake's promised "intellectual War." In the Blakean Last Judgment that seems to deliver us to any morning, Urthona's "golden armour" may equally describe dewy droplets glistening in the early sunlight of the passage's first lines: as though "sweet Science" simply "rains," even as it "reigns."[11]

Epicurean materialism as *De rerum natura* teaches it is hard medicine to swallow because it denies any transcendent privilege (immortality, immateriality) to the human soul, any teleological plan behind the chance workings of the natural universe, and any super-natural sources of meaning, form, or truth. Instead, Lucretius offers the Epicurean conjecture that the universe begins, like the "reign" of Blake's "sweet Science," in elemental "rain": indestructible, indivisible, insensate atoms falling "Like Drops of Rain," through the void.[12] The atoms' apparent tendency to swerve without cause—to "Decline / Tho' Very Little, from the exactest Line"—is the scandalous philosophical posit by which anything else comes into existence in this aleatory materialist view.[13] Chancing into contact, atoms entangle to produce the intricate figures (*perplexis figuris*) that constitute the apprehensible world.[14] That is, in the classical atomist tradition to which Blake unexpectedly returns us, all extant things—animal, vegetable, mineral, and as I shall stress, mental, figurative, textual, and spectral—are transient congeries of elements, composed of, and decomposing into, the myriad little bodies in motion that are nature's only permanent parts. From worn paving stones, to thinning finger rings, to wasting elders, Lucretius writes, "we see each thing . . . as it were ebbing through length of time, and age withdrawing them from our eyes [*ex oculisque vetustatem subducere nostris*]" (2.68-70). He is voicing a commitment to mortality and transience that, Epicurus teaches, is, paradoxically, the only reliable consolation against the fear of death that motivates people to seek and submit to damaging accumulations of power.

Hence Lucretius's concern that his Epicurean materialist doctrine is bitter, stern, or sad: it promises immortality, but for atoms, in their "single solidness," not for us (in our plural softness). Life, self, sentience, and memory in this philosophy are effects of the lifeless atoms' fantastically intricate collocations, and they are irretrievably lost at death, when the configuration ceases to hold, and persons, as Percy Shelley puts it, "die in rain."[15] "To make them take it," Lucretius suggests, he'll "tinge the Cup with Sweet" (Creech, 1.946). But the apparently modest role set out for poetry in this topos—a "tinge" of sweetness that makes the mortal materialist medicine go down—already intimates, etymologically, poetry's indispensible role in the kind of contact that constitutes the con*tinge*nt beginning of any thing.[16]

In fact, *De rerum natura* makes a strong case for poetry as intrinsic to any empirical knowledge of nature, and this because, not in spite, of its sweetness, contingency, and tendency to trope or turn "from the exactest Line." I argue in this book that *De rerum natura*'s case for poetic science was revived with fresh urgency in the Romantic period as a means to resist the reorganization of knowledge that was increasingly invalidating—we might say *fictionalizing*—poetic connections to the natural and social world. Moreover, thinkers from Erasmus Darwin to Goethe, Percy Shelley, and the young Karl Marx rediscovered in Lucretius's "sweet Science" a theory, lexicon, genre, and imaginary—a poetics—fit to conjoin two defining fixations of Romantic modernity: the problem of biological life, and the period's pressing new sense of its own historicity. Like multiple Romantics before him, Marx seized on the ancient materialism as a means of grasping sensuous life as "embodied time"—shaped and shot through with wanted and unwanted bequests from elsewhere and before that both riddle and enable the present making of history.[17]

Grasping the poem's Romantic era utility and fascination means attending to a technical feature frequently overshadowed by the more glamorous scandals of its aleatory *clinamen*, its libertine sensualism, and its proto-Enlightenment advocacy of knowledge of nature against religious orthodoxy. *De rerum natura* presumes figuration to be central to the reality of things (*res*) in their being and appearing alike: the figural improprieties of perceptual and linguistic representation are taken to originate in and index this general reality, rather than to betray an insuperable gap between human knowledge and other things. The Lucretian case for poetry intrinsic to the (extravagant) real begins in the last words of the "honeyed cup" topos to which Blake has already pointed us:

> volui tibi suaviloquenti
> carmine Pierio rationem exponere nostram
> et quasi musaeo dulci contingere melle,
> si tibi forte animum tali ratione tenere
> versibus in nostris possem, dum perspicis omnem
> naturam rerum qua constet compta figura
>
> (1.945–50)

[I have chosen to set forth my doctrine to you in sweet-speaking Pierian song, and as it were to touch it with the Muses' delicious honey, if by chance in such a way I might engage your mind in my verses, while you are learning to see in what shape is framed the whole nature of things]

The passage's surprising last word—what the speaker hopes a person will discern (*perspicio, -spicere*) while detained in the sticky sweetness of the verse—is not *ratio* but *figura*: "in what *shape* is framed the whole nature of things," or "by what woven figure [*compta figura*] the whole nature of things holds together [*constet*]."[18] The whole nature of things seems especially, flamboyantly figurative because Lucretius calls its figure *compta*, "arrayed" or "adorned," as in braided hair, ornamented rhetoric, or a dressed-up person. It is tempting to read the line, then, as a textually self-reflexive moment: just where we expect to see through (*per-spicere*) the poem's embellishments into its natural philosophical content, we are instead deflected back upon the lavish surfaces of its own rhetorical shape.

But in fact the phrase *qua constet compta figura* makes a good deal of scientific sense, too, according to Epicurean physics. There, to exist as anything other than an atom, which is the only thing that this natural philosophy allows to exist indivisible and alone, is to stand together (*consto, -stare*), interwoven (*compta*) of many elements and the void between them. In this way, *figura* connotes for Lucretius not only an extended analogy for the relationship between poetry and knowledge, but also the indispensible contexture by which any compound body, whether a universe, dust mote, or poem, consists and persists. *De rerum natura* is the kind of text, the kind of poetry and science that results from declining to conceive of figuration exclusively as a strategy of consciousness or a linguistic effect. The Romantics in my study revived its precedent not only as an emphatically plural logic of organic and inorganic formation and transformation, but also as a monumental and gorgeous case for poetry's role in the perception and communication of empirical realities.

For the items of the universe not only take their composite shapes by (con)figuration in this materialist perspective; they also unwittingly collaborate to produce a sensuously available world by physically figural means. According to Lucretius's astonishing semiotics of sensation—which Goethe made central to his life science of "morphology," Shelley to his poetics of "Shape," and Marx to his notion of lived history as "embodied time"—not only are bodies inescapably transient, they transpire into perceptibility, losing themselves in atom-thin layers that disperse their likenesses into others' senses. Sentient beings perceive the world by means of these airborne exfoliations, whose touch occasions sensation and thought as they enter and move another body with floating traces (*vestigia*) of near or distant objects (4.87).

Epicurus's name for these extravagant material films, which give sense perception both its veracity and its belatedness in his philosophy, is *eidola*, which Lucretius justly rendered in terms that belong as much to rhetoric as to physics: *simulacra, figurae, imagines*. Offering no separate ontology for the poem's verbal figures, the physical ones it investigates, and the sensible apparitions that disclose them, Lucretian poetry and science each presume and constantly rediscover the material continuity between physical and poetic shapes.[19] Here figures are fractions of the real estranged from their sources, and all bodies, not just poems, have the capacity, indeed the necessity, of producing them. Being a body in time means shedding figures of self—involuntarily re-presenting oneself—and weathering others' particulate bombardment: "bending within each other's atmosphere," as Percy Shelley would put it in *The Triumph of Life*. And as Shelley and some of his contemporaries were particularly motivated to realize, to think with Lucretian materialism was to accept figuration as an indispensible object, medium, and means of empirical investigation, without invalidating any of these phases as "mere" metaphor, anthropomorphism, or linguistic construction.[20]

Sweet Science is a sustained experiment in taking this estranging possibility as seriously as earlier moderns did, and in tracing out the now unfamiliar implications for the semiotic and somatic actions and passions of human and nonhuman bodies, for the experimental arts and sciences that sought to know them, and for the ethics and, ultimately, the politics of that relation. I argue in this book that Lucretian materialism—a materialism that granted substance to tropes and tropic activity to nonverbal things—afforded Romantic poetry a strong claim upon the real at an early nineteenth-century moment when disciplinary consolidation was rendering that claim incredible. Available in a fresh wave of editions, translations, and imitations that qualify the Romantic period as "the second British Lucretian moment," *De rerum natura* was summoned as a kind of license to resist the prevailing epistemic compromise that would make prose the presumptive medium for communicating social and natural realities and secure poetry its enduring place at the prestigious margins of cultural life—as a discourse expert in the *im*material reaches of subjectivity and the special materiality of language itself.[21] By the same token, Romantic neo-Lucretianism poses a challenge to the ongoing presumption, common to the most diverse among present-day theoretical orientations, that "matter" and "the body" are what suffer, resist, or elude—rather than produce, demand, elaborate, or perpetuate—signification and the imposition of discursive, cognitive, social, and political form.[22]

Leaving us, then, after the Last Judgment, in what would appear to be the least Blakean, and least Romantic, of places—at the beginning of a notoriously atheistic, materialist poem precious to the radical French Enlightenment—Blake's "sweet Science" invites us to unfold a lapsed possibility near the nineteenth-century beginnings of the now familiar division of labor between literary and scientific representation. Against the pressure, then and now, to treat the culture of science as context or antithesis to literary production, this book uncovers a countervailing epistemology that cast poetry as a privileged technique of empirical inquiry: a knowledgeable practice whose *figurative* work brought it closer to, not farther from, the physical nature of things.

From Blakean sweet science to Goethe's call for a "tender Empiricism" (*zarte Empirie*) and Wordsworth's for a "science of feelings," I track a set of related, Romantic-era attempts to reinvent inherited empirical and experimental methods—and to do so with the resources of verse and figure.[23] The chapters that follow put Blake's poetic embryology, Goethe's journals *On Morphology*, and Percy Shelley's "poetry of life" back into conversation with materialist life sciences then coming under fire for being overly *poetic* and *literary*: those of Erasmus Darwin, Johann Gottfried Herder, Jean-Baptiste Lamarck, and Étienne Geoffroy Saint-Hilaire. Together, this minoritarian corpus of life writings declines either to conform the task of scientific representation to emergent professional standards or to valorize literariness as a critical escape from the confines of empirical representation or contextual determination.

Romantic, revisionary poetic sciences, I argue, challenged emergent life-scientific *and* aesthetic protocols to understand "raw" sensation itself as susceptible and generative of social and rhetorical transformation; to countenance the mutual, material influence between the subjects and objects of experiment; and to position vulnerability—to impression, influence, and decay—as central, not inimical, to biological life.[24] Against the vitalist ideal of self-generating "organic form" with which Romantic biology is frequently identified, sweet sciences conceive of animation as a relational effect of contact and context; against the twinned, post-Kantian scientific and aesthetic ideals of impartial observation, they pose an ethos of ineluctable participation in that which is felt or known.[25] Two chapters of this book unfold Goethean "tender empiricism" as the explicit method of this counterdisciplinary entanglement. "Every new object, well seen," the morphologist maintains, "opens up a new organ in us."[26] Mischievously resignifying "objectivity" to mean an observer's vulnerability to transformation by the objects under view, Goethe recasts the sensations that ground empirical enquiry as occasions of bilateral transformation, shifts biology's question from the life force *within* beings to the metamorphic relations *between* them, and renders experiment irreversibly historical, rather than infinitely replicable. Yet might such complications be fortuitous, as well as inevitable, for empirical knowledge?

Erasmus Darwin, Goethe, Herder, Blake, Shelley, and Marx variously suggest that such complexities must be confronted, and indeed *cultivated*, by any science out to investigate living activity through the testimony of the senses. Rejecting prior empiricist and experimentalist strictures against figurative language, moreover, the poets among them seized upon developments in the physiology of sensation to present poetic tropes as a crucial register—perhaps the only adequate register—for actual but fugitive aspects of the entities knowable through sensuous experience. "Our whole life," as Johann Gottfried Herder put it, "is to a certain extent poetics," and the assertion was physiologically meant. In his account, bodily nerves and fibers enact forms of nonverbal figuration—"allegorizing," "translating," "schematizing," "image-making"—such that sensory experience might be consciously borne.[27] And yet, in contrast to Nietzsche's famous appropriation of Herder's insight in "On Truth and Lie in an Extra-Moral Sense" ("the stimulation of the nerve is first translated into an image: first metaphor!"), Herder's understanding of sensation as "poetics" does not work primarily to reveal and valorize our role as "*artistically creative subjects*" who inadvertently construct the world that our science purports to discover.[28] At stake instead is an unfamiliar understanding of physical bodies—that of the human knower and the perceived object alike—as really, fundamentally tropic, physically enmeshed in multilateral figurative transactions of which human poetry of words is but one index and instance.

If the sensory basis of empirical knowledge is poetic down to its very nerves and fibers, the epistemic status of poetic *language* changes: the tropes, figures, images, and metaphors formerly thought to adorn or obscure the simple evidence of sense may instead index fidelity to it—may constitute the best possible linguistic register for aspects of subjective and objective activity sidelined by the philosophical pursuit of "clear and distinct ideas."[29] In this respect, Charles Darwin's once-renowned naturalist grandfather, Erasmus, was being empirically thorough (as well as delightfully fanciful) in systematically doubling his prose medical, botanical, geological, and zoological writings in verse. Blake was more sincere than we have supposed when, playing on the shared etymology of *experience* and *experiment*, he claimed poetry as the faculty to which the empirical method belongs and by which it ought to be redisciplined: "As the true meth-/od of knowledge/is experiment/the true faculty/of knowing must/be the faculty which/experiences. This/faculty I treat of."[30] Herder praised poetic language in terms we frequently reserve for the scientific and journalistic prose genres that now monopolize objective description: "accuracy," "faithfulness," "truth," "efficacy," and fidelity to "the minor details" that the "frigid understanding . . . omits as superfluous."[31] As in the first, seventeenth-century wave of experimentalist literature Joanna Picciotto brilliantly reconstructs, such poems do not offer themselves primarily as fictions but rather as instruments for displacing the fictionalizing force of perceptual routine—for registering "the given" otherwise.[32]

And though it would indeed be strange, now, to look to poems for scientifically valid information, it wasn't always so. We can hear, in Herder, the eighteenth-century figural materialist tradition that Natania Meeker and Monique Allewaert have begun to elaborate in French and American contexts, respectively: here "figures are key conduits for the exchange between human and non-human forces, which together act on and produce the real."[33] The possibility dovetails with Robert Mitchell's searching work on sensation in Keats and Shelley as "a corporeal power that bound the body in the world," linking it with "other forms and states of matter and motion."[34] As Noah Heringman has long since argued, to explore the literariness of Romantic science in historical context is less "to tip the balance further toward 'social construction'" than to "enrich and complicate the sense in which discourses may be understood as 'reliable representation[s] of the natural world.'"[35] What has become of poetry's physical role in the induction and communication of objective and empirical knowledge, of its frank claim to tell "truth, not individual and local, but general, and operative," as Wordsworth put in the 1802 Preface to *Lyrical Ballads*?[36]

b. Undisciplined Romantics

The poetic science in Romantic writing is kept fictional through methodological, generic, and institutional divisions between the natural, human, and social sciences that were not only unfinished but also brilliantly contested in Romantic period projects. At work in the period were certainly arguments and impulses that contributed to the now naturalized division between literary and scientific production: Kant's division of his third *Critique* into aesthetic and natural-philosophical halves, for instance, and Georges Cuvier's influential scorn for his rivals' poetically and philosophically expansive style.[37] But there were also attitudes that contested, ignored, did not accept, or did not know that division—positions of which the period's many retrospectively "hybrid," "interdisciplinary," "amorphous," and "monstrous" projects are the sign. As Bruno Latour's attempts at an anthropology of Western modernity have shown, conceptual divisions between nonhuman nature and human culture (and between the forms of expertise devoted to their study) have, in practice, proved extraordinarily productive of such illicit "monsters."[38] However, in this study I take the sheer normalcy of "scientific" investments among the writings later canonized as Romantic poetry as a clue that such projects were not always or only products of cognitive dissonance, or of unfallen, predisciplinary naïveté. Some were supported by positive ontological, ethical, and epistemological premises, even if those supports have since slipped from view. These include the philosophical vitalism, Spinozism, and natural theologies that Romanticists have begun to reconstruct, as well as the form of neo-Lucretian poetic materialism unfolded in the following pages.[39]

The task of thinking along this gray spectrum of un- and otherwise-disciplined writings has been impeded, ironically, by the very Foucauldian paradigm that makes it possible and desirable in the first place: the work of puncturing intellectual historiography with the "epistemic ruptures" that render prior knowledge formations perceptible in their alterity has also served to emphasize new and unprecedented utterance in ways that inadvertently collude with high Romantic strategies of self-definition.[40] After the breaks and innovations "around 1800" have been obsessively told, then, comes the privilege of attending to minor and untimely positions, instead: to attitudes that wittingly or unwittingly fail to endorse the epistemic reconfigurations that will have been, in retrospect, epochal—attitudes that fill out the heterogeneity of a modernity that Keats, Shelley, Cowper, Wordsworth, and Coleridge, frequently with the help of Lucretius, tended to visualize less as a rupture than as a "mist."[41]

Hoping to pluralize our account of these attitudes and to do better justice to what Simon Schaffer and Jon Klancher call the period's *"indisciplines*," Luisa Calè, Adriana Craciun, and Heringman its predisciplinary *"disorder* of things," this book focuses on poets and naturalists who declined to cede the task of empirical representation to rapidly specializing genres of scientific discourse.[42] They did so, I suggest, less out of a concern to expand the purview of subjective imagination (as might be expected from Romantics) than out of concern for what of the world would be lost to representation if empirical realities were constrained to manifest in antifigural, antipoetical genres and codes. As Robin Valenza's important research on Romantic disciplinarity has made clear, the crisis of vocation that catalyzed the most familiar manifestos of Romantic poetics arose in no small part from the tension between poetry's historical role as a general medium of knowledge—a major means, with history and philosophy, of communicating virtually anything to a broad readership—and the modern sense of a *discipline*, devoted to and defined by a specific object of expertise.[43]

Romanticists know Wordsworth's expression of this distinction virtually by heart. In the 1802 Preface to *Lyrical Ballads*, "the Man of Science" (whether "Chemist," "Mathematician," "Anatomist," "Botanist," or "Mineralogist") is found "conversing with those particular parts of nature which are the objects of his studies," while "the Poet" carries on a versatile conversation with "objects . . . everywhere."[44] While much attention has been given to the ways Wordsworth justifies poetry's generality by claiming suspect custody of the "real language of men" and the "inherent and indestructible qualities of the human mind," the aim of poetic attunement to "human nature" in this area of the Preface is less to isolate what is essentially human or to naturalize poetic craft than to keep increasingly "remote" men of science and pretentiously "celestial" poets in physical contact with other natures:

> What then does the Poet? He considers man and the objects that surround
> him as acting and re-acting upon each other, so as to produce an infinite
> complexity of pain and pleasure. (258)

In a way that the new men of science are increasingly unable to do, Wordsworth argues, the poet "converses with *general nature*" in this surprisingly materialist sense of action and reaction, registering the "complex scene of ideas and sensations" undergone by participant-observers exposed to "the goings-on of the Universe" (259, my emphasis, 258, 256).

In other words, Wordsworth's surprising worry in 1802 is less that the specializing sciences will not do justice to the intangible reaches of subjective individuality or universal humanity than that they are becoming insensible and inadequate to natural phenomena: to "operations of the elements and the appearances of the visible universe," of which our "animal sensations" and "moral sentiments" are effects and signs (261). When Wordsworth writes, as he does incessantly in the Prefaces, of *passions*, he is thus intoning the now archaic sense of *patior*, to bear or suffer: a sense of *being acted upon*, as well as acting, in complex corporeal processes in which, startlingly, the "loss of friends and kindred" takes place on a continuum with "cold and heat" (261).[45]

In fact, one argument I offer here is that the burden of the brilliant, long-eighteenth-century experimental and theoretical work on animation and sensation captured in expressions like *irritability, sensibility, excitability*, and *associability* was to articulate a nuanced spectrum *between* autonomous action and passive determination as the zone proper to biological life. The period's myriad, vital "-abilities" not only afford immanent and active powers to living bodies, as our histories never tire of affirming, but also cast those very "powers" as specific and specifying modalities *of being moved*.[46] Romantic poetic sciences challenge us to grasp biological life in subtle combinations of power *and* liability, activity *and* passivity that have eluded literary historians' reconstruction of the discourse as one of "vital power" and "self-organization."[47] Throughout this study, I attempt to deploy these various passions and passivities in ways that restore their peculiar determination to name capacities of receptivity, liability, and susceptibility that are neither action nor nothing.[48]

Writing that poetry "is the impassioned expression which is in the countenance of all Science," then, Wordsworth not only positions poetry as intrinsic to knowledge formation but also implies, as Herder did, that the *dis*passionate form of looking valorized in scientific objectivity after Kant in fact endangers empirical induction and representation (259). Impartial looking sidelines not just psychic experience but features of actions and objects generally, "goings-on of the Universe" that ought to be the Man of Science's first concern. A similar assessment motivates Goethe's "*tender* Empiricism" as a revisionary experimental practice open to being passively, passionately *moved* by objects, for the sake not of subjective fulfillment but of a fuller record of objects' efficacies and effects. The professionalizing habitus of the "Man of Science," Wordsworth thinks, needs poetic "transfiguration" for the sake of full and accurate knowledge. And though the Prefaces are famous for their denunciations of "poetic diction," Wordsworth in fact argues not for the elimination of "metaphors and figures," but for keeping them closely "fitted" to the "actual pressure of those passions" that excite them, so as to avoid any "falsehood of description" (255–56, 251). *Lyrical Ballads* makes many claims for Poetry, but one that remains tellingly, provocatively difficult to accept is the claim that poetry might be refined into an empirically accurate figural practice: "the history or science of feelings."[49]

The accumulated evidence of decades of groundbreaking historical studies proves that intensive engagements with themes we would now call "scientific" were the rule, rather than the exception, among the writers later canonized as Romantics.[50] And though Romantic literature and science has become a robust subfield in its own right, too rare are the moments when we follow through on Noel Jackson's important suggestion that "literature might be construed in this period as a form of scientific practice."[51] The tentative formulation, shortly followed by the reassurance that Jackson's pioneering study nonetheless "represents a chapter in the history of the aesthetic's emergence as, in theory, an autonomous affective and cognitive domain," indexes the hermeneutic and political difficulty of dwelling with that deceptively simple thought, which still smacks of magic, oxymoron, or apostasy.[52] As Maureen McLane wittily reflects, in the midst of a slow-burning "crisis" in the humanities, a student of literature and science wants to take the right side.[53] And, I would add, she wants to answer the serious objection that stressing the science in poetry capitulates to indiscriminately scientistic norms of cultural valuation. In our own historical and institutional position, it has seemed more imaginable and more conscionable (and both accurate and important) to speak of the "permeability" of prior disciplinary bounds, of the aesthetic critique of positive science and instrumental reason, of metaphoric "borrowing" between scientific and literary discourses, and of the scientific "contexts" of literary texts.

But on the issue of capitulation, such formulations presume that science is extrinsic to poetry and thereby concede too much. They concede *scientia* in the beautiful old sense of "knowledge," the *Wissen* in *Wissenschaft*, evacuating Wordsworth's frank aspiration to "truth . . . general, and operative" of its surprisingly material, physical, and natural-descriptive dimensions.[54] They thereby reinscribe the terms of ill-fated nineteenth-century epistemic settlements around poetic subjectivity and scientific objectivity, aesthetic autonomy and scientific facticity, that many texts did not endorse or know.[55] For as Lorraine Daston and Peter Galison show in a magisterial history, the aspiration to scientific "objectivity" as we still know it—representation that "bears no trace of the knower," defined in constitutive opposition to the expressive "subjectivity" of creative art—took broad hold only in the middle of the nineteenth century, despite its prescient articulation and justification in Kantian critical philosophy.[56] Only decades into the nineteenth century was the natural and physical sciences' capture of the word "science," complete.[57]

A further motive for exploring alternatives to aesthetic defenses of Romantic literary autonomy is their unintended consequences for poetry's modern social value. As Raymond Williams concluded with palpable regret more than half a century ago, Romantic arguments for aesthetic autonomy that began as forms of social protest also scripted poetry's subsequent "practical exile."[58] If (some) Romantics (sometimes) produced and valorized a sphere of literary culture critically detached from society—a timeless "court of appeal," as Williams put it, where one could object to "the kind of civilization that was being inaugurated"—they thereby conceptualized "society" as symmetrically autonomous, capable of getting on without them.[59] Hence what Jon Klancher identifies as the double-edged "illusion about reading" that (some) Romantic literary theory helped engender: "it frees us from a materially intolerable social world."[60] This promise instates not only a crucial, critical and emancipatory distance from immiserating material realities but also the imperviousness of those realities to literary intervention. "The very definition of 'the literary' as an autonomous domain," Natania Meeker concludes in a study of French Enlightenment materialism that deeply informs and inspires this one, "may ironically be dependent upon the simultaneous concretization of a material world that remains fully immune to its effects."[61] And yet luckily, as poets well understand, this process of "definition" and "concretization" may never be absolute.

Romantic critical theory, in a profound post-Kantian vein that stretches from Theodor Adorno on lyric poetry and society to an ongoing tradition of negative aesthetic theory, has elaborated poignant and permanently convincing means of appreciating the paradoxical social value of aesthetic turnings from the world—not least the suspension of social value, up to and including the impulse to make a value of suspending it.[62] This book goes a different way, drawing inspiration from Renaissance and eighteenth-century studies, where the early modern fact of the physical force of rhetoric has become a critical commonplace, and where Lucretian materialism, in particular, is enjoying a full-fledged renaissance.[63] And yet, as I have already hinted, the specifically Romantic inflections of the material conduit between *res* and *verba* under investigation here somewhat reverse its usual polemical direction in literary criticism and theory. For in our case, the continuum between physical and rhetorical figures works to assert less the poetic power to "refashion" the world (though the poets in question certainly assert this) than that world's power to fashion a poem into a partial image of its complexity, with a degree of nuance that other verbal forms cannot bear.

This very difference owes something to rhetoric's devaluation over the course of the early modern period from the technical apex of humanist education to the dross from which seventeenth-century "plain style" might be purified—a devaluation that left rhetorical tropes, Jenny Mann shows, closer to given nature than to technical erudition.[64] The names "Rousseau" and "Herder" are sufficient to remind us that at or near the "natural" origins of language is indeed where tropes are situated in a major current of Romantic linguistic theory. As we will see, though, in an almost exact reversal of present-day critical expectations, such Romantic "naturalizations" frequently worked to submerge tropes in, rather than exempt them from, social practice and historical development. As I will continue to stress, this is because "natural" origination was not yet defined in opposition to cultural convention or social construction: instead, these mundane processes together composed a domain of naturalist explanation still pitted against super-natural creativity as the source of sublunary meanings, purposes, and forms.

The chapters that follow read more and less familiar writings in Romantic poetry and science as a reservoir of neglected alternatives to disciplinary and generic alignments normalized over the course of the nineteenth century. The "science" in question, no less than the "poetry," looks quite different from the ultimately triumphal version we have come to know. At stake in Romantic poetic empiricisms are what Gernot Böhme has called "alternatives *of*" (rather than *to*) "science": historical forms of knowledge that no longer qualify as such, but prove the "actual possibility" of science *otherwise*—making visible "the specificity, functionality . . . and also the costs, of our reigning types of science."[65] Such alternatives renew their relevance as it becomes unmistakably obvious that the organization of knowledge inherited from the later nineteenth century not only fails to resolve but also positively helps to produce the crises in ecological, biopolitical, and social justice that chronically evade its grasp.

c. The "Poetry of Life"

The present-day strangeness of the proposition that the empirical, scientific method might benefit from poetic renovation reveals the extent to which it drew force from a forgotten understanding of physical nature. As I show in the first chapter (before delving into specifically neo-Lucretian viewpoints), the Romantic-era credibility of the notion that corporeality in general (like poetic language, in particular) was productive of tropic and figurative transports owed much to the emergent sciences of life. Directing experimental scrutiny toward the specific properties and propensities of *living* matter, eighteenth-century physiology, medicine, and the incipient disciplines of zoology and biology had shifted cultural focus from the inert, passive matter exemplary for classical physics to living, active, and susceptible bodies endowed with specific propensities toward form. As Theresa Kelley shows vividly in the case of Romantic botany, the category transgressions and cross-species mimicry evinced by nonhuman natures "invite[d] poets to perform their own figural mimicry."[66]

In an era before the master metaphor of inherited information ("code") would seem to moot the question of how living bodies get (in)formed, the eighteenth century's favored objects of life-scientific experiment—headlessly reproducing polyps, "spontaneously" proliferating animalcules, "irritable" and "sensible" fibers and nerves, "mere" seminal matter perpetuating family resemblance in embryogenesis—impressed researchers with the likelihood that *matter* might be productive of figures, copies, and likenesses.[67] The formation of specifically biological sciences at the turn to the nineteenth century was motivated by questions about how unorganized seminal and amniotic matter could produce the intricate and diversified form of an embryonic body, how the sensitive and irritable nerves and fibers might transfigure and re-present stimuli, and how static taxonomies might be inadequate to the dynamics of animal and geological nature in motion and with a history.

This "vitalization of nature" and the conduit between science and poetry that it opened are widely known to have been authorized, in the Romantic period, by vitalist philosophies that construed life as an ontologically and causally unique force or power: a power "not analagous," in Kant's influential assessment, "with any causality we know."[68] In contradistinction to mere material bodies, which Kant construed as obedient to "blind mechanism" and "in the highest degree contingent," the living organism, according to the seminal theory of the third *Critique*, was an autonomous "causal nexus" whose exceptional "self-propagating formative power" enabled it to constitute "both cause and effect of itself."[69] Coleridge's generative reception of Kant's theory of "organized and self-organizing being" capitalized on the remarkable echo, in the third *Critique*, between the self-sufficient objects of aesthetic judgment that concern its first part, the self-sufficient organisms of natural-philosophical judgment that populate its second, and the capacity of both to induce experiences of human self-sufficiency.[70]

Coleridge's 1812–13 lectures on Shakespeare render "organic form" an explicit criterion of art, capable of distinguishing an author's "living power" from "lifeless mechanism," his "free and rival originality . . . from servile imitation."[71] Here Kant's precise definition of the exceptional means-ends circularity at work in organic life returns as the "living power" of *poetry*:

> The spirit of poetry, like all other living powers. . . . must embody in order to reveal itself; but a living body is of necessity an organized one—and what is organization but the connection of parts to a whole so that each part is at once end and means![72]

It is in light of assertions such as these that Romantic aesthetics and biology are so readily understood to converge in the ideal of "organic form": the organism or artwork as teleologically autonomous whole, sustained by the "vital power" of life or imaginative genius. As Denise Gigante sums up an important recent study that presents organic form as the unifying aspiration of Romantic science, aesthetics, and politics: "once life was viewed vitalistically as power, science and aesthetics confronted the same formal problems."[73] But what problems (formal and otherwise) did science and aesthetics confront when life was *not* viewed vitalistically as "power"?

In choosing organic form as Romanticism's defining innovation in biological nature and linguistic culture alike, Gigante's vitalist and idealist Romanticism comes into ideologically unlikely concurrence with critical methods as powerful, and as distinct, as Foucault's archeology of knowledge and Paul de Man's mode of deconstructive rhetorical reading. For de Man, organic form and its rhetorical correlate, the symbol, epitomize the powerful illusion of unity, integrity, life, and meaning that Romantic texts cannot but produce in figuration and dismember at the level of their arbitrary linguistic materiality. In Foucault's *The Order of Things*, meanwhile, the organism's debut around 1800 not only typifies the "break" between classical and modern epistemes that is the book's major thesis but also pioneers the very possibility of *breaking* crucial to Foucault's own iconoclastic historiography. Out of Georges Cuvier's organicism, stirringly narrated as life's retreat from the domain of representation into the inscrutable depths of the animal interior, which shatters the visible continuum of natural history, Foucault educes not only the paradigm for the fresh fascination with "life" as a distinct epistemic challenge around 1800, but also, implicitly but unmistakably, the allegory and heroic precedent for the radical shattering to which *The Order of Things* subjects the continuum of traditional intellectual history.[74] The point is not to restart the tired game of catching Foucault out in historical inaccuracy, but rather to mark as overdetermined, in his thinking, an organicist concept of "life" that is historically coincident with Romanticism and that will continue to undergird his coming theorization of "biopolitics."[75]

That so many otherwise diverse approaches to Romanticism are staked (even negatively) on Kantian and Coleridgean organicism—from readings in the tradition of M. H. Abrams and Harold Bloom, to post-Kantian critical theory, de Manian deconstruction, Freud beyond the pleasure principle, and the Foucauldian archeology of "1800"—explains its enduring ability to dominate critical reconstructions of what the Romantic poetics of life may have sought to do. Marking this permits me to ask, very much in the spirit of the archeological method, after other possible grounds for the trifold agreement between natures, signs, and exchanges that Foucault teaches us to seek in a coherent episteme.[76] It is possible that the neo-Epicurean composite bodies (natural, political, textual) that populate this minority position, in their incessant traffic in signs that disclose the invisible in the visible and connect otherwise irreconcilable scales and orders of analysis, constitute such an alternative. And it is certain that tugging at the thread of Blake's phrase "sweet Science" reveals a hidden counterpoint of atomist materialism and didactic poetry in a period best known for organic and lyric form. Incidentally, tracing this counterpoint begins an alternative comparative romanticism that runs athwart its usual basis in the English reception of Kantian and post-Kantian idealist philosophy.

At the conjunction between rhetoric and biopolitics that Sara Guyer terms "biopoetics" were in fact multiple, competing articulations of living form, with polemically different political-economic and semiotic entailments.[77] Indeed, Coleridge understood his own early nineteenth-century organicism to be an embattled, minority position: a system deliberately elaborated, like Burke's before it, to combat the "mechanic philosophy" that each diagnosed as the "general contagion" of his age.[78] Coleridge's *The Statesman's Manual* (1816) tracks a "system of disguised and decorous *epicureanism*," transmitted—or, more malignantly *"transvenomed"*—from seventeenth-century revolutionary England, with its enthusiastic New Scientific rediscovery of classical atomism, through Locke's and Hume's empiricist philosophies of mind, to the recent French "madhouse of Jacobinism."[79]

Even now, Coleridge warns, this empirico-materialism pervades the English ruling classes that ought best to recognize and expel its irreligious and leveling threat. Gifting the *Manual* and his second *Lay Sermon* to the then prime minister, Lord Liverpool, in 1817, Coleridge wittily but earnestly stressed the real-political exigency of inoculating the national metaphysics against the threat of Epicurean physics: "An atom, is an Atom," he observed in the enclosed letter, ". . . and by the pure Attribute of his atomy has an equal right with all other Atoms to be constituent & Demiurgic on all occasions."[80] The *Lay Sermons* offer "the Higher Classes of Society" their emphatically Christian and immaterialist "vital philosophy" not merely (or even primarily) as a philosophy of biological life, but rather as a systematic philosophical counterweight—complete with metaphysics, epistemology, aesthetics, semiotics, and moral and political philosophies—to the subtilized *"epicureanism"* Coleridge felt to be pervading nineteenth-century life and thought.[81]

Rather than take his openly partisan perspective as representative of the Romantic logic and aesthetics of life, then, my study credits Coleridge as a hostile witness to subtle, post-Revolutionary combinations of empiricist epistemology, materialist ontology, and atomist poetics that literary critics of the period have yet to reconstruct. (In fact, moderns like Coleridge may owe the very idea of "pervasive influence" to the partial recoveries and disseminations of *De rerum natura* in the Renaissance, when the illicit text itself, Gerard Passannante shows, epitomized the possibility of "an influence that was unseen and everywhere" that its theory of subliminal particles advanced.)[82] But following through on Coleridge's unheeded testament to the pervasion of atomist materialism decades after its radical Enlightenment heyday will also mean challenging his (much better heeded) caricature of that philosophical tradition: as naively ocularcentric in its epistemology, individualist in its ethics and politics, and inadequate, in its natural philosophy, to the dynamic notions of complexity and emergence necessary to a theory of life. "This is the philosophy of death," Coleridge concluded a passage on the differences between "mechanic" and "vital" philosophy, "and only of a dead nature can it hold good."[83]

But, in fact, to assert a polar opposition between vital and mechanical systems, organic and inorganic being, was the defining task of specifically *vitalist* philosophies at the turn to the century, and not, as too often assumed, the obligatory premise of any theory, philosophy, or science of life. And the same must be said for Coleridge's accusation that atomist ontology endorses a liberal individualist politics of "Demiurgic" citizen-atoms "miraculously invested with certain individual rights."[84] Vitalism and organicism exhausted neither the field of philosophical and experimental enquiry into living form, nor the political theorization of interdependency. (Not for nothing does the Lucretian materialist tradition strictly constrain the category of "individual" to a very specific and bare kind of being that quite technically can never appear among us.) Challenging the vitalist premise that life "is to be considered as a *distinct* and *independent* essence," a spiritual "something" interfusing the gross matter of the body, other thinkers—here the unapologetically Jacobin and materialist writer John Thelwall—argued for life as a "state" of the organism in context rather than an essence "superadded" to it: a compound *"modification* or *effect"* produced by perfectly earthly causes in complex "co-operation."[85]

Leaving aside, for the moment, the elective affinities between this position and currents in radical democratic theory (explored in chapter 5), Thelwall's *Essay towards a Definition of Animal Vitality* (1793) helpfully typifies the contrapuntal materialist possibility that is my focus in this study. First, Thelwall's *Essay* theorizes life as an emergent property of nonliving materials in "co-operation," the second-order *"result of a co-operation of other causes,* [none] of which need, in themselves, of necessity be alive."[86] This thought of the emergence of qualitative difference through the cooperation of simple elements, rather than the introduction of ontologically distinct kinds or forces, is a hallmark of Lucretian materialist reasoning and its primary response to the recurrent charge (here levied by Coleridge) that a philosophy of "composition" and "decomposition" among "unproductive particles" can only think quantitative change, "the exact sum of the component quantities, as in arithmetical addition."[87] For such an accusation rather exactly (if not purposely) misses the point of an ontology that grants virtually nothing to individual particles in order to attribute every form of productivity, variety, and sophistication to their combinations—the outcomes of which, never belonging to "component quantities" in the first place, is never fully calculable in advance.

"Matter is denominated inert and brutish," complained Lucretius's early nineteenth-century editor and translator, John Mason Good (1764-1827), yet only a single atom, "placed at an infinite distance from all other atoms," could possibly "deserve such an appellation." ("I am like an atom," Blake laments from this theoretical position in *The Four Zoas*, "A Nothing left in darkness yet I am an identity / . . . Ah terrible terrible.") Let more atoms into the picture, and matter's "plastic and prolific" dimension comes into view: "they will begin to act with reciprocity," such that matter "immediately evinces other powers or attractions—and these will be perpetually and almost infinitely varied in proportion as we vary its combinations."[88] So decisive, and so generative of irreducible, reputedly ontological or immaterial difference, are the "position" in which primordia "are held together, and what motions they give and receive mutually" (*quali positura contineantur/et quos inter se dent motus accipiantque*).[89] *De rerum natura* repeats the point four times. As I argue at some length in chapters 1 and 3, one effect of this position is to dramatically diminish the role of "necessity" without wishing away the tenacity of existing structures, norms, and forms in a general affirmation of "flux." In this view, the stubborn regularity with which biological, political, and artistic forms perpetuate themselves has to be traced to continuities in context and practice, forces that are no less intransigent for being contingent, yet cannot be explained (away) as natural or supernatural "law."

As this book's chapters explore in detail, the logic of associative emergence—debated under the heading of "spontaneous generation" in the eighteenth and nineteenth centuries—accounts, in the Lucretian tradition, for several *un*quantifiable relays: between a life and the separately nonliving parts upon which it depends, between a word's meaning and the separately insignificant letters that make it up, between atoms qualitatively "destitute" in their singular being and sensuously splendid in their collective appearing, between personal feelings and the astonishingly powerful force Shelley recognizes as "national sentiment."[90] How Romantics put this logic of (re)combinatory mutualism to use in minoritarian theories of life that render it morphologically inseparable from contextual experience and collective history is the cumulative subject of chapters 1, 2, and 4; chapter 3 delves into a corollary theory of language as a material semiotics natural in its very dependency upon convention, and chapter 5 explores a frighteningly vulnerable practice of political association that does not, as atomism is too often supposed to do, substantiate a liberal individualist picture of political sovereignty.

But it suffices to observe for the moment that Coleridge and Kant approach life as a question of the parts-whole relation *internal* to an individual organism, and still more, as the very power of individuation and autonomous integrity: "I define life as *the principle of individuation*, or the power that unites a given *all* into a *whole* that is presupposed by its parts."[91] Thelwall and like-minded materialists, meanwhile, advocated the remarkable, but largely unremarked thought that *life is not attributable to an individual organism at all*.[92] Organic form, that "particular harmony and correspondence of the whole or aggregate combination [of animal parts]," Thelwall explains, is necessary but not sufficient to "Life." Such organization amounts to "the *susceptibility*" of life, which must yet be "*induced by the application of proper stimuli*."[93] Proper neither to the organism nor to extrinsic stimuli, then—let alone to any immaterial principle or power—"life itself," argues Thelwall, is a "state" induced and sustained by "the intimate combination" of a susceptible body and a surround that keeps exciting its activity. An ontologically relational affair that arises between an "animal frame" and its stimulating surround, "life itself" issues from the "perfect contact" between an amenable form and that to which it responds, arising or expiring according to "the susceptibility and presence, or the non-susceptibility and absence" of these "cooperating causes."[94]

Against organicism's teleologically insular ideal, then, other Romantic thinkers cast life as dependent upon context, contact, and combination: a contingent *susceptibility* rather than an autonomous *power*. As I argue in chapter 1, for Lamarck and Erasmus Darwin, as well as for William Blake, a being's morphological viability was a question not (or not only) of the formative "force" or "drive" emphasized in vitalist discourse, but also of its "maximum suppleness": its receptive capacity to con-form with the social and material surround, such that its plastic, forming organs made up a physical archive of successive interactions with the milieu.[95] The ancient conceptual resource of atomist science, combined with a host of new experimental evidence,[96] furnished motives and means for describing animate bodies less as self-generating causal circuits—"Self-closd" and "self-begotten," as Blake caricatures Urizen's body-type—than as what Goethe called "Being-Complexes": open-textured, irreducibly plural bodies, dependent for their very life, and for good or ill, upon the "co-operation" and "intimate combination" of causes that are not their own. Indeed, to hold at bay the false alternative between a kind of scientific positivism not yet in existence and the idealism and spiritualism still expected from any "Romantic science" is to allow sophisticated and theoretically challenging logics of life to come into view. Late eighteenth- and early nineteenth-century biologies offer surprising, scientific corroborations of the freshest present-day revisions of Romantic subjectivity: the "impersonal, asystematic, and nonintentional" expressions of "saturation *in* the world" that Jacques Khalip calls "anonymous life," the paradoxical forms of recessive action and passive disclosure that Anne-Lise François charts in her unforgettable revision of poetic agency and address, and the romantic "phenomenophelia" that Rei Terada discerns through her own, post-Kantian aesthetic lens.[97]

In these and other ways, then, "the poetry of life" (Percy Shelley's phrase) under examination in this study differs dramatically, even polemically, from what is now the most familiar account of Romanticism's role in the modern history of "life."[98] I focus on the antivitalist counterpoint that worked to uncouple professionalizing aesthetics *and* biology from their shared rhetoric of autonomy, impartiality, and power. And I pursue the once-palpable stakes—an experimental ethics and a biology of transience, a rival image of the body politic, an alternative organization of knowledge, a fresh analysis of state violence and record of the historical present—of the at times abstrusely philosophical Romantic attempts to keep the substance and motion of life materially continuous with other natural, social, and linguistic processes. For to reject the vitalist antimony between life and "mere" matter in the late eighteenth and early nineteenth centuries, I argue, was in fact to refuse to dissociate the problem of biological vitality from other material histories, social, environmental, and political. Such were the shaping pressures, resources, and powers once enfolded in Lamarck's capacious plural, *milieux*, whose biological significance, I argue in chapter 1, both nineteenth-century organicism and twentieth-century neo-Darwinism worked, until recently, to discredit.

Goethe and Shelley are the principal avatars of this book's unrecognized biopoetic perspective, and each takes up two chapters in its middle. This is because they elaborate a poetic science and a scientific poetry, respectively, that make "Shape" the fundamental unit of biopoetical (and, ultimately, biopolitical) inquiry, recognizing in its physical and poetic double meanings a real continuity between rhetorical and corporeal figures, and by extension, between "natural" bodies and the structures of power and signification that culture is still frequently thought to impose upon them. In complementary life writings of the 1810s and 1820s, when, as the sociologists have it, a professionalizing "revolution" meant that such poetic polymathy required explicit "Defense," the Goethean science of *morph*ology and the Shelleyan poetics of "Shape" together illuminate the connections between individual, social, and natural histories that stood to be lost with the peculiar figural realism of poetry. In so doing, they illuminate an occluded, Romantic-poetic through-line between radical Enlightenment materialisms of nature and the coming historical materialism.

In fact, as Shelley and Marx were particularly keen to notice, to be a body in Lucretian and Epicurean materialist philosophy is to participate in an ineluctable collectivity of decay and transformation that produces, at once, temporal succession and the order of the visible. That collectivity is named "nature" in this materialist view, and as it will be the further task of this book to show, it was conceived in a way that gave Romantic era thinkers—up to and including Marx on the cusp of writing something recognizably Marxian—extraordinary resources for *historical* thinking. This would be historicity in the sense that

Kevis Goodman, following Raymond Williams, has taught us to see: "that immanent, collective perception of any moment as a seething mix of unsettled elements." The phrase itself demonstrates the felicity of an atomist idiom for describing a quintessentially Romantic sense of "history in solution."[99]

Said to travel long-distance and long term, the capacity of Lucretian figures to nuance the perceived present with distant and prior happenings—"in one moment of time perceived by us . . . many times are lurking [*tempora multa latent*]" (4.794–96)—was historiographically suited to the period in which, in Reinhart Koselleck's teaching, "the contemporaneity of the non-contemporaneous became a fundamental datum of all history."[100] Constituents of a mixed, material atmosphere of cast-off images and affects, *eidola* were summoned to articulate the way bodies produce and suffer what "is in the air" of a shared, historical interval: how singular lives produce and integrate passages of historical time (chapter 4); how the exertions of the milieu shape the viability of forms (chapter 1); how formal causation might be conceived contingently and relationally, rather than according to an organicist logic of autotelic power (chapters 2–3); how any "present" figure obtrudes particulate histories of elsewhere and before (chapter 5); and how beings body forth their times, though not just as they would like (Coda).

In each instance, Romantic neo-Lucretian figures hold social, biological, and poetic materiality together, making historical experience felt as what Wordsworth called an "atmosphere of sensation." And though this book is not written under the sign of Marxian materialism, by the time we reach its Coda, the young, notoriously "romantic" Marx's insistence on "an historical nature and a natural history," his recourse to ancient poetic materialism to suffuse the biological "production of life" with the force of sociohistorical circumstance, and his use of emphatically figurative discourse to do so will appear thoroughly *romantic*, albeit in an unfamiliar sense that I hope the study makes visible. This would be an ancient and Romantic recourse to materialism as the common idiom between corporealities of the text, the body, and history that later specializations of materialist criticism have tended to hold apart.

d. Matter Figures Back

Like the bumper crop of "new materialist" ontologies in critical and political theory, this project arose in part out of frustration with rigid and hard to reconcile senses of "materialism" invoked in Marxian ideology critique and deconstruction: with the way, in Karen Barad's words, "Language matters.

Discourse matters. Culture matters . . . the only thing that does not seem to matter any more is matter."[101] And yet it has proved more worthwhile to engage these traditions—and to reconstruct their divergent genealogies—than to proceed as though their tasks were finished. New materialisms (as "old" Marx and Engels were well aware) reach a limit as long as they reactively attend to portions of being neglected by discursively turned theory: doing so continues to cede language, culture, representation, ideation, fantasy, and policy to immaterialist logics of explanation.

This is where Lucretian materialism, being also and irreducibly poetry, has something strange and specific to offer. Which is why, in (sympathetic) contrast to Jane Bennett's broad and enthusiastic sense of "vibrant matter," I hew to the particularity of Lucretian materialism and its specific historical appropriations, taking care not to fuse it with other unfamiliar, non-Kantian and nondualist ontologies.[102] I say this not only because what now seem to be obscure distinctions were politically loaded ones, but also because I am not sure a suitably "*political* ecology of things," in Bennett's phrase, will be attained until the ecologically renovated materialisms elaborate alternative accounts of the themes that have traditionally driven accounts of the political (representation, ideology, rights, policy, protest). I thus attempt to illuminate Romantic Lucretius with the specificity, rigor, and imagination that Marjorie Levinson has afforded to Romantic Spinoza; and in fact, reviving the period's *other* heretic ontology demands kindred but distinctive suspensions of the paradigm of representation and the "seemingly axiomatic resistance of matter to mind."[103] And attempting to treat this Lucretian materialist tradition not only philosophically, but also as a set of formal and generic technologies in competition with other instruments of perception and mediation, this study owes debts of inspiration to Kevis Goodman's unsurpassed treatment of Romanticism's georgic mode.[104] Like Goodman's Romantic Virgil and Levinson's Romantic Spinoza, Romantic Lucretius in this study works to make familiarly "ideological" poems strange and to interrogate the ongoing project of historical materialist criticism.

Goethe, Shelley, Marx, and other contemporary appropriators point us toward something unique to Lucretius among the welter of vitalist, materialist, Spinozist, hylozoist, and animist positions then in circulation. *De rerum natura* presents figuration as the basic action and passion of matter, without thereby casting matter either as alive or as produced by (the poet's) words. The first caveat, as I have begun to argue and will elaborate throughout this book, has rare and critical consequences for philosophies of life: Lucretian materialism, as Goethe and Shelley show, could counteract the valorization of youth, power, force, and autonomy that attends vitalist rhetoric from Romantic anthropology through to fascist biopolitics, opening instead onto epistemologies of relation, senescence, and dependency, and onto the questions Judith Butler has posed of which lives and which body morphologies are socially livable, visible, and grievable.[105] The second caveat generatively estranges Romanticists' most practiced strategies of deconstructive rhetorical reading.

For to argue, as I do, that Lucretius's mode of materialist, didactic poiesis remained available to Romantic writing—writing deeply and lastingly identified, in the criticism, with the performative feats and fiats of lyric address—is to accept an unfamiliar possibility: that the faces, forms, and figures that Nature puts on in Romantic poetry attest to something other than anthropomorphic rhetorical effects.[106] Like the "not myself" that "goes home" to Keats when "the identity of every one in the room begins to to [*sic*] press upon me," Romantic poetic figures can also index attempts to cede poiesis to something that is neither the self nor the alien force of pure language, including the action-at-a-distance of things excluded from nonpoetic norms of perception and representation.[107] Yet the materialist tropes that Romantics sometimes mobilized to do so have been difficult to appreciate through the de Manian model of rhetorical reading: there, the identification of natural and linguistic materiality has been the very definition of ideological mystification, while the materiality of language, properly grasped, can be relied upon to disrupt this illusion.[108] And yet, as I take pains to show in chapter 3, at stake in neo-Lucretian figuration is a kind of material semiosis, neither symbolic nor allegorical in de Man's sense, for which the material commonality between sign and referent accentuates, rather than masks, the "temporal predicament" that divides them, as well as the arbitrary origins of *both* words and things.

Lucretian materialism of nature thus also opens onto, but dramatically re-inflects, arguments in gender, science, and cultural studies about the discursive production of (corpo)reality. Does the fact that *De rerum natura* posits an inexplicable trope or turn at the origin of every physical entity and process, the fact that its materialism is inexorably and unabashedly figural, make Lucretian poetic science a form of pure social constructivism, linguistic monism, or "textualization of nature"?[109] The unique force and manner with which *De rerum natura* answers "no" to this question—positing a reality that is figural not because representation fails to capture things-in-themselves, but because things inadvertently collaborate in their perception and representation—is the basic interest and ongoing provocation of this text. We are confronted, in Meeker's perfect formulation, with "a classical materialism that does not find its own origins in the split between representation and things-in-themselves."[110] No lesser aficionado of *figura* in the Western tradition than Erich Auerbach credits Lucretius with the "extremely individual, free, and significant" transfer of the term from physical, plastic form to that form's *imitation*, "from model to copy"—a leveling of the hierarchy of representation that Deleuze would celebrate as "the most innocent of all destructions, the destruction of Platonism."[111] Unlike Platonic simulacra—for the target of Epicurean philosophy, it is thought, was indeed Plato's theory of forms—the Epicurean and Lucretian kind are separated from their referents by lapses of time, rather than reality.[112]

These observations permit us to mark *De rerum natura*'s affinities with the answers carefully elaborated by Butler and Bruno Latour in the 1990s, when each report being confronted with varieties of the same question regarding their purported constructivisms: "What about the materiality of the body, *Judy*?" and "Do you believe in reality, [*Bruno*]?"[113] Butler's answer, that there is certainly an "outside" to discourse and power, but one that we can grasp only in and through them, "at and as" their "most tenuous borders," brings her to advocate "a return to the notion of matter," not as a blank site or surface for the inscription of discursive power, but as "*a process of materialization that stabilizes over time to produce the effect of boundary, fixity, and surface we call matter.*"[114] And it is notable, in our context, that one way of acknowledging bodily matter as processually enacted, rather than given, in *Bodies That Matter*, is to notice that "matter never appears without its *schema*"— its "form, shape, figure, appearance, dress"—morphological contours that attest, for Butler, to the apparently given body as a shaped and shaping instantiation of power. Lucretius, too, uses figuration to make a boundary effect of that "which, in discussing philosophy, we are accustomed to call matter," construing *figura* as something that speculatively simple, single, quality-less atoms acquire when, entering into plural con-*figurations*, they "come forth into the borders of the light," establishing a realm of phenomenal perceptibility and intelligibility that instances, without exhausting, the extant.[115]

With Butler, but perhaps more emphatically, Lucretian poetic philosophy enjoins readers to resist ascribing the figurative *action* within this event of figural materialization wholly to the exercise of discourse, and the material *passion* to its bodily, natural, inhuman, or unspeaking "outside." As we have begun to see, matter figures too in Lucretius, which means that the schema a body bears in any scene of perception or knowledge production has to be understood as the joint work of knower and known alike. This is where Bruno Latour's attempts to think around the stalemate between realism and constructivism (and corollary fact / artifact, discovery / invention binaries) in the "science wars" help. For as Latour's science and technology studies argue in a strikingly Goethean vein, "*all*" parties to an experiment, human and non-human "actants" alike, "leave their meeting in a different state from the one in which they entered." This alteration, Latour thinks, ought to be taken as nonparadoxical proof of the reality *and* the historicity of a scientific *fact*, as "that which is made and that which is not made up."[116]

More than two centuries earlier, Herder offered a brilliant theory of real-and-figural facticity by revising the inherited Enlightenment notion of a simple sense "impression" printed on the passive human mind. Herder redefines this "impression" as a "stamp of *analogy*" through which the knower "leave[s] his own mark" on what he perceives: "For everything that we call image in Nature becomes such only through the reception and operation of his perceiving, separating, composing, and designating soul."[117] The "stamp of *analogy*" works first in this passage to defamiliarize the sensory image of the object, to reveal its appearance "in Nature" as the outcome of a prior assimilation, a "reception and operation" between observer and observed for which the sensory image must be understood not as the raw or simple datum but as an appropriative "mark." The wrought or figured contours of a phenomenon in Herder—its "outline, dimension, and form"—bear witness to a bilateral process of figuration that not only conforms an object to the needs and habits of the perceiver's acculturated sensorium, but attests undeniably to the impressive force of something "alien which imprints itself in us."[118]

Remarkably, then, Herder treats the frankly figural stamp that marks embodied sensuous perception not as a disqualifying blemish but as a "seal of truth": a testament to the fact of a real-physical transaction between observer and observed that does not conceal the biased, power-imbued, species-specific shape of the record that the observer will bear. In this vein, we likewise find Goethe and Blake advancing in improbable anticipation of Donna Haraway a form of knowledge whose objectivity draws force from its partiality and situatedness.[119] Here the figurative improprieties that testify to the contingent limits of human knowledge also mark sites of contact with other natures, effects of common materiality rather than unilateral projection or inscription. That empirically, sensuously derived matters of fact are figural does not make them invalid, and certainly not invented: "poetic" is the perfect name for their impropriety and their truth.[120]

Thus, Romantic calls for sweetness and tenderness renovate the empirical habitus from two sides: they entail both an attitude toward the object that is conscious of its fragility, and a concomitant consciousness, indeed, a cultivation, of the plasticity and permeability of the observing self. Lucretius's "honeyed cup" topos emphasizes that one effect of the poet's *dulce labor* is to induce the auditor *to open* (mouth, mind) to a kind of knowledge that is at once potentially frightening and supportive of life. In *De rerum natura*, the knowledge to which one must open turns out to be a lesson in ineluctable openness, in the physical permeability of a self dependent for its life on constant interchange with other things. "[H]owever solid things may be thought to be," Lucretius repeatedly insists, they are in fact "rare" (*raro*), tenuous, open-textured, and loosely bound (1.346–47). The politics of such poetic receptivity come into relief in Lily Gurton-Wachter's study of Romantic attention, which reveals that attitudes "invit[ing] the strange or foreign without seeking to protect against it" could amount to radical counter-conduct in a militarized culture pervaded by calls to sensory vigilance.[121] "Nothing exists unless loosely bound in body" (*nil est nisi raro corpore nexum*), Lucretius insists in the mnemonic, alliterative form in which the poem delivers Epicurus's most fundamental axioms (1.346–47, 6.958).[122]

The permeability of bodies that appear bounded and circumscribed is indeed an issue of extraordinary importance to *De rerum natura*, which constantly directs attention to the matter (from food to figures) that passes between them: looking at a rural scene, for instance, we are asked to watch pastures, rivers, and leaves transform themselves into cattle.[123] And unlike the highest and better-known Epicurean philosophical pleasure of *ataraxia*—a negative freedom *from* anxiety that arguably does not survive its translation into poetry intact—sweetness in *De rerum natura* tends to describe pleasures of in- and ex-corporation:[124] the smooth particles of juice, for instance, that "sweetly touch and sweetly stroke" (*suaviter attingunt et suaviter omnia tractant*) as they pass through "the pores of the palate and the winding passageways of the open-textured tongue [*rarae per flexa foramina linguae*]."[125] In its sweetness, then, this didactic poetic modality induces the listener to experience with pleasure rather than fear this basic tenet of atomist corporeality: the open, vulnerable texture of his body and its dependence for every kind of experience upon the touch and motion of other things.[126] "Nothing can touch or be touched, except body" (*tangere enim et tangi, nisi corpus, nulla potest res*), Lucretius sums up, restoring the particular risk and privilege of being a body, upon which every kind of knowledge and experience depends (1.304).

De rerum natura's (still unlearned) lesson is that you can't wage war on "terror," because fears cannot be frightened: "fears and haunting cares fear neither the clang of arms nor wild weapons." Instead, in a passage repeated four times in the epic:

> Dark fears of the mind, then banish quite away
> Not with the Sun-beams, or the light of day,
> But by such *species*, as from Nature flow,
> And what from right informed reason grow.

> [Hunc igitur terrorem animi tenebrasque necessest
> non radii solis neque lucida tela diei
> discutiant, sed naturae species ratioque.]
> (1.146–48; Evelyn [1656], pp. 21–23)

For radical, high Enlightenment culture, the appeal of *De rerum natura* is traditionally said to have lodged in that last note, *ratio*: Lucretius as the classical avatar of reason's power to free humankind from the dark thrall of superstition.[127] Leaving aside the accuracy of that assessment in its context, the passage's subtle insistence that light is *not* terror's antidote—"this terror of mind" (*terrorem animi*), Lucretius writes, will *not* be banished by "Sun-beams" and "light"—must have rung out with welcome sophistication to Romantic generations reared in the knowledge that Terror could be Enlightenment's consequence as well as its casualty.[128]

The chapters on Romantic neo-Lucretianism that follow also track a shift in focus to *ratio*'s pendant here, *species*. This multifarious term, which John Evelyn wisely left untranslated and which many translations simply elide, gives terror-dispersing power to the visible outsides of things. *Naturae species* connotes the "view," "sight," or "spectacle" of nature, nature's "outward appearance," "shape," "figure," "aspect," or "mien."[129] Evelyn treats it as a plural, and his rendering, "such *species* as from nature flow," highlights a medieval and early modern usage that glossed a shared aspect of both atomist and Aristotelian sciences of sensation. Here *species* denotes not only a subject's "view" of an object, or that object's outward "look," but the subtle material media that, "flowing" between them, enable one body to affect another in perception.[130] Lucretius's leveling of nature's flowing face and its ratio—*naturae species ratioque*—constitutes a rare and radical validation of sensuous appearance whose consequences for poetic science will also be explored below.

Recognizing the Romantic Lucretian tradition means reawakening the nineteenth-century possibility of an empirical science that Marx called "sensuous," Goethe "tender," and Blake "Sweet": one once suited to think vital, poetic, and historical materialisms together and maybe still to open what poems might know about our own moment of ecological and biopolitical precarity.

e. Chapters and Scope

Building on Marilyn Butler's seminal reconstruction of a Romantic culture war between Coleridge's "right wing cult of the Germanic" and a "left wing cult of the classical," Martin Priestman has revealed the scope of neo-Lucretian culture in the Romantic period by exploring the ongoing risk, shock, and generativity of atheism in the neglected milieu of Richard Payne Knight, Sir William Jones, and Erasmus Darwin.[131] This study contributes to this growing understanding of Romantic materialist countercultures less by extending the survey of Lucretian influence than by opening out a series of Lucretian "moments" in more and less canonical Romantic texts—moments that exhibit the minute workings and far consequences of what I take to be an insufficiently examined biopoetic mode.

Reading into and out of these moments permits me to trace Lucretian materialist effects on the theory and representation of natural life, semiotics and rhetoric, historical experience, and political consciousness, roughly in that order. My intention is not to claim the texts at issue, let alone the authors or oeuvres to which they belong, as certifiably "Lucretian" in any systematic sense, but rather to demonstrate that this materialist mode was a live possibility for Romantic era writers, and an available (if risky, Terror-tinged) technique in the poetic and epistemic repertoire of writers as divergent as Coleridge and Marx. I try to develop a mode of reading that allows this live possibility to reconfigure perennial cruxes in the criticism of Romantic poetry and science. The chapters differ in the intensity and proximity of their relation to *De rerum natura*. Chapter 1 does not specifically concern Lucretius, but sketches, through Blake and his interlocutors, the greater biological field in its amenability to figural materialism; chapters 2, 4, and 5 home in on intimate relations of translation, adaptation, and strategic redeployment of *De rerum natura* in Goethe, Shelley, and others; in chapter 3, Lucretianism surfaces principally as a practice of rhetorical reading. As mentioned, Goethe's and Shelley's life writings receive disproportionate attention in this series because their poetic sciences most flagrantly and self-consciously stage the entrance of Lucretian figures and the generative category confusion that ensues.

Chapter 1, "Blake's Mundane Egg: Epigenesis and Milieux," introduces the problem of organic form in Romantic philosophy of biology, reading Blake's *The First Book of Urizen* as a strong critique of the vitalist organicism that depicts life as the power to self-organize. Pinpointing self-reflexivity as the lynchpin of the ideal of organic form, Blake parodies that idiom with a string of reflexives—"Self-closd," "self-contemplating," "self balanc'd," "self-begotten," "In anguish dividing & dividing"—aligning Urizen's political and theological despotism with the uniquely autotelic logic of organicist theory. But self-generation, I argue, is but the best-canonized of the period's theories of life's emergent causality (*epigenesis*). In this chapter, I recruit Jean-Baptiste Lamarck and Erasmus Darwin for materialist theories of embryogenesis that stressed receptivity to the ambient pressure of the social and material surround, construe viable morphology as an archive of interactive con-figurations with those milieux, and ultimately challenge the consensus that bodily life resists symbolic power. It is this biological and poetic tradition, I argue, that Blake advances when he depicts livable form as a product of industrial and labor relations operative even within the "Mundane Egg(s)" that were life scientists' favored objects of experimental vivisection. In "Epigenesis, an Epilogue," I sketch how formerly heretical Romantic and Lamarckian ideas are enjoying a revival in the "epigenetic" correctives to the excesses of twentieth-century genocentrism.

My second chapter, "Equivocal Life: Goethe's Journals on Morphology," looks to Goethe's microscopy logs (1785–86) and *On Morphology* periodicals (1817–24) for their exemplary neo-Lucretian treatment of living beings as composite, rather than strictly organic, forms. For Goethe, each "seeming individual"—not least the observer—is a fractious "assemblage of independent beings," a plural "Being-Complex" whose incessant metamorphoses the morphologist ought to match with every available mimetic, generic, and figurative resource. The *Morphology* project, I argue moreover, shifts biology's focus from the problem of embryogenesis that drove its emergence as a science, toward neglected, late-life processes of senescence and decomposition: "Going to Dust, Vapor, Droplets," as Goethe names one late essay. What, this essay begins to ask, might life look like from the perspective of the nonreproductive, but communicative, effluvia that mediate *between* beings in their transience? What arts of discomposure would be adequate to this view? Focusing on an experiment in which a mushroom "draws" its own image in spores, I argue for the credibility in the period of nonhuman acts of representation that do not take verbal form.

I next follow Goethe's "natural simulacrum" into a thorough discussion of neo-Lucretian semiotics and poetics, showing how this kind of material trope has eluded both structural linguistic and poststructuralist modes of reading poetry. Chapter 3, "Tender Semiosis: Reading Goethe with Lucretius and Paul de Man," rereads Goethe's neo-Lucretian didactic poem "Dauer im Wechsel" (Permanence in Change) from the perspective of objective figuration, centering on a neo-Lucretian simulacrum that, I argue, eluded Paul de Man's seminal taxonomy of Romantic allegorical and symbolic figuration. At issue is neither the bad-faith simultaneity of subject-object, mind-nature, word-world that marks symbolic diction for de Man, nor the salutary renunciation of such illusions in favor of responsibly intralinguistic allegorical self-referentiality. Goethe's poem instead stages the strange and estranging arrival of an atomist sign in its joint capacity as verbal figure and datum of perception, and I draw on Charles Peirce's understanding of indexicality, Thomas Sebeok's biosemiotics, and Michel Serres's information theory to explain its efficacy. For Goethe, I argue, such semiotics entail a sly critique of the Kantian epistemology so frequently said to undergird Romantic practices of aesthetic and scientific observation. Cultivating his own vulnerability to the object under view, Goethe revalues the feminized quality of "tenderness" as an epistemic virtue, permitting the ways in which the self is (also) an object to reenter natural and aesthetic philosophy.

Chapter 4 examines how, in an English context, Lucretian poetics could be summoned to rebuke a kind of triumphalism shared by both vitalist life science and post-Waterloo historical consciousness. "Growing Old Together: Lucretian Materialism in Shelley's *The Triumph of Life*" shows how Shelley mobilizes figural materialism to articulate the way personal bodies produce and integrate historical eventfulness, especially at a distance. Offering a new account of the face-giving trope of prosopopoeia in the poem, I argue that Shelley adapts *De rerum natura* to depict wrinkling faces as mutable registers of the "living air" of a post-Napoleonic interval. Casting personal senescence as the unintended work of multitudes, *The Triumph* presses toward a biology and epistemology of senescence—yet in this case, we open our inquiry outward from the intimate scene of experimental trial toward the collective experience of social history, in league with Walter Benjamin's historical-materialist practice of "weak messianism."

My fifth chapter, "A Natural History of Violence: Allegory and Atomism in Shelley's *The Mask of Anarchy*," rounds off the book's increasingly historical trajectory by putting its valorization of poetic sweetness and tenderness to the test of the quick and slow violence of material exploitation. Chronicling the violent suppression of a mass demonstration for parliamentary reform in 1819, Shelley's *Mask* has been a lightning rod for debates about aesthetics and politics in Romantic literary criticism. My chapter explores why the poem Shelley called "wholly political" elects to treat the event as a matter of natural history, obsessively chronicling the recirculation of wrongly spilled blood through geological and meteorological cycles. Far from "naturalizing" the damage, I argue, this discourse enables *The Mask* to navigate between affective communication and structural analysis, subjecting the kind of felt, historical "atmosphere of sensation" that has built up over the course of the book to didactic breakdown (as if in answer to Lauren Berlant's point that "we can over-respect the work of emotional justice").[132] Sketching a history of related moments in eighteenth-century neo-Lucretian poetry, I track the proto-Marxian and proto-Foucauldian features of Shelley's poem as affordances of its genre, arguing that the notoriously "lyrical" *Mask* needs to be seen as part of a *didactic* poetry that speaks polemically for long-traveled, bloodstained materials and whose attempted politics are those of radical pedagogy, rather than lyric performance. Such didactic materialist passions, neither subjective nor divine, motivated significant blowback from John Stuart Mill and Thomas De Quincey in early nineteenth-century genre theory. Shelley, however, puts this tradition to use in a call to assembly that capitalizes on the difference between that embodied power and parliamentary freedoms of speech.

Chapter 5 discovers, in the "accent unwithstood" that Shelley attributes to the Earth in *The Mask of Anarchy*, an extreme of natural-historical materialism at which the posthuman, ecologically motivated "new materialisms" and the socioeconomically focused, Marxian "old materialism" might circle back and rejoin one another. And indeed, each of the book's chapters differently argues for the Romantic compatibility of earlier modern materialisms of *nature* and new, nineteenth-century forms of materialist *history*—a conjunction held open by the problem of natural "life." My Coda, "Old Materialism, or Romantic Marx," resituates Marx in this Romantic tradition by returning to the "sensuous science" he elaborated reading Epicurus and Lucretius, for its ecological, historical, and poetic intelligence. While the newness in "new materialist" ontologies is meant to break from the Marxian materiality of history and the deconstructive materiality of the letter, I argue that Marx was operating in a mode of counterdisciplinary materialism that deployed figural matter as a substance fit to articulate terrestrial rhetoric, history, and biology together. Attaining to such materialist semiotics might help historical materialist criticism rise to the again pressing task of writing for social justice in the idiom of natural history.

CHAPTER 1

Blake's Mundane Egg: Epigenesis and Milieux

Voyez-vous cet œuf? C'est avec cela qu'on renverse toutes les écoles de théologie et tous les temples de la terre.

[Do you see this egg? It is with this that one overturns all the schools of theology and all the temples of the earth.]

DENIS DIDEROT, *Le rêve de d'Alembert*

a. Into the Egg

When William Blake, in *Milton* and *Jerusalem*, began to call *this* world, the world disclosed by our fallen senses, "the Mundane Egg," he tucked the theater of human history within the most disputed experimental object of Enlightened and Romantic life science. For the longest conceivable eighteenth -century, controversies in the study of animal "Generation"—discipline-forging controversies for what would become experimental physiology, biology, embryology, and comparative anatomy—hinged on events of conception, animation, and organ formation under way within mundane eggs. In this chapter, Blake's books serve as our surprisingly informative, unsurprisingly polemical guide to the mundane egg's politically and theologically fraught territory, introducing problems in the early life sciences that will prove central to each subsequent chapter. In this introductory vein, the Blakean poetic lines that aspire to "vary" in "every Word & Every Character/ . . . According to the Expansion or Contraction. the Translucence or/Opakeness of Nervous fibres" divulge the special status of the new sciences of life as invitations and incitements to poetic art: the enabling idiom, rather than the target, of Romantics' critique of previous empiricist and experimentalist knowledge formations ("Bacon & Newton & Locke") and a license for poetic renovation of what the "senses five" purport to know.[1]

Before zeroing in on the particularly neo-Lucretian strain of figural materialism followed in detail over the next four chapters, then, this chapter lays out the wider field of interaction and indistinction between biological and poetic approaches to living form in the period, paying special attention to arguments against and beside the organicist ideal that still dominates literary-critical reconstructions of this terrain. Reading scenes of embryogenesis from *The First Book of Urizen* (1794), *Milton* (c. 1804–1811), and *Jerusalem* (1804–c. 1820) with related poetic and prose accounts from Erasmus Darwin and Jean-Baptiste Lamarck, this chapter presents a now unfamiliar premise that, I argue, held contemporary experimental life science open to poetic participation: an understanding of biological life as dependent, for its very development and viability, upon actions and passions of lineation, imitation, and figuration, of which poetry is one. At issue in recent life science was a body tropologically active and reactive even at the level of its basic biological needs and functions, a body whose sensory means of "converting objects" into "symbols," "signs," and "images" seemed to furnish the pattern, prehistory, and present condition for more obviously linguistic and social means of production and instruction.[2] And given the pre-Darwinian conceptions of biological time that I attempt to restore in this chapter, this logic rendered "forms / Of life" intrinsically historical and poetic in several time signatures at once: not only in the now-familiar, geological "deep time" of species evolution (phylogeny), but also at the quicker tempo of individual development and life course (ontogeny)—not to mention in every apparently immediate event of sensation.

Blake's books brilliantly illustrate the potential of the experimental philosophy of *epigenesis*—the science, as we will see, of real-time embryonic embodiment—to represent living bodies as sociohistorically elaborated and collaborative "forms / Of life," without thereby alleging their unilateral construction or inscription by discursive power.[3] In Blake's and related biologies, I argue in this chapter, beings owe their very life to the social and symbolic processes that give them viable and recognizable form, and they collaborate, down to their nerves and fibers, in the production and perpetuation of signifying systems we now tend to figure as imposed upon bodily life by force of language or law. By virtue of this congruence and complicity with the social forms of art, I argue, the life of the body in Blake and like-minded life sciences cannot fulfill the emancipatory role that Romantic vitalist criticism assigns to it—cannot be trusted, that is, as a source of pure, creative, ontological power to resist, elude, or break through what Blake calls the "mighty Incrustation" of discursive, epistemic, political, and historical forms.[4] Instead, as Jean-Baptiste Lamarck argued in texts that helped coin the term and define the science of *biologie*, the problem of natural life in fact delivers inquirers directly to the intersection of "the *physical* and the *moral*": biology names the logic by which the "two orders of things" we might now call culture or discourse and nature or the body "have one common source [and] react upon one another, above all when they seem most separated."[5] In Lamarck and Blake's context, I argue here, "life" frequently names matter's movement toward, its reception, co-production, and perpetuation of, structures of power and meaning. This also means that bodily "life" cannot be hailed as "totipotent," discursively innocent, or free.[6]

There is no denying that Blake's poetry frequently aspires to absolutely unfettered vitality: a life in which "words of man to man" redound "in fury of Poetic Inspiration/To build the Universe stupendous," in an activity Denise Gigante has justly termed "ontopoietic," and Saree Makdisi, "a life of endlessly proliferating creation, a life of ontological power."[7] Yet unlike his more recent, presumptively secular, vitalist advocates, Blake names this mode "*Eternal* Life" and the "Life of *Im*mortality" (my emphasis), in frankly millenarian invocation of a "Life" available once the pathos and finitude of terrestrial life "Are," as one Eternal Prophet puts it bluntly, "finishd."[8] That is, Blake's poetic theology makes clear that "Life" as vital power unbound is not synonymous with actually existing "forms/Of Life," and may even come at their expense. This chapter proceeds from the observation that exhilarating, vitalist visions of "Eternal Life" in Blake's oeuvre are frequently shadowed by a different style of demand for life, a demand that voices marked as finite, terrestrial, feminine, and infantile issue against an ideal of Life "*too* exceeding unbounded."[9] Even *Jerusalem*'s final, tour-de-force vision of Humanity redeemed is punctured by a "Cry" of protest from "all the Earth from the Living Creatures of the Earth," a contingent legitimately concerned that Life without limits spells the end of the specific bodies in which their finite lives inhere.[10] As if amending the vision of infinite ontopoetic vitality in response to this cry, Blake's magnum opus ends by picturing participants in the "Life of Immortality" taking restful dips into the transience and finitude of "Planetary lives"—"reposing," that is, in "lives of Years Months Days & Hours."[11]

Heeding such compromised and compromising voices, I want to suggest, is the exact, if somewhat deflationary task of thinking through "biology" with Blake, a logic of life he and his contemporaries called "Generation" and situated within mundane eggs. The very term "Generation" keeps the problems of contemporary experimental life science rigorously coterminous with mundane history in Blake's lexicon, where "Generation" not only names what we have come to call developmental biology but also, more broadly and basically, creatures' (ongoing) fall into sexual difference and corollary systems of (re) production and moral and natural law. In the "Looms of Generation" within the Mundane Egg, Blake writes, "Men take their Sexual texture Woven."[12] Sexual "Generation" weaves the span of fallen history for Blake, a "Woof of Six Thousand Years" he expects to be wrapped up and discarded at Jesus's second coming.[13] There is thus an important sense in which the science of "Generation"—a science brandishing, since the seventeenth century, the motto *ex ovo omnia* (everything from an egg)—is, from a Blakean point of view, curiously preoccupied with the processes by which fallen being replicates itself.[14] In what follows, then, I take seriously both Blake's minute engagement with generation science and his delimitation of any such science to the "Finite & Temporal" set of perspectives confined to "The Mundane Egg."[15] Scenes of generation in Blake pertain precisely and modestly to works of inadvertent repetition and minor transformation within a fallen world, opening a technically mundane dimension in the work of an artist best known for radical, revolutionary, and prophetic aspirations.[16] Thus, after examining the rival theories of epigenesis at work in *The First Book of Urizen*, I turn to a quiet passage in *Milton* that explores how generated beings might discover their bodies' negotiable lineaments, were they but to recognize the histories casually "imbodied" there (fig. 1).[17]

For in fact, if "Generation" in Blake invites us to curb our critical enthusiasm for the ideal of life unbounded, then accepting this curb allows rival Romantic theories of Generation to come into view as rival theories of *this-*worldly historical process.[18] Thanks to the groundbreaking research of numerous scholars over the last two decades, it is no longer news that Blake, with the majority of more and less canonical Romantic poets, intimately engaged the emergent, scientific study of life we now know as biology.[19] And yet as literary critics have rediscovered that science's foundational controversy between "epigenesis" and "preformation" theories of animal generation, we have too quickly seized upon *epigenesis*—loosely construed as the vital, dynamic refutation of deistic, mechanical predetermination—as Romanticism's fit, provocative, and progressive life science.[20] It is not that preformation, which was anyway on the wane in the late eighteenth century, should be rehabilitated instead. Rather, Blake's "epigenesist poetics," in Denise Gigante's fine phrase, in fact demands a more rigorous grasp of competing variants within that epoch-making idiom, not least because rival theories of generation were then understood to carry competing systems of moral and political order to term.[21]

I argue in this chapter that Blake's specific mischief on the mundane egg

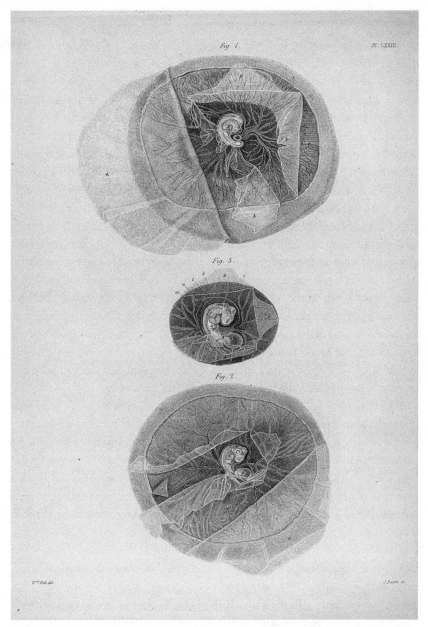

FIGURE 1. Illustration of the progressive development of the allantois membrane in a goose embryo, prepared by John Hunter. From Richard Owen, *Descriptive and Illustrated Catalogue of the Physiological Series of Comparative Anatomy Contained in the Museum of the Royal College of Surgeons in London*, Volume 5, Plate LXXIII, Products of Generation (London: R. and J. E. Taylor, 1840). Compare Blake's Mundane Egg of "labyrinthine intricacy twenty-seven folds of opakeness" (*Milton* 16.26). Photograph: Courtesy of the Division of Rare and Manuscript Collections, Cornell University Library.

satirizes one powerful model of *epigenesis*—incidentally, the one most frequently recovered by Romantic literary historians—and invites us to reconstruct another. Rejecting the rhetoric of autonomous "self-organization" from *within* a broader epigenesist idiom, Blake's books undermine the persistent tendency to equate epigenesis with organicism, as well as to frame the question of Romanticism and biology as a question of the fate and merit of a particular, organicist ideal: the Kantian and Coleridgean ideal of living form as autonomous "cause and effect of itself."[22] In Blake, as in the contemporary zoologies of Lamarck and Erasmus Darwin, the theory and lexicon of epigenesis casts animal formation as a work of acute circumstantial dependency, rather than of autotelic power. Here living forms are those that tend, for better or worse, to make an organ of experience, their developing bodies presenting a compounding archive of prior interactions with their social and material surrounds. The epigenesist idiom these writers share permits them to depict living forms—even at the level of the bodies they take for natural—as bearers of contemporary circumstance, exquisitely susceptible to incorporating the material relations of their multilevel *milieux*. By the lights of this lesser-known logic of life, the living bring circumstantial histories to bear on any present interaction and owe their very capacity for improvisation to the gap between present exigencies and the pasts embodied in their very organs.

When it comes to biological timekeeping, then, Blake's form of "incarnate history," as Noel Jackson perfectly puts it, challenges that persistent commonplace whereby it is first on the grand scale of species evolution across geological time that biology is said to have granted a history to nature; or whereby the organism's teleological *Bildung* is a microcosm for phylogenetic and world-historical progress.[23] Blakean epigenesis, I argue, instead casts the interval of ontogeny as historical time: operating, like Erasmus Darwin's and Lamarck's, on the micrological scale of a being's mutability over the span of its own life, this biological outlook chronicles the susceptibility and resourcefulness of the living vis-à-vis contemporary environments and practices, the way the pressures of inheritance, need, habit, and contingency interact to mold their organs from conception until death. And, as I hope to show, it is in concert with this neglected biological tradition that Blake depicts livable form as a fabric of many hands whose psychobiological nerves and fibers link individual selves to supra-individual contexts and forces.[24] Here Blake, Lamarck, and Darwin depart from a powerful, Romantic organicist conception of life, but also, as I will close suggesting, anticipate what "epigenesis" has come to mean for biology now, in the aftermath of "The Century of the Gene."[25]

b. The Missing *Baumeister*

In the English context, it was William Harvey's groundbreaking *Exercitationes de generatione animalium* (*Anatomical Exercitations concerning the Generation of Living Creatures*) (1651, trans. 1653) that reignited experimental research into embryogenesis and positioned mundane hens' eggs—"because *Egges* are a cheap merchandize, and are at hand at all times, and in all places"—as *the* paradigmatic experimental object for life scientific enquiry.[26] Blakeans who read Harvey discussing, for instance, the embryonic substance "in whose *Center* is born . . . the *Bloody panting point*, together with the *Ramifications* of the *Veins*," cannot help hearing Urizen's forming heart: "Panting: Conglobing, Trembling / Shooting out ten thousand branches" from "a red / Round globe hot burning."[27] And, indeed, Blake arguably modeled his seven-stage cosmogony in *The First Book of Urizen* not only on Genesis, *Paradise Lost*, and the Gnostic gospels, but also on Harvey's by then classic embryological "History of an Egge."[28]

Harvey's opening chapter, "The Reason why we begin with a Henns Egge," defended his choice of a model object by famously arguing that "all *Animals* may in some sort be said to be born out of *Egges*," the apparently "*Viviparous*" (live-birthing) among us being merely those who keep their eggs "so long within their bowels, till the *foetus* come forth shaped, and alive."[29] Harvey sealed his point about the comprehensive exemplarity of egg histories by pronouncing "*man* himself to be made of an *Egge*"—a conviction glossed equally, as I mentioned, in *De generatione*'s ovi-centric motto, *ex ovo omnia*, and in Blake's mischievous equation of "Generation" with the "Egg form'd World."[30] Subjecting mundane eggs to the New Scientific rigors of experimental vivisection and "ocular inspection," Harvey's series of replicable Exercitations promised that in the common egg, "it is an easie matter to observe . . . the first evident, and distinct ground-works of *Generation*; what progress nature makes in *formation*, and with what wonderfull providence shee governs the whole worke."[31] But deep into the book, the disheartened title of Exercitation 49 confesses that "*The Efficient Cause of the Chicken, is hard to be found out*," and Harvey's research in fact unleashed one hundred and fifty years of controversy around his suggestion that embryonic formation occurred by *epigenesis*. Generation works "*per Epigenisen*," he wrote: "by a post-generation or after-production; that is to say, by degrees, part after part."[32]

By *epi*-genesis as postproduction, Harvey meant that the simple materials enclosed in the egg ("namely, the White and Yolk") differentiate *serially* into the splendidly intricate body of an embryo, and crucially, that they do so in the observer's presence and in his present tense. Beginning from a mere "speck" that "jets forth" in the yolk, fetal complexity arises through a series of observable *post*generations of subsequent organs and parts. Neither always-already present in the fertilized egg, nor "constituted at once" by the conjunction of "heterogeneous *Elements*," Harvey argues, the "future Body" instead "grows distinct as it grows," visibly "making out of a *similar Matter . . . dissimilar Organs*" until "at the last the *Foetus* doth result." Nor, Harvey emphasizes, are we to mistake that "first particle of the Chicken" for some dedicated kind of matter or privileged site of insemination, as if there were "some little portion of the *egge* out of which the *foetus* should assume his body," a control center from which an agent would direct the "delineation of the parts." *Epigenesis* instead names the possibility that the embryonic animal is "formed" and "imbodyed" by the same stuff that merely feeds it: "constituted," in Harvey's words, "out of the same *matter*, by which it is sustained and augmented." When the doctor hazards the thought in the active voice, the subversion of figure-ground and form-matter binary priorities becomes unmistakable: "the *Matter . . .* grows also, and takes some shape; so soon as there is a *Nutriment*, there is a creature to be nourished."[33]

The early life sciences are known for the panoply of formative principles and powers with which they attempted to answer the question of the *cause* driving epigenesis. But it is worth holding this question and its answers at bay for a moment in order to recognize the more basic provocation of epigenesis as an account of living body formation. This provocation was *temporal*. Epigenesis concerned *when* matter attained to form, generation's tense, and it amounted to what Jacques Roger has called "the reintroduction of duration . . . into the science of life."[34] The prefix *epi* that Harvey conjoined to his theory of generation denotes "upon, at, or close upon (a point of space or time)" (*OED*), and in describing his intervention Harvey incessantly stressed the term's *temporal* dimension. An embryo's "parts are not fashioned all together at the same time, but emerge successively," "some parts are born after other parts," "this thing coming after that."[35] Marking out fetal development in a numbered series of "Inspections," his book serialized the embryo into a succession of forms, a "dayly progress" under way in a present shared by observer and observed. In so doing, Harvey's experiments and others like them turned the interval of individual embryonic development now known as "ontogeny" into the standard preoccupation of early life science. And it was this fundamentally temporal provocation that sparked epigenesis's host of controversial conjectures regarding the causal powers and principles responsible for embryonic formation: John Turberville Need-ham's *vis plastica*, Pierre Louis Moreau de Maupertuis's particulate attrac-tion, Georges-Louis Leclerc de Buffon's *moule intérieur*, Caspar Friedrich Wolff's *Bildungskraft*, Johann Friedrich Blumenbach's *Bildungstrieb*, and so on. From Harvey to Blake, the basic provocation of any theory of epigen-esis was to represent living forms as form*ing* in the present progressive, in mundane natural-historical time, against opponents who represented form as fixed since Creation.

Indeed, in the seventeenth and eighteenth centuries, research such as Har-vey's elicited a powerful counterreaction from natural philosophers sensitive to the threat epigenesis posed to the divine prerogative of conferring specific form and skeptical of the epigenesists' groping attempts to define the shaping agency immanent to living matter. In response, naturalists such as Jan Swam-merdam, Lazzaro Spallanzani, Charles Bonnet, and (ambivalently) Albrecht von Haller put forward theories of "pre-existence" or "pre-formation" that were more compatible with both the prevailing deistic understanding of a cosmos set into lawful motion by a (now) aloof Prime Mover and the intel-lectual allure of infinitesimal calculus.[36] Preformation theory responded to the provocation of epigenesis's processual present tense with a rival, past-perfect temporality: the finished bodies of each member of each generation of each species had been formed by God at Creation. In the most memorable ver-sions, these forms were stowed like nesting dolls in progressively smaller min-iature within the sperm or ovaries of each species' first parents—"like the cups of a conjurer," Erasmus Darwin would scoff—ready to unfold and expand at the proper time.[37]

As Richard Blackmore put their case in *The Creation: A Philosophical Poem Demonstrating the Existence and Providence of a God* (1712): "ev'ry Foetus bears a secret Hoard,/With sleeping, unexpanded Issue stor'd."[38] Whereas Harvey chronicled the serial emergence of the "whole edifice" of an animal from a single point, Blackmore's didactic epic recast that "vital Speck" as an animal already whole, if a little "rumpled":

> And as each vital Speck, in which remains
> Th'entire, but rumpled Animal, contains
> Organs perplext, and Clues of twining Veins;
> So ev'ry Foetus bears a secret Hoard,
> With sleeping, unexpanded Issue stor'd;
> Which num'rous, but unquicken'd Progeny,
> Clasp'd and inwrap'd within each other lye:[39]

Preexistence theory was then frequently known as "evolution," and Blackmore's "Clasp'd and inwrap'd" progeny, progeny that need only "unravel and untwist/Th'invelop'd Limbs, that previous there exist," play on the close kinship the theory retained with *evolution*'s etymological sense of "unrolling."[40] "Evolution" in this earlier modern, preformist sense described the way insensibly minute and transparent organs would gradually "unravel" and "expand" over the threshold of *visibility*, rather than existence. "Some people," as Charles Bonnet put it in terms that recall Blake's own sensitivity to the problems of "Minute Particulars" and of "Translucence or Opakeness," "do not reflect that minuteness and transparency alone can make these parts invisible to us although they really exist all the time."[41] Just because we cannot see organization doesn't mean it isn't there.

When Blake, on a commercial engraving commission in the early 1780s, produced a tiny portrait of Albrecht von Haller (1708–77), he drew out the lineaments of one of the most influential protagonists in the eighteenth-century epigenesis debates and in experimental physiology more generally (fig. 2). Haller's prodigious midcentury research on the "sensibility" and "irritability" of animal organs, nerves, and fibers had fueled pan-cultural interest in the distinctive properties of living matter, furnished a physiological dimension to the cult of sensibility, and provided fresh impetus to epigenesist theories of generation.[42] In the words of the biography Blake illustrated, Thomas Henry's *Memoirs of Albert de Haller, M.D.* (1783), Haller "traced the formation of the chicken from the instant in which the first change in the egg is perceived, and the vital speck begins to dilate, to that when the little animal quits the shell in which it has been formed."[43] Attesting that Haller "saw the organs successively spring up before his eyes, acquire life and motion; saw them transformed and perfected," the text effectively glossed and endorsed epigenesis, not least the theory's characteristic stress on the contemporaneity of this process and its observation.

Over the course of his career, though, Haller (the subject of Blake's portrait) in fact embodied the generation controversy, ultimately repudiating his epigenesist stance on the conviction that mere physical and chemical laws could not explain "by what means the rude and shapeless mass of the first embryo is fashioned into the beautiful shape of the human body"; a body "so exquisitely fitted for its proper and distinct functions" that "no cause can be assigned for it below the infinite wisdom of the Creator himself."[44] In a challenge to Buffon at the height of the midcentury debate, Haller argued that the theory of epigenesis lacked a "Building Master" ("Es fehlet ein Baumeister"),

ALBERT DE HALLER.

FIGURE 2. William Blake, portrait of Albrecht von Haller [Albert de Haller] (1783). Frontispiece to *Memoirs of Albert de Haller, M.D.* by Thomas Henry (Warrington, Printed by W. Eyres, for Joseph Johnson, St. Paul's Church Yard, London, 1783). Photograph: Courtesy of Widener Library, Harvard University.

an intelligent supervisor to "construct the scattered microscopical parts of the body according to the wonderful plan of the human body, who would prevent an eye from ever sticking to a knee, or an ear to a forehead."[45] But let's pose the question of generation's "missing Building Master" to Blake.

"*We* form the Mundane Egg," he responds decisively in *Milton*: "And every Generated Body in its inward form,/Is a garden of delight & a building of magnificence,/Built by the Sons of Los."[46] Indeed, in *Milton* (c. 1804–11) and *Jerusalem* (1804–c. 1820), "Los," Blake's avatar of poetry and prophecy, "continual[ly] builds the Mundane Shell."[47] He, his estranged partner Enitharmon, and their myriad sons and daughters are pictured tirelessly hammering, beating, blowing, weaving, spinning, knotting, binding, fabricating, molding, carving, and ornamenting tissues and organs within the egg into shape. In *Milton*, "Thousands & thousands labour" to industrially produce its embryonic organs: "The Bellows are the Animal Lungs: the Hammers the Animal Heart / The Furnaces the Stomach for digestion."[48]

Blake's illuminated books, that is, turn the experimentally embroiled egg into an ongoing social construction site, as though for all the numerous competing accounts of the "forces," "drives," and "powers" controlling generation within mundane eggs, what is really "missing" is an account of the force of historically contingent labor, social, and sexual relations in the formation of new bodies and their sensoria:

> Let Cambel[49] and her Sisters sit within the Mundane Shell:
> Forming the fluctuating Globe according to their will.
> According as they weave the little embryon nerves & veins
> The Eye, the little Nostrils, & the delicate Tongue & Ears
> Of labyrinthine intricacy: so shall they fold the World[50]

In scenes of generation such as this, Blake magnifies the mundane egg into an inhabited world, suggesting that the shape of a new being's body *in ovo* is neither preformed by God nor the product of its own self-making. Rather, the "little embryon" is a fabric of many hands, "Forming" under the aegis of numerous social actors who precede it in a "World" that its "delicate" senses are folded to fit and verify.

Indeed, it is a hallmark of Blakean scenes of generation to show the embryonic rudiments of sensation—epigenetic in so far as they are "forming" and "fluctuating" in the present progressive—taking shape from industrial materials and under social pressure. Here resemblance between generations and the exquisite fit between form and function are assured, that is, not because the embryo preexists, but because its circumstances do.[51] This now counterintuitive point is worth lingering over, given the present association of biology with determinism, with "nature" as the inflexible (hard, even "hard-wired") side of the nature-culture divide or continuum.[52] In Blake's texts, the body natural—a "fluctuating," nervous, and sensitive body furnished by epigenesist life science—is more tractable in its organization than the social surround, which has extraordinarily consistent and effective means of perpetuating itself across generations. (As examples of such means, the above scene invokes the gendered division of labor, the force of parental instruction, and the system of industrial textile production.) The scene's two implicit, presiding puns, "looms" for "wombs," and factory labor for the labor of childbirth, drive home the point that a host of historical circumstances enforce the biological regularity by which "like produces like."

In the Blakean inflection of development by epigenesis, animal life archives and perpetuates these amniotic, sexual, and industrial relations, quite literally incorporating them in the shapes of its organs. Blake thereby exercises an early aptitude of biology not for "naturalizing" historical contingencies but for conceptualizing and narrating processes of "naturalization." One would be hard pressed to find a better exhibit of Pierre Bourdieu's notion of body *hexis* than Blake's "little embryon": its development rather exactly chronicles the "apprenticeship which leads to the em-bodying of the structures of the world, that is, the appropriating by the world of a body thus enabled to appropriate the world."[53] And, in fact, this strain of Romantic epigenesis was requisite not only for the deep historicism of later evolutionary theory but also, less familiarly, for the coming sciences of sociology and historical materialism.[54]

Blake glosses this chicken-and-egg problem by punning on "fluctuating Globe" in the above passage, which means first the "embryon" yolk gestating "within the Mundane Shell," and then the "World" this body is formed to fit and discover. In the Blakean corporeal "apprenticeship" that begins (as embodiment does) in the womb, sensorium and sensible world con-figure each other, their congruity guaranteed by practices that knit bodies and structures into mutually appropriate and appropriative relation: "According as they [the Sisters] weave the little embryon nerves & veins / The Eye, the little Nostrils, & the delicate Tongue & Ears . . . so shall they fold the World."[55] Thus Blake cheekily points out that the world proper to epi-generated bodies will be an "Egg form'd World": in his poetry the "Mundane Shell" is also our terrestrial sky, described alternately as "lovely blue & shining" and as a "mighty Incrustation" of institutional norms and forms (for sexual, political, economic, and spiritual life). At stake is the kind of confining *and* indispensable Shell-ter that the living continually demand, build, and require for their survival in Blake, a structure no less given, stubborn, and factual for being exposed as also "built" ("constructed").[56] In the next chapter, Goethe will insist that "the whole activity of life requires a covering to shelter it . . . Whether this envelope appears as bark, skin, or shell, everything that emerges into life, everything that acts vitally, must be enveloped."[57] In Blake, too, liability to this kind of sheltering limitation may be essential to biological viability.

As Blake's scenes of embryogenesis stress by focusing tightly on the social (indeed, industrial) development of the organs of sense—"The Eye, the little Nostrils, & the delicate Tongue & Ears" that will "fold the World" as they have been folded—it is in sensation that the figural activity and susceptibility of living bodies are at their most undeniable, and the risk and potential of complicity between biology and poetry at their most acute. Any poetry of the senses will be "politically ambivalent" work, as Noel Jackson points out: though Blake famously calls for the emancipatory unbinding of sensuous energy, he also tends to depict sensory formation as disturbingly "coeval" with "the enslaving experience of modern historical time."[58] If, for Goethe, the act of seeing a new object promises to "open a new organ in us," in Blake the novelty and benignity of this transaction are never secure: just as frequently, sense organs overactively "fold the world" to fit their contours and over-passively "become what they behold."[59] *The First Book of Urizen*, to which I now turn, shows how such extremes of biopoetic power and vulnerability arise from the failure to admit sensuous life and its science as collaborative effects that transpire, wittingly and unwittingly, between the poles of autonomous action and passive determination.

c. From Epi- to Autogenesis

Incessant the falling Mind labour'd / Organizing itself

BLAKE, *The Book of Los* 8.49–50

The synthesis of epigenetic theory best known to Romanticists derives, appropriately, from Kant. In the "Critique of the Teleological Power of Judgment," pendant to that of "Aesthetic Judgment," Kant argues that natural philosophy was lately delineating a specific type of being for its object: the "organized and self-organizing being," whose "self-propagating formative power," he set out to show, was "not analogous with any causality we know."[60] Unlike material beings—defined, in Kant's view, by contingent obedience to "blind mechanism"—the organism appears "related to itself reciprocally as both cause and effect." The finished whole seems to function as the governing *telos* behind the parts' formation, yet also first to come into existence as their consequence.[61] As an autonomous "causal nexus" unparalleled in the domain of inert material bodies, the organism merits investigation under the rubric of its special, autotelic insularity. And organisms thereby serve to authorize teleological thinking in the epistemology of science, a regulative "way of judging" not otherwise justifiable.[62]

Kant's argument for the autonomy and singularity of organic causation proved highly useful for establishing the disciplinary legitimacy of an autonomous science of "life" in the early decades of the nineteenth century. His ginger, heuristic notion of autotelic organic form was emboldened in the research of Göttingen school naturalists, extended to the edges of earth and consciousness in idealist *Naturphilosophie*, and richly appropriated in the natural and aesthetic philosophy of Coleridge and his disciples.[63] But the proposition that organisms are constituted by their own "self-organizing" power needs to be understood as but one answer, among many, to the challenge raised by epigenesist biology. "Kant has a role in that historical constellation," as John Zammito rightly insists, "but not as a coherent master model."[64] Historians of science tend to represent strong-form organicism as one theory among others; moreover, they point out, as did Goethe at the time, that Kant in effect converted the *problem* motivating early biology (whence organization?) into its *premise* (organisms are self-organizing).[65] In this clever organicist reformulation, *auto-*, the prefix of self, consequentially displaces *epi-*, the prefix of sequence.

But despite Frederick Burwick's early admonition that "organicism" was a partial and polemical "ism" at the turn to the nineteenth century, among Romanticists the Kantian notion of organic form—as form "without recourse to extraneous causes"—has retained pride of place.[66] Deftly identifying in organicism the logic of "the literary absolute," Helmut Müller-Sievers criticized epigenesis (synonymous, for him, with "self-generation") as the very condition of possibility for Romanticism and its ideological resilience: "only organically can the interminable chain of causes and effects be bent back onto its own origin, and only as organic can a discourse claim to contain all the reasons for its own existence."[67] And Denise Gigante recently celebrated the aspiration toward organic form—natural, national, and textual—as the hallmark of "Romanticism as a shared intellectual project." In Blake's case and in general, she contends, this project seeks the "pluripotent, even totipotent" vital power to regenerate fresh wholes "from the ground up in the manner of epigenetic particulars," rather than by "tinkering with" the "existing structures and organizations that constitute society."[68]

But neither assessment of epigenesis as "self-generation" accounts for the version of epigenesis toward which Blake has already begun to point us: one in which the language of epigenetic embodiment—the language of bodily tissues and lines taking shape in the present tense—permits the poet to describe a being intricately enmeshed in, rather than triumphantly free from, "existing structures and organizations." That is, Blake's texts remind us that in releasing the embryo from the divine predetermination defended by preformist theory, epigenesist discourse did not only or always deliver it cleanly into autonomous self-determination. Here, epigenesis turns the forming body over to the numerous powers operating in a given historical present, articulating a kind of being exquisitely, even dangerously susceptible to manipulation: "helpless it lay like a Worm / In the trembling womb / To be moulded into existence."[69] In these lines, "existence" follows, rather than precedes, the embryo's conformation to circumambient pressure, as though liability to physical and social "moulding" were the very condition of its life.

The heroic, idealist myth of autonomous self-generation has a name in Blake's books: "It is Urizen." Such, at least, is the assessment of the spectators in *The First Book of Urizen* who observe Urizen's "Self-closed, all-repelling" generation of a body and world with "horror."[70] Pinpointing self-reflexivity as the rhetorical and philosophical lynchpin of the powerful organicist conception of life, Blake's text parodies that idiom with a flurry of reflexives: "Self-closd," "self-contemplating," "self balanc'd," "self-begotten," "In anguish dividing & dividing."[71] The book simultaneously establishes Urizen as the uncompromising, lawgiving, paternal God—"no flesh nor spirit could keep / His iron laws one moment"—and aligns his oppressive and life-denying political theology with the uniquely "Self-closed" causal circuit of organicist self-generation.[72] Vaunting his autonomy and singularity—"I alone, even I!"— Urizen speaks an organicist rhetoric: "Life is I myself I!," Coleridge would enthuse in a letter soon afterward.[73] Yet in *The Book of Urizen*, self-reflexivity redounds in totalitarian simplicity, rather than the infinite complexities of the Romantic absolute: "One command, one joy, one desire, / One curse, one weight, one measure / One King. one God. one Law."[74]

Working "ceaseless round & round," Urizen produces a "dark globe" that is at once our "pendulous earth" and an epigenetic "globe of life blood trembling."[75] Blake's equation of "striving, struggling," "beating," "surgeing," "Heaving" vitality *with* Urizen's moral and physical laws of "iron" is a pointed intervention.[76] It neatly undermines the shared strategy of vitalist discourses from Xavier Bichat to Bergson: recourse to "unfathomable life," as Donna Jones puts it, that pits "raw, unverbalized, lived experience" against "the petrification of social forms" and "sedimented categories and schema."[77] Though Blake scholars frequently repeat the gesture, Blake's refutation thereof is exact.[78] His books instead picture organs that simultaneously self-generate *and* petrify, as when Urizen's ears "Shot spiring out and petrified / As they grew"; or when, in perfect parody of organic autonomy, Urizen builds his own "petrific" womb.[79] One falls into law-bound, sociocultural constraint and the raw, pulsing, self-organizing body at the same time. And against the persistent view that vitalism is mechanism's opposite, *The Book of Urizen* captures a suspicious isomorphism—a "Conglobing"—between quivering "Globule[s]" of organic "life blood" and the solid particles of a mechanical universe.[80] From this perspective, the two ontologies alike reduce to an individualist and identitarian logic, whether of the (etymologically in-dividual) atom or the "self-begotten" organic life.

But for all his triumphal crowing, Urizen's *Book* makes clear that he did not, in fact, begin alone. Much of his story is told from the perspective of "myriads of Eternity" who surround him, commenting with a mixture of horror, grief, and disdain on his curious form of self-inflicted torment. Lest Urizenic *self*-organization be mistaken for life, generation, or organization in general, their voices declare Urizen "unprolific!," "Unorganiz'd," and "Disorganiz'd."[81] They recontextualize Urizen's flesh as torn from a prior social and somatic fabric: "Disorganiz'd, rent from Eternity." His "Sund'ring" sets a pattern in the book for tissues whose injured loose ends (nerves, fibers, spines) "writh[e] in torment / Upon the winds."[82] Meanwhile, the "tearing" that Urizen is philosophically and constitutionally unfit to acknowledge revisits his world in "tears." An unfelt wound weeps from his back, "Like a spiders web, moist, cold, & dim"; it forms a net that entraps the denizens of the world he engenders.[83]

The Book of Urizen reveals Urizen's heroic form of Romantic autonomy— "Urizen so nam'd / That solitary one in Immensity"—to have been "self"-generated at high cost.[84] The "vacuum" in which this act transpires was once an inhabited space, less a physical fact than a forcible expulsion that the poem laments in sad, simple present progressive:

> Departing; departing; departing:
> Leaving ruinous fragments of life
> Hanging frowning cliffs & all between
> An ocean of voidness unfathomable[85]

Here Urizenic life, the life of autonomous self-organization, is achieved at the expense of lives that took shape differently. The lines register this predicament in a strange prosopopoeia that allegorizes Blake's work of animation over and against the Urizenic kind: the enjambed middle lines render

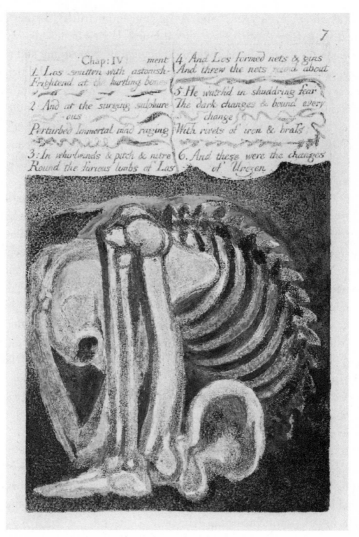

FIGURE 3. William Blake, *The First Book of Urizen* (1794), Copy D, Plate 7. Typical attitude of Urizen; examples continue in figs. 4 and 5. Contrast with fig. 6, below. Photograph: © The Trustees of the British Museum.

the cliffs' frown an effect of inadmissible forms of life glimpsed "hanging" at the margins of Urizen's world(view). Here, Blakean figuration purports to harbor "fragments" that are outcast rather than unreal: "Portions of life; similitudes / Of a foot, or a hand, or a head" that are less imaginative projections than injured forms rendered scarcely perceptible, livable, or tolerable ("Frightning; faithless; fawning") under a given epistemic regime.[86]

F I G U R E   4 . Typical attitudes of Urizen (*continued*). William Blake, *The First Book of Urizen*, Copy D, Plate 9. Photograph: © The Trustees of the British Museum.

That regime, I have argued, enforces a literally monomaniacal love of "one" that stretches from atomist physics to organicist physiology, and from subject formation to authoritarian rule ("One King, one God, one Law"). And yet *The Book of Urizen* leaves the success of this project vividly uncertain. Even in the above lines, the marginalization of dependent forms of "life" is incomplete, productive not quite of void, but of "void*ness*." In one of the book's most stirring illuminations, Urizen—usually curled, bound, or muscling raw elements around (figs. 3, 4, and 5)—floats passively, arms open, through a plate awash in

F I G U R E 5 . Typical attitudes of Urizen (*continued*). William Blake, *The First Book of Urizen*, Copy D, Plate 29. Photograph: © The Trustees of the British Museum.

blue (fig. 6). If this is Urizen's "ocean of voidness," then it seems to buoy and caress his body, lifting and parting his trademark heavy beard and blurring the edges of his "self-clos'd" shape.

W. J. T. Mitchell correlates the image to lines on the next plate, where, despite the anguished din of physiological "changes of Urizen," we learn that he also floats in a kind of sustaining sea:

All the myriads of Eternity
All the wisdom & joy of life
Roll like a sea around him
Except what his little orbs
Of sight by degrees unfold[87]

In this perspective on Urizen's self-generation, the myriads "of life" return—
were they ever absent?—as his vastly underfelt medium or surround, the fluid

FIGURE 6. William Blake, *The First Book of Urizen* (1794), Copy D, Plate 6. (Compare with figs. 3, 4, and 5 above). Photograph: © The Trustees of the British Museum.

milieux sidelined from the strong organicist version of epigenesis. To glance ahead to the end of this chapter, when Blake draws attention, within Urizen's own *Book*, to the myriads implicated in his purported *self*-generation, he comes into long-distance agreement with another philosopher of epigenesist biology, in her present-day attempt "to speak for the background—the mute, manipulated materials, the featureless surround":

> Sometimes, the peripheral is the political. Ultimately it may seem less appropriate to speak of entities' replicating themselves (thus marginalizing and instrumentalizing everything else) than to say that they may be assembled or constructed again and again as a result of processes in which they play a part but do not 'control,' except when one isolates a part of the process for analysis.[88]

But there were epigenesists who spoke for the surround, material and social, in Blake's time, too.[89]

d. Epigenesis and Milieux

> Many sorrows and dismal throes
> Many forms of fish, bird & beast
> Brought forth an Infant form
> Where was a worm before.
>
> BLAKE, *The First Book of Urizen*, 17.33–36

In setting out his version of epigenesis in *Zoonomia; Or the Laws of Organic Life* (1794), the polymathic naturalist (and sometime Blake employer) Erasmus Darwin argued that embryonic life begins with a "simple living filament," endowed not with self-organizing power but, like Blake's malleable worms, with an exquisite, passive-voiced receptivity to influence. Darwin likens the embryonic filament to the kind of fibril that "compose[s] the mouth of an absorbent vessel," supposing it to be "endued with the capability of being excited into action by certain kinds of stimulus."[90] The surrounding fluids bend and stimulate this receptive "rudiment," already likened to a mouth, into "a living ring" suited to "embrace or absorb a nutritive particle of the fluid in which it swims." Each such absorption, Darwin explains, reshapes the embryo's body, and each different shape entails different desires and capacities for interaction: with every "new form," he insists, come "new powers, new sensations, and new desires," themselves fresh goads to further diversifying engagements between animal and surround.[91]

At stake is a process of recursive conformation in which every interaction leaves a morphological alteration in its wake. In Darwin's theory of epigenesis, living bodies literally gain and sophisticate their organs by experience, their organization instantiating less a self-fulfilling prophecy than a living archive of successive interactions with material milieux. The technical term that arises in *Zoonomia* for the formative relation between living tissue and stimulus is "configuration," and this term is rigorously and capaciously meant. It denotes lifelong, epigenetic processes of *shaping-with* that compound to form not only complex morphological structures during embryogenesis and postpartum development, but also the complex psychological capacities for sensibility, volition, ideation, and social sympathy that accompany those structures in Darwin's thoroughly corporeal psychology. Animal perceptions of motion, time, space, and number, *Zoonomia* explains, are felt changes in the "configurations" between the shape of sensory organs and the shapes that variously impinge upon them.[92] An idea itself, as Darwin defines it, is one such momentary "configuration," one shape assumed by the tips of the nerves of an organ of sense within their constant succession of interactive motions.[93] Epigenesis as "configuration" thus captures the real-physical coincidence of physiology and psychology in Darwin's understanding of animal development from fibril to person: a sweet scientific relay between bodily *figures* and *figures* of thought and speech that Darwin frequently and fittingly chooses to represent in poetic verse.[94]

At nearly the same time, Jean-Baptiste Lamarck (1744–1829), tasked in 1793 by the French Revolutionary Convention with inaugurating the study of zoology at the new Muséum d'Histoire Naturelle, offered a remarkably similar theory of epigenesis just as irreconcilable with vitalist organicism.[95] Setting out to define animal life as an object of study in his influential *Recherches sur l'organisation des corps vivans* (1802), Lamarck began by pointedly repudiating the ontological distinction and narrative agon between bounded organism and inorganic surround posited in vitalist and organicist discourse. With the recent claims of Xavier Bichat in mind, Lamarck asserted: "It is not true, as has been said, that all that surrounds living bodies strives to destroy them, such that if these bodies did not possess in themselves a *principle of reaction*, they would quickly succumb to the effects of the action that the inorganic bodies that environ them exert upon them."[96] In many animals, he continued, life "does not exist except by an external influence," and "this influence, far from tending to destroy them, is on the contrary the sole cause of their conservation."[97] In fact, according to Lamarck, it is through the action of its "environing milieux" (*milieux environnans*) that a body obtains "the first sketches of the order of things and the internal movement that constitutes life."[98] Even more than for Erasmus Darwin, what enables "a little mass of materials" to attain to life and organization in Lamarck's text is a susceptibility rather than a power: an "aptitude to receive," from surrounding circumstances, the "impression *that traces . . . the first lines of organization.*"[99] To be disposed toward life, a body must be of "maximum suppleness"—just consistent enough, Lamarck argues, to detain the fluid *milieux* that course through its permeable mass, sculpting and diversifying its shape, and rendering it capable, at each turn, of new and different absorptions. "With time and all the right circumstances," Lamarck concludes, these processes "complicate and perfect" the simple mass into a viable animal delicately suited to the circumstances that continue to sustain it.[100]

It is at once clear that our later, colloquial, and singular sense of "environment" as an organism's outside is inadequate to the fluid *milieux environnans* at work in Lamarck's and Erasmus Darwin's biologies. Both take interest in the processes by which an embryo's surround constitutes its body, contributing substance, shape, and line in a manner true to William Harvey's initial gambit for the science of epigenesis: that a life might be "formed" by "the same matter by which it is sustained and augmented."[101] In such versions of ontogeny, to distinguish between organism and surround is a temporally contingent gesture, complicated by the processual temporality of epigenesis and the "suppleness" of the body it proposes: in Lamarck's account of organization, even "the internal movement that constitutes life" is a kind of eddy in the surrounding movement, an "intussusception" that waylays circumambient motion into a vital tension that cannot persist without it.[102]

In the work of both naturalists, grasping the literally constitutive, stimulating, and sculpting pressure of material "circumstances" in the production of animal lives—construing animal organs and capacities as responsive records of this impress—supports a critical understanding of the (corpo)real force of social, educational, and cultural circumstances and practices over the courses of animal life. However gratifying it might have been for the humanist camp, then, we cannot credit Blake's grasp of epigenesis as social process "imbodied" to his brilliant, poetic subversion of naively positivist biology. As I mentioned at the outset, Lamarck, who coined the very word *biologie* in the text we have been examining, gave that science the express task of investigating the *c*ooperation and "common source" of "the *physical* and the *moral*": how an animal's "habits, its way of life, and the circumstances in which those individuals, from whom he derives, found themselves" have literally "constituted the form of its body . . . and the faculties it enjoys."[103] The now notoriously mutable animal body posited in Lamarckian transformism lives by way of its susceptibility to circumstance, its capacity to make an organ out of repeated experience, and to archive this history as a physical organization that it transmits to its (equally plastic) offspring.

Lamarck ends the *Recherches* indicting the inequitable distribution of "favourable circumstances" that consigned the "multitude" of modern citizens to "conserve and propagate a multitude of prejudices that subjugate them," much as Darwin felt himself perfectly on topic in extolling abolition and the American and French Revolutions in his own natural-historical texts.[104] As Alan Bewell, Noel Jackson, Noah Heringman, Kevis Goodman, Martin Priestman, and Catherine Packham have variously and trenchantly shown, Darwin's synthetic and wide-ranging literary science consistently foregrounds "the constitutive role of literary, political, and other cultural elements in natural history and philosophy," presenting "a conception of nature . . . inescapably bound up with global commerce, industry and consumption."[105] It did so in a mode made scarce by the process of disciplinary differentiation in the early decades of the nineteenth century.[106] Darwin and Lamarck's tradition of unapologetically political and philosophical life science—unapologetic, perhaps, because it saw politics and philosophy as biologically consequential—was an eventual casualty not only of anti-Jacobin reaction, but also of biology's disciplinary modernization in the purportedly modest, apolitical, antiphilosophical, and antiliterary style advocated by Lamarck's colleague, Georges Cuvier.[107] But not before "radical morphology" took hold among a broad cadre of London medical men, fueling the popular vitality controversy that Romanticists know as the background to *Frankenstein*, and provoking concerted theological and political counterattacks long before Erasmus's grandson published *The Origin of Species*.[108]

Yet Lamarck's stress on the constitutive power of "circumstances," on the one hand, and of "actions," on the other, makes his approach to viable form, like that of Blake and Erasmus Darwin, more interesting than the mere inversion of the organicist line that would substitute environmental determination for vital autonomy. For recall that Lamarckian zoology—whose famous giraffes have gradually lengthened their necks by straining to reach acacia trees—is usually ridiculed for excessive voluntarism, not determinism: for maintaining that, striving to satisfy their needs under changed circumstances, organisms can effect heritable changes in their own morphology.[109] In fact, such logics of life will not sort into autonomy or contingency, power or passivity, "the physical" or "the moral," because they theorized life as transpiring between or athwart such alternatives in a domain of *use, disuse, needs, habit, practice, circumstance* and *desire* articulated in their stead.[110] In Blake, Erasmus Darwin, and Lamarck, this science integrates the stubborn corporeality of social and historical patterning *and* the ongoing capacity of lives so formed to be transformed—through pragmatic changes and improvisations that leap the gap between the pasts congealed in their body parts and the exigency of present circumstance.

In this tradition, Blake's "sweet Science" challenges readers to conceive of biological life in subtle combinations of power *and* liability, spontaneity *and* structuration, that have eluded literary historians' reconstruction of the discourse as one of "vital power" and "self-organization," whether idealistically or materialistically construed. Since, in Blake's books and related zoologies, beings *live* by recursive configuration with relations that precede and condition their coming, the biological life of the body cannot be summoned to adjudicate between given and made, or between natural and social expressions of power. For if, as Saree Makdisi has brilliantly shown, the Blakean logic of "organization" posits "a seamless continuity among the social, legal, economic, and political organizations and the organisms inhabiting the world defined by them," the full, multidirectional force of this continuity eludes its characterization as a problem of "the law's" imposition upon "life," of the law's "attempt to appropriate life's endlessly proliferating creative energies and to confine being into limited forms."[111] The contest between discursively turned ideology critique and "positive, materialist ontology"—where Blake's poetry, Makdisi suggests, would express the Deleuzian-Spinozist promise of "a life of creative *power*" against the Foucauldian actuality of a life "regulated, controlled, constituted, directed, and hence stunted" by discursive, political, and juridical institutions—misses the most interesting lexical and theoretical resources of earlier modern logics of life (not to mention the subtleties of subjection *as* subjectification in Foucault's and Spinoza's serious appreciation for determinacy).

In his poignant essay "The Living and Its Milieu" (1946–47), Georges Canguilhem emphasizes that the oft-forgotten, original value of the notion of *milieu*, which Lamarck was the first to import into biology from physics, was its "fundamentally relative" character.[112] In the eighteenth-century French, *milieu* translated the notion of a rare ethereal medium that supplemented the Newtonian mechanics of central forces: the fluid, penetrating vehicle of gravity's action-at-a-distance between bodies separated in space. For Canguilhem, this background illuminates the "relative" dimension frequently elided in later usage:

> The fluid is the intermediary between two bodies, it is their milieu; and to the extent that it penetrates these bodies, they are situated within it [in the middle of it, *au milieu*] . . . it is only because there are centers of force that we can speak of environment. . . . To the extent that we consider separately the body on which the action, transmitted through the medium, is exercised, we may forget that the milieu is *a between two centers*. . . . [M]ilieu tends to lose its relative meaning and takes on an absolute one. It becomes a reality in itself.[113]

It is helpful to recall that biological environments could be conceived in this way, as penetrating *milieux* that communicate relations of force between multiple "centers," of which the developing body in question is one. With their "subtle fluids" working across each animal lifespan and skin, Lamarck and Darwin posit a transcorporeal substance of relation through which diverse and dispersed orders and scales of force act and react upon one another in the production of viable forms.[114]

For Canguilhem, the notion of *milieu* also opens onto something that distinguishes biology from physics and links it, perhaps counterintuitively, to subjectivity. While in physics the *milieu* contributes to a "general theory of the real"—an aspiration to universal and impartial truth—Canguilhem notices that biological appropriations inflected the term into a kind of circle, associating it with words like *circum*stances and *environner* that express an "intuition of a centered formation."[115] In biology, that is, the concept of *milieu* instantiates a view from and toward a living center for whom *this* set of factors, among the "anonymity of elements and universal movements," compose *its* milieu.[116] The point recalls the Blakean twist on the ancient correspondence between micro- and macrocosm that we read above, in which the World was less the body's analogue than its ineluctably partial world*view*: the "Eye, the little Nostrils, & the delicate Tongue & Ears" were said to "fold the World" just as they themselves had been folded. Biology, Canguilhem argues, is dignified rather than diminished by this partiality of perspective and by the inability to exempt the knower from the problem. Suspecting that science "is rooted in life," biology "perpetuates a permanent and necessary relationship with perception," with the problem of looking out as a subject of need and a center of force.[117]

As though accentuating the feature Canguilhem describes, Blake's scenes of epigenesis obsessively thematize the generation of organs of perception in particular, as well as the transformative (configurative) presence of an observer. And in Blake, observation frequently entails involuntary and morphologically consequential forms of sympathetic mimicry: in an archetypal scene, Los witnesses Urizen's generation and "became what he beheld," sprouting organs to match those of the forming body under view.[118] As Blakean infants from the *Songs of Innocence and of Experience* forward testify, vital "becoming" in this logic of life depends upon "beholding" in at least two senses. Beings develop ("become") amid relations of sensing and being sensed that, in Erasmus Darwin's terms, "confound perception with imitation" and make basic physiological formation contingent upon sensory stimulus and mimicry (cf. "Infant Joy").[119] Moreover, life and language cannot be abstracted from their dependent beginnings in what Herder called "the husks of human need, in a cradle of childhood, in swaddling clothes"; put simply, nobody can "become" without *being held* (cf. "Infant Sorrow").[120] In these senses and more, a certain strain of Romantic life science—the epigenesist science of *milieux*—facilitates Blake's vehement critiques of the organicist logic of life that substantiates Urizenic natural and political theology. In this, Blake, like Lamarck and Erasmus Darwin, draws on a particular risk and resource of biological thinking that Canguilhem identified at the inception of that science: constitutionally unable to exclude normative questions of value and meaning, biology sometimes becomes a study of the dependent partiality of both living forms and the project of knowing them.[121]

e. Beholding

If we take Urizen as organicist physiology's paradigmatic experimental object—the self-organizing life form—then this object finds its scientific subject in Blake's early parody of his comparative anatomist acquaintance John Hunter, who makes a cameo as "Jack Tearguts" in "An Island in the Moon" (1784).[122] *The First Book of Urizen* shows Urizen's self-organization to be a forcible denial of social form, his weeping organs a tearful product of unacknowledged tearing. In this context, "Tearguts" is legible as Urizen's perfect counterpart, an anatomist who evinces a truly Urizenic insensibility to organ tearing (and tears):

> [H]e understands anatomy better than any of the Ancients hell plunge his knife up to the hilt in a single drive and thrust his fist in, and all in the space of a Quarter of an hour. he does not mind their crying—tho they cry ever

so hell Swear at them & keep them down with his fist & tell them that hell
scrape their bones if they dont lay still & be quiet—[123]

Juxtaposed, Jack Tearguts and Urizen typify two poles of a damaging dynamic between the subjects and objects of life-scientific experiment: the co-
constitutive dynamic between "self-clos'd" living form (the Urizenic body)
and the aggressive experimentalist who seeks to break or cut it open (the
Teargutsian physiologist). This relation drives Blake's most famous anatomical scene, in which Jerusalem tries to explain that violently exposing a body's
interior will harvest only despair and melancholy for its knowledge:

> Why wilt thou number every little fibre of my Soul
> Spreading them out before the Sun like stalks of flax to dry?
> The Infant Joy is beautiful, but its anatomy
> Horrible ghast & deadly! nought shalt thou find in it
> But dark despair & everlasting brooding melancholy![124]

The "everlasting brooding melancholy" that issues from this anatomy is am-
biguously attributable to the anatomist and to the being he dissects in the passage: it belongs to them both in the science of living form as originally closed
and violently disclosed. Thus the anatomist discovers in his object ("*in* it")
the same set of adjectives that characterized Urizen: "Horrible," "deadly,"
"dark," and "brooding." Urizen curls his body into a globe; Jack Tearguts
curls his hand into a fist.

But there is a minor life scientist in *Milton* whose experimental habitus,
whose way of holding his hands, finds something fortuitous in the inevitable
relation of mutual "configuration" between experimental subject and object.
Antamon, whose name anagrammatically recalls and (slightly) reforms anatomy and its atomism, is not an apocalyptic figure, nor does he redemptively
reintegrate Urizenic bodies and anatomized Infants back into the prelapsarian social fabric. Rather, like that of many of the beings laboring busily in
Milton, Antamon's activity is a compromised and compromising work of minor rehabilitation. Blake, whose works have noisy apocalyptic pretenses ("Six
Thousand Years / Are finishd. I return! both Time & Space obey my will"),
is not typically thought to give much time to mundane and palliative labor.[125]
And yet, in *Milton*, the poet Los and his progeny are out less to shatter the
mundane egg than to repair, maintain, and refurbish its organs and structures.
In language that recalls the "ruinous fragments of life" exiled from Urizenic
genesis, they set to work giving "a name and a habitation" to amorphous and

fragile entities "With neither lineament nor form but like to watry clouds."[126] These marginally real beings would pass for "nothing," without the shaping attention that establishes "form & beauty around the dark regions of sorrow"; they otherwise stand "on the threshold of Death / Eternal."[127]

What follows is a form of experimental contact, contouring, and lineation that allows these beings to collaborate in the biopoetic animation and figuration that affords them livable, perceptible form. Privileging the bilateral sense of touch over "ocular inspection," Antamon's "sweet Science" cultivates the "holding" in "beholding" against the violent pursuit of life in the organic depths of a body's interior and the melancholy inability to recognize knowers as shaped by the object they seek to know.[128] Notice how the passage begins with the melancholy affective hallmarks of anatomy in the Teargutsian style and transfers them into a form of sweetness:

> Others [of the Sons of Los]; Cabinets richly fabricate of gold & ivory;
> For Doubts & fears unform'd & wretched & melancholy
> The little weeping Spectre stands on the threshold of Death
> Eternal; and sometimes two Spectres like lamps quivering
> And often malignant they combat (heart-breaking sorrowful & piteous)
> Antamon takes them into his beautiful flexible hands,
> As the Sower takes the seed, or as the Artist his clay
> Or fine wax, to mould artful a model for golden ornaments.
> The soft hands of Antamon draw the indelible line:
> Form immortal with golden pen; such as the Spectre admiring
> Puts on the sweet form; then smiles Antamon bright thro his windows
> The Daughters of beauty look up from their Loom & prepare.
> The integument soft for its clothing with joy & delight.[129]

Antamon, of the "beautiful flexible hands," appears only three times in Blake's work, notably in *Europe* where his mother, Enitharmon, calls out to him: "I see thee crystal form, / Floting upon the bosomd air; / With lineaments of gratified desire."[130] S. Foster Damon's *Blake Dictionary* defines Antamon as "the male seed," and it is indeed wonderful how many models of generation Antamon's brief appearances collate: the materialist position that looked to inorganic crystallization as an analogue for life; Harvey's famous image of the forming body as its own clay *and* potter; the preformist topos of an embryo unfurling from a tightly compressed seed; Buffon's mind-bending notion of "internal molds" for each of the new being's parts; and, in the Spectre's sudden duplication, the scandal of asexual reproduction without differentiation.[131]

What Antamon takes in hand with the Spectre above is, then, among other things, a series of metaphors for ontogeny, bringing these models into touch with Blake's characteristic insistence upon the sociability of achieving embodiment through generation. But the scene is rare in Blake's oeuvre for representing this interaction as reciprocity and pleasure. Unlike Cambel and her sisters, whom we saw "Forming the fluctuating Globe according to their will," here the fluctuating Spectre consentingly "puts on the sweet form." And there are numerous ways in which Antamon receives as much "sweet form" as he bestows: to begin with, it is first in conjunction with this Spectre that Antamon assumes organized body at all—in *Europe*, he was a cloud, among the form*less* and endangered class that *Milton* worries over.

Readings of the figure are rare, but Leo Damrosch has called Antamon "a sexualized expression of the Imagination," since "to draw outlines at all, in a symbolic system in which outline is male and body or matter female, is to be a progenitor."[132] Though the gender argument does not hold, given the delineating power we have seen Blake concede (however spitefully) to women in matters of generation, Damrosch's emphasis on "drawing outlines" is deft. Antamon is legible not only as a "progenitor" here, but also as an anatomist or experimental physiologist, thanks to the interior-exterior ambiguity of the passage's central act of lineation. The "line" Antamon draws could also mean "a 'cord' of the body," a nerve or fiber (as in Cowper's "She pours sensibility divine / Along the nerve of every feeling line" [*OED*]). A similar ambiguity in the word *draw* means that to the inky lines left by Antamon's pen we can add the kind of drawing that pulls or induces something *outward*: "drawing out-lines," as Damrosch has it, would become "drawing *out* lines," and Antamon's "golden pen" can be read as the more incisive instrument of an anatomist or engraver.[133]

That "line" also tended to connote spun flax and its fibers heightens the status of Antamon's drawing as a counterpart to the anatomy of cruel exposure: "Why wilt thou number every little fibre of my Soul / Spreading them out before the Sun like stalks of flax to dry?" The opposition is precise: whereas Albion seeks to reveal (internal, self-closed) life by prying the body open and finding "nought," Antamon's drawing covers as it dis-covers, drawing out a bodily fiber in a way that also produces a new garment, a soft "integument" (membrane) that the Spectre puts on. The scene thus manages to dramatize knowledge of the body natural as both found and made, drawn out in a way that does not reduce to either pure construct or pure fact. The line originates both within the Spectre's body and within Antamon's pen, belonging to observer and observed as a negotiable boundary that determines the shape of each. Here both livable form and its science are, between fact and construct, outcomes of bilateral *figuration* that is anything but "mere": the indispensible, integumental "clothing" by which anything comes to be viable or graspable, to and among others, physically and cognitively, at all.

Antamon's uniqueness as an anatom(ist) resides in his hands, "beautiful, flexible" and "soft" in contrast to Jack Tearguts's driving fist. Maximally plastic, they not only *form* but *are formed* in the experimental process pictured here: hands so soft that they conform to the object of their touch. In this susceptibility to impression, Antamon risks object-like passivity, and, indeed, scientific subject and spectral object converge repeatedly in the passage. Each one's attributes are echoed in the other: in addition to the line that is their shared boundary, the Spectre's similarity to "wax" and "clay" finds its likeness in "soft" and "flexible" Antamon; the Spectre's "golden ornaments" echo in Antamon's "golden pen," its lamp-like eyes reappear in the "bright" look Antamon casts "thro his windows," and its admiration rebounds in Antamon's smile. In an antitype of that scene of compulsory group-embryo-weaving cited earlier, the joint softness and contentment of Antamon and Spectre ripple past them to affect the waiting weavers, who "look up from their Loom & prepare / The integument soft for its clothing with joy & delight." If the melancholy Spectre was a form of life rendered deathlike, unlivable by the (Urizenic) reduction of generation to autogenesis, its new "integument" is a livable figure fashioned in concertedly nonviolent acknowledgment of the relational labor in generation and its science. In such a "sweet Science," the pretense of impartial and unperturbed inspection would give way to a practice that cultivates epigenesis as a multivalent configuration from which no party can emerge unchanged.[134]

I mentioned at the outset of this chapter that there are contrapuntal choruses in *Milton* and *Jerusalem* who insist that the ideal of totipotent, vital creativity advanced by their high-prophetic "Fathers & Brothers" can in fact pose "terrible" threats to life. Those who understand finitude as the condition of their life—"we who are but for a time, & who pass away in winter"—hear, in the ideal of "unbounded" life, a negation of the "earthly lineaments" that bind and define their specific bodies, and thus justly recognize, in the ideals of "Eternity" and "Immortality," a life-threatening hostility to mortality and transience. I want to close this part by suggesting that when these minor and terrestrial voices, voices rather derisively called "weak & weary / Like Women & Children,"[135] demand "bounding," sheltering limitations on the ideal of Life—for the sake of the living—they also recall Blake to his "great and golden rule of art." Art and life alike subsist "by the bounding line," Blake argues in one of his few direct aesthetic statements in prose—"leave out this l[i]ne and you leave out life itself":

> The great and golden rule of art, as well as life, is this: That the more distinct, sharp, and wiry the bounding line, the more perfect the work of art. . . . How do we distinguish the oak from the beech, the horse from the ox, but by the bounding outline? How do we distinguish one face or countenance from another, but by the bounding line and its infinite inflexions and movements? What is it that builds a house and plants a garden, but the definite and determinate? . . . Leave out this l[i]ne and you leave out life itself; all is chaos again, and the line of the almighty must be drawn out upon it before man or beast can exist.[136]

The actions and passions of lineation in this passage migrate from drawing, painting, and engraving (the immediate context of the excerpt) to building and gardening, from epistemology (distinguishing) to ontology (existing), from genus, species, and character to the act of *Genesis* itself.[137] Suffice it to focus for a moment on the "face or countenance," in which "bounding" and infinity suddenly work in concert, rather than opposition: "How do we distinguish one face or countenance from another, but by *the bounding line and its infinite inflexions and movements?*" The phrase makes clear why an aspiration to life "too exceeding unbounded" would be an overreach in Blake: by one of the most basic and generative puns in his oeuvre, to "bound" is not only to bind but already to leap into motion.[138]

Coming into "definite and determinate" form and into "infinite inflexions and movements" at once, the delineated "face or countenance" in this passage exemplifies Blake's rule about life and art as joint effects of line: equivocating between a "work of art," a face "drawn out," and a living being ("oak" or "ox"), the face personifies Blake's point that both varieties of animation are effects of bilateral delineations that meaningfully confuse "be[i]ng drawn out" with being self-moving. Nor is Blake taking purely poetic license by putting the problem of life this way: William Harvey had used much the same language as he inspected the emergence of organized form from the material chaos within his mundane eggs. "The parts are at first *delineated* by an obscure division, and afterwards become separate and distinct organs," wrote Harvey, who was unable to avoid making his own analogy to Genesis, even as he groped toward a theory of immanent, material causation: "As if by the Omnipotent's command, or fiat," the chick was created, except stepwise, by successive "secretion and delineation of parts."[139]

f. Epigenesis, an Epilogue

In this chapter, I have attempted a Blakean pluralization of our understanding of the Romantic era science of living form, arguing that epigenesist science was not limited to theories of internal purposiveness and self-organization, but encompassed theories in which life and shape were drawn from the configurative pressure (damaging or sweet) of the beings and forces that made up an organism's physical, moral, and historical milieu—theories of embodiment for which viability was conceived as the ongoing work of "myriads." But reconceptualizing epigenesis in this way has contemporary effects, too, making visible this Romantic science's recurrence in the turn from genetics to "*epi*-genetics*" in present-day developmental and evolutionary biology.

In a fresh, self-conscious revival of biology's founding controversy, the molecular genetic perspective that was the prolific achievement of twentieth-century life science has, since the millennium, come under criticism for "preformationism" and been subjected to wide-ranging epigenetic revision.[140] At stake are the explanatory limits of the biological paradigm known as the "Modern Evolutionary Synthesis," which drew (Charles) Darwinian natural selection theory, Mendelian population genetics, and molecular biology together under the notion of the "gene" as the fundamental unit of biological identity and heredity. Nobel-winning molecular biologist François Jacob explained the "genetic programme" this way in the early 1970s, as both the vehicle of evolutionary history and the signature of individual life:

> In the chromosomes received from its parents, each egg therefore contains its entire future: the stages of its development, the shape and the properties of the living being which will emerge. The organism thus becomes the realization of a programme prescribed by heredity.[141]

Jacob's echo of classical preformism is unmistakable in our context, above all in the prescriptive temporality that makes living form a fait accompli, secured through an inheritance invulnerable to present processes and powers.[142] Here questions about the shape of the living refer temporally to a formal endowment inherited from the deep past, spatially to the subvisible interiors of germs, and causally to the in-formative power of an anterior logic. ("Modern preformationism," observe Susan Oyama, Paul E. Griffiths, and Russel D. Gray, is just "subtler" than its classical precedent: "The organism is not preformed in the egg, but the information that programs its development is preformed in the genes.")[143] And classical organicism is audible, too, when "each egg" is said to "contain its entire future" in perfect, autotelic sufficiency. In fact, the twentieth-century doctrine of "Weismann's Barrier" worked to guarantee the immunity of genetic information not only from the influence of external environments but from genes' own cellular and somatic contexts. Francis Crick's "Central Dogma" of molecular biology stipulated that "living things are created by an *outward flow* of causality and form from the nucleus," impervious to contextual feedback.[144] And in Jacob's account, the history of biological progress beats a trajectory inward from surface morphology toward that microscopic control center: toward the genetic perspective capable of "distinguishing between what is seen, the character, and something else underlying the character," between "phenotype and genotype."[145]

In other words, the midcentury answer to biology's two-hundred-year-old question regarding the "missing Building Master" behind living forms was the univocal authority of genetic in-formation over the body's material "building blocks." Defined in explicit contrast to cultural and personal memory, the genetic inheritance, Jacob stressed, "does not learn from experience." Redefining living beings as pre-scripted materializations of inherited code, recasting their "shape" and "properties" as consequences of this "underlying" program, molecular genetics effectively mooted the question of ontogenetic development that had motivated biology since Romanticism. Institutionally, the rise of genetics meant the decline of once-formidable disciplines devoted to that question and less dismissive of life's phenomenal appearing—developmental biology, embryology, and comparative morphology—and, with them, of arguments for environmental interaction and learning as causally relevant factors in the physical formation and evolution of living beings.[146] At its limits, molecular genetics tended to press bodily development and life course to the peripheries of biology: Richard Dawkins's *The Selfish Gene* (1976), for instance, cast mortal bodies as mere "survival machines" for immortal genetic "replicators."[147]

"The Century of the Gene" was set to culminate in the triumphal completion of the Human Genome Project; yet as Evelyn Fox Keller has reflected, this 2003 event came as something of an anticlimax. "For decades," she writes, "we were confident that if we could only decode the message in DNA's sequence of nucleotides, we would understand the 'program' that makes an organism what it is." Instead, the project's most significant legacy may be the prolific puzzlement it generated regarding DNA's predictive and explanatory power.[148] For in many areas, genetic sequencing fell instructively short of establishing the expected causal links between the code and particular developmental and medical outcomes. Within molecular biology, this discrepancy has catalyzed the precipitous rise of "*epi*genetics," dedicated to taking on the contextual and environmental processes "in addition to genes" that influence the expression or suppression of genetic information. The name deliberately harks back to classical *epigenesis* as an alternative fit to succeed, by preceding, the Modern Synthesis.[149]

From quarters as diverse as developmental psychology, experimental embryology, immunology, ethology, ecology, "evo-devo," biological philosophy, and feminist science studies, a once-heretical consensus is gaining ground: patterns of gene expression can indeed be modified by environments (intracellular, somatic, amniotic, ecological, cultural), and even by learned behaviors. Such modifications are even *heritable*, prompting some to frankly hail evolution's "Lamarckian dimension."[150] In ecological developmental biology, "developmental plasticity" names the ways environmental cues elicit diverse morphological outcomes from an invariant genotype, keying newborn bodies to their circumstances in both beneficial and pathological ways.[151] Such emphasis on organisms' plastic receptivity proceeds in tandem with new attention to their ability to actively modify their environments and bequeath an "ecological inheritance" to their offspring.[152]

In groundbreaking news from the mundane egg, meanwhile, Gilbert Gottlieb's research on the "psychobiological" embryogenesis of ducklings showed that full auditory development required the external stimulus of birdsong upon the forming ears *in ovo*.[153] The notion that "experience" could "call forth" gene expression and determine physiological outcomes undermined the Central Dogma as well as ethology's traditional division between instinct and learning.[154] As Susan Oyama, preeminent philosopher of the epigenetic approaches that coalesced as "developmental systems theory" in the 1990s, remarked in incidental rebuke of Urizen: "no egg can develop in a vacuum."[155] Her critique of "internalist" (Urizenic) organicism—of development as "self-contained, self-guiding, and self-maintaining," as she puts it—goes further.[156] DST attempts a "new epigenesis" attentive to the constitutive action of multiple *milieux*; in place of the organism or its genome, its researchers take an inclusive "developmental system" as their unit of analysis. What passes between generations, they argue, is not an immaterial code but a "wide range of resources"—including "other people's ideas, actions, values, habits, and beliefs"—"that are 'passed on' and are thus available to reconstruct the organism's life cycle."[157]

As theoretical geneticists Eva Jablonka, Marion Lamb, and Eytan Avital have been arguing for more than a decade, behavioral traditions and symbolic systems are causally crucial for biology when it comes to the transmission of "information" during development and the morphological repetition between generations known as "inheritance."[158] We are witnessing, in other words, a revival of the surprising claim central to Blakean, Lamarckian, and *Erasmus* Darwinian biologies: animal cultures can act as agents of the regularity we associate with their natures, while natural patterns and structures are more variable than we have assumed. And as if to substantiate the Blakean insistence that any being's organs are literally bustling with the activities of these other lives, the Human *Micro*biome Project made headlines in 2010 by announcing that within a "human" body, other organisms outnumber human cells by a factor of ten to one.[159]

I have argued in this chapter that the basic provocation of epigenesis as a recurrent conceptual formation is less the notion of autotelic self-production with which it sometimes coincides than an insistence upon the temporality of ontogeny—"the process by which the sequence of forms that comprise an individual's life history come into being," in Richard Lewontin's words—as biology's irrepressible explanandum. Stressing livable form as an unfinished product of mundane powers and purposes, epigenesists' continuous present tense has counterintuitively historicizing effects: less for thinking phylogenetic evolution across deep, geological time (for this better-documented question has not been my focus), than for representing particular animal life courses—in the shorter, mortal interval of their overlap with contemporaries, parents, and progeny—as mutable registers of material interchanges and powers. This interval raises challenging questions about the production, incorporation, naturalization, and transmission of social forms. Such epigenesis theory cannot properly be called "deterministic" in the way humanists sometimes type knowledge furnished by biology; but neither does the epigenetic body offer a locus of pure indeterminacy or self-authorizing power, inherently resistant to preexisting strictures, as neovitalist Romanticists would like it to do. Taking embodiment as their problem, not their premise, epigenesis, early and late, instead theorizes a being's active and passive shaping into observable fact and viable form.

But perhaps more provocative for our present moment than the redescription of eggs as social construction sites—for who, lately, defends the nature-culture divide?—is the observation that, at the near-simultaneous beginning of Romanticism and biology, science and poetry could be allied and not opposed in this description. Notions of *need, disposition, habit, milieu,* and *circumstance* were the common property of the natural, human, and social sciences in the epoch of their modern differentiation—and they take precedence again in our later modern mode of calling this differentiation into question.

CHAPTER 2

Equivocal Life: Goethe's Journals on Morphology

"Up until the end of the eighteenth century, in fact, life does not exist: only living beings." Or so, at least, Michel Foucault unequivocally and influentially proclaimed in *The Order of Things* (1966).[1] François Jacob, fresh from winning the Nobel Prize for his own research in cellular genetics, amplified the thesis this way in *La logique du vivant* (1970), the first of his many important theoretical histories of biology:

> From that time onwards, there were only two classes of bodies: the inorganic was non-living, inanimate, inert; the organic was what breathes, feeds, and reproduces; it was what lives and is 'inevitably doomed to die.' Organization became identified with the living. Beings were definitively separated from things.[2]

Nearly two decades after the epistemic rupture repeatedly—almost compulsively—alleged to have produced "life" as a discrete and sovereign issue "around 1800," Goethe gave his late attempt to curate and publish three decades of his scientific writings a divided, double structure that seems, at first glance, to exemplify the thesis.[3] This scientific periodical project, published between 1817 and 1824, dutifully divides each journal number into separately bound parts, one for living, the other for nonliving things: *On Natural Science* (*Zur Naturwissenschaft*) comprises mostly meteorological, geological, and optical studies, while its counterpart, *On Morphology* (*Zur Morphologie*) collects pieces concerning "the formation and transformation of organic natures." But an unwieldy, rarely cited umbrella title stretches over this clear distinction between organic and inorganic science, tipping them into indistinct and asymmetrical relation.

The double journal's full name reads: *On Natural Science in General, Particularly on Morphology* (*Zur Naturwissenschaft überhaupt, besonders zur Morphologie*). This chapter is about the subtle maneuver contained in that "particularly," that is, about the stakes of Goethe's apparently anachronistic attempt to recontain the science of life within a more general study of nature. The periodical project gently demotes living natures to an "especially" engaging, but by no means ontologically distinct, set of objects for experimental investigation. Complementarily, the shape-shifting science of *morphology* that living beings particularly elicit and exemplify for Goethe extends past them to encompass many of life's purported opposites: the mineral, the textual, the pathological, and the dead. In the following pages, I examine the way a neo-Lucretian notion of equivocal corporeality licenses this unusually permissive biology, both before and after the much-attested epochal shift to an organicist conception of life. For from his 1780s microscopy experiments on infusoria to the composite life forms (including the morphologist) inhabiting the demi-journal *On Morphology*, Goethe's life-scientific writings, like Blake's by different means, tend subtly to contest the ascendant biological and aesthetic virtue of autotelic "organic" form, as well as its associated protocols of objective and subjective looking. Goethe's genre-bending life writings exemplify the way Lucretian materialism provided an alternative, minoritarian license for the period's prolific indistinction between science and literature, through notions of life and form radically different from the better-known, Kantian and *naturphilosophischen* grounds for this commonality.[4]

On Morphology's opening essay apologizes for that science's composite and provisional form, for the fact that "what I, in youthful spirits, dreamt of as one work, now comes forward as a draft, indeed as a fragmentary collection [*fragmentarische Sammlung*]" of the "sketches of many years" [*vieljährige Skizzen*].[5] The items in this collection not only range from gnomic epigram to osteological treatise but also tend to erode generic bounds internally: between personal essay and scientific trial, between lyric and didactic poem, between the methods and conclusions of experiments.[6] Nor are the journals temporally sequential: the many-year-old sketches of which the apology warns are not just aged, but variously so, since Goethe sprinkles his own and others' recent writings among three decades of previously published and unpublished work. The resulting tissue of multiple times and voices, marked and unmarked, body forth what turns out to be an emphatically late and heterogenous science of life: *morphology* both actively antiquates and disperses the "youthful spirit" of Romantic genius that a younger Goethe helped invent and shifts focus away from the problem of new, embryonic life that, as chapter 1 showed, motivated biology's initial disciplinary formation.[7]

For what that first "apology" frames as a formal failure—*On Morphology*'s status as a disorderly assemblage of disparate pieces and times—emerges in the periodical's course as acutely expressive of its life-scientific contribution. As Eva Geulen has argued in the first studies to take seriously the journal's "disturbingly heterogenous" form as a significant statement about seriality, in biology and narrative alike, the notion of "form" (*morphe*) as the "unifying, closing, and enclosing shape of things" does not survive its encounter with Goethean morphology intact.[8] The science Goethe names "morphology" takes up each entity (morphologist included) as a fractious "assemblage" (*Versammlung*) of "independent beings." Its collective status as living or no—as animal, plant, or picture, Goethe suggests—remains "determinable" rather than ontologically endowed: a matter neither of inexorable determination nor of pure indeterminacy, life in the journals issues from circumstantial affiliations and an indirectly, figuratively apprehensible politics of parts.[9]

I show in this chapter how this neo-Lucretian perspective enables Goethe to practice and theorize a nonvitalist biology: a logic of life that declines to define itself against mere matter, decay, vulnerability, or death, and declines to take organic autonomy as its ideal. Morphology manages to represent life as a condition rather than a power, to turn from self-sufficient integrity toward a proto-ecological notion of contingency and interrelation, and to experiment in theorizing and writing "life" from the perspective of senescent extravagance rather than self-production and procreation.[10] Likewise, Goethe's equivocal approach to matter sanctions a nonvitalist mode of life writing sensitive to the relational, contingent, abnormal, and obsolescent aspects of embodied science. In this kind of life writing, I argue, Lucretian composite form conjoins biological life and post-Revolutionary historical experience: for Reinhart Koselleck's diagnosis of that experience as distinguished by the palpable "contemporaneity of the noncontemporaneous" in fact deftly glosses the most provocative feature of both *On Morphology*'s neo-Lucretian living forms and the form of representation it models as adequate to their metamorphoses.[11]

The first part of this chapter, "Goethe and the Equivocal Matter of *De rerum natura*," situates Goethe in the contemporary life-scientific field via the controversy over "equivocal generation," a controversy whose relevance for the relation between life and teleological explanation in fact stretches from Lucretius to Marx. From the microscopic entities that equivocate between life and nonlife in Goethe's early experimental logs to the complex forms that inhabit the journals *On Morphology*, I argue, Goethe studies beings as collectivities whose vitality depends on timing and who contribute materially to the medium through which they are perceived. Part 2, "Obsolescent Life," shows how Goethe's late life science deliberately revises his celebrated early theory of metamorphosis to carve out a space for nonprocreative activity within the botany of life and to redirect experimental enquiry toward the processes and poetics of senescence and decomposition.[12] What, the essay "Going to Dust, Vapor, Droplets" begins to ask, might life look like from the perspective of the nongenerative, but perhaps communicative, effluvia that mediate *between* beings? What arts of discomposure would be adequate to this view? Focusing on an experiment in which a cut mushroom "draws" its own image in spores, I argue for the credibility within this science of nonhuman acts of representation: that is, of material, neo-Lucretian signs that emanate not just from verbal agents, but from things. This sets the stage for chapter 3, which delves into Goethe's neo-Lucretian poetics and the epistemic attitude he names "tender empiricism": taken together, these dimensions constitute a method of poetic empiricism capable of registering and representing the transfigurative activity (*metamorphosis*) among *res* against the grain of an intellectual culture Goethe diagnosed as "shy of the real."[13]

1. GOETHE AND THE EQUIVOCAL MATTER OF *DE RERUM NATURA*

a. Endlessly Small Points, 1785–86

The roughly three decades of writings Goethe published in the double-journal project *Zur Naturwissenschaft überhaupt, besonders zur Morphologie* are not incidentally coincident with another fraught, thirty-year project in which Goethe was intimately involved: the first, full German verse translation of Lucretius's *De rerum natura*, undertaken by Goethe's classicist friend Karl Ludwig von Knebel, at Goethe's and Herder's prompting, in the1780s, and published in 1821.[14] Now, Goethean science tends to inhabit both sides of the epistemic shifts historians of literature and science place "around 1800": it turns toward "life," but keeps it in the company of other natures; it employs the literary forms of natural-historical representation Wolf Lepenies diagnosed as dying out with Buffon's reputation, but does so in the periodical form emblematic of the new models of scientific authority; it implements techniques of observation that, for Jonathan Crary, pioneer optics' modern entry into "the unstable physiology and temporality of the human body" but, for Lorraine Daston and Peter Galison, typify the prior century's mode of "truth-to-nature" observation.[15] Indeed, this very mediating, sophisticated, and nondogmatic character would seem to exemplify the way of knowing Peter Hanns Reill christened "Enlightenment Vitalism," and yet to do so *after* that attitude's purported "destruction" by "scientism, positivism, and romantic *Naturphilosophie.*"[16]

This resistance to periodization, between Enlightenment and Romanticism, romanticism and classicism, suggestively parallels *De rerum natura*'s untimely (but enthusiastic) reception in Goethe's circle and oeuvre: in effect, one of the materialisms that had fueled radical, French, high Enlightenment culture preoccupied Weimar-Jena during the term of that Enlightenment's palpable obsolescence—during its complex aftermath of terror, restoration, and imperial modernity.[17] At stake in this later context, of course, was a different *De rerum natura*, answering to a different set of needs: less a humanist polemic advocating the power of natural knowledge against the tyranny of superstition than a poetic science so attuned to the historical complexity of the present that it affords a distinctive past to its every particle.[18] This second text exhibits a kind of poetic materialism nuanced enough to undermine any easy antitheses between German and British Romanticism and all things materialistic, atheistic, Enlightened, and French, as well as between radical Enlightenment and post-Revolutionary reaction.

In 1785 and 1786, Goethe conducted a spate of microscopic experiments touching the then-controversial phenomenon of "spontaneous" or "equivocal generation"—the possibility that animal life could generate from "mere *brute matter,*" or categorically different matter, without parents or their specific seed. Like many contemporary amateur microscopists, Goethe prepared a variety of vegetal infusions (cactus, truffle, boletus mushroom, lentil, rye, peppercorn, potato, morel, beer, banana) and exposed them to sunlight in order to observe the resulting proliferation of microscopic forms—"the dance of the infusion-animalcules," as he described it to Charlotte von Stein.[19] Goethe, Knebel, and von Stein were at this time so preoccupied with such research that Herder reportedly joked that they risked "becom[ing] infusoria," warning them not to "let the large-scale world become fatal to [them] just because the little seed- and tree-world is so adorable."[20]

To read Goethe's and other eighteenth-century infusion experiments, with their language of "compounding" and "disintegrating elements" and their irrepressible delight in chronicling the proliferation of endlessly variegated *Körperchen* (little bodies, corpuscles)—"Clouds of moving Atoms . . . Particles of various forms, oval, oblong, and cylindrical, which advanced in all Directions spontaneously"[21]—is to detect the currency of experimental philosophy's older, atomist imaginary in the heart of the eighteenth-century turn to life. Historians of literature and of science typically date and designate that "corpuscular" imaginary as a *pre*biological, Renaissance and New Scientific way of looking: fueled by the recent rediscovery of classical atomisms (especially *De rerum natura*) and advancements in microscopy, seventeenth-century experimental philosophy and poetry alike were fascinated by the alien, motile substructure of the visible world and keen to attribute all manner of actions-at-a-distance to the subsensible interactions of particulate effluvia.[22] But Goethe's infusoria observations, like those of his contemporaries cited above, still bristle with the little bodies that made up corpuscularian natural philosophy—particles (*Teilchen*), globules (*Kügelchen*), bubbles (*Bläschen*), and rodlets (*Stäbchen*) "of great beauty."[23] Above all, Goethe observes "innumerable," "endlessly small" points, looking to these "pointlets" (*Pünktchen*) with increased, though not exclusive or polemical attention to their role as constituents of animation.[24] The "points" (*Punkte*) that are the minimal observable products of his cloudy infusions range in character from those with "the most beautiful round form . . . without the least trace of life," to those "lacking only movement to be considered infusoria," to those "in lively motion," to "Point-animals" (*Punkt-Tiere*).[25]

All matter is not alive in these papers, nor does life inhere in certain types of material as its special, ontological principle or privilege. Instead, basic material minima acquire prefixes, suffixes, and lengthy descriptors that express diverse degrees of liveliness, animality, and motion in diverse contexts and diverse moments of observation. "Points" and "pointlets," for example, are inflected in some of the following ways in the text: points "with no trace of life"; "inanimate," "immobile," and "resting" points; points "whose movement I could not perceive"; points "full of life" and "moving themselves livelily"; points "that moved slowly"; "point-animalcules"; and, rather beautifully, "a few bright, motile Life-points."[26] The notes document the successive, present-tense emergence of increasingly complex animal structures (threads, fibers, tissues, plants, and ultimately animals) from the minimal points at hand, testifying without anxiety to the epigenetic account of generation discussed in the previous chapter. What is striking, though, is that the infusoria notes do so without mention of a governing force, power, or drive, distributing the adjective "lively" with a significantly casual—rather than vehemently *causal*—logic.[27] Written a few years after Blumenbach coined the catching phrase *Bildungstrieb* (formative drive) (1780) and a few years before Kant gave the notion his influential endorsement and epistemological justification in his third *Critique* (1790), Goethe's infusoria notebooks are "especially" interested in sighting animal "Life," but not concerned to assert its state of exception, police its bounds, or attribute its activity to unique forces or powers.[28] The experimental log bears witness to a liveliness gradient rather than a stark antinomy between organic and inorganic being.[29]

b. Life Is Not a Power

This permissive logic of life appears to have endured through to a moment when the limitations of "vitalism" could be explicitly identified and rejected. In 1830, Goethe received, bound, and saved an article from the *Gazette médicale de Paris*, for instance, in which the comparative anatomist (and Lamarck protégé) Étienne Geoffroy Saint-Hilaire rather theatrically broke his long-standing public "neutrality" to come out against "la proposition absolue des vitalistes." Saint-Hilaire glosses vitalism's "absolute proposition" this way: "living beings resist the laws of affinity of brute bodies, and the composites that they form are indebted to other laws than the ones that operate upon chemical mixtures." What Saint-Hilaire diagnoses and denounces above all, in fact, is the *absolutism* of vitalist expression, which proclaims its "vital forces" with such "certainty," "conviction," and "positiv[ity]," he observes, that contributions from "physico-chemical researches" to the study of life are frequently silenced by fiat.[30]

Meanwhile, at the time of Goethe's infusoria experiments, Johann Friedrich Blumenbach's popular treatises on the *Bildungstrieb*, or formative drive, exemplified the vitalist approach, and did so in the rhetoric of absolute conviction that Saint-Hilaire would diagnose as its hallmark: "One cannot be more inwardly convinced of something than I am," Blumenbach wrote in his *Über den Bildungstrieb und das Zeugungsgeschäfte* (1781), "of the powerful gulf that nature has fixed between the animate and inanimate creation, between the organized and inorganic creatures."[31] For Blumenbach, "organic" designates those entities dignified by their drive to self-organize and procreate, their *Bildungstrieb*. Though other, "inorganic" (*unorganisch*) beings may indeed exhibit "extremely regular shapes formed out of previously unformed stuff," such matter is not to be credited with a *Bildungstrieb*. This is because, Blumenbach explains with a circular logic indissociable from the causal circularity of the organisms he seeks to describe, the *Bildungstrieb* "is a life force and as such is consequently unthinkable in the inanimate creation" (*denn der ist eine Lebenskraft und folglich als solche in der unbelebten Schöpfung nicht denkbar*).[32] It is such pronouncements that have supported the thesis that the period's defining epistemic gesture was to radicalize the distinction between living and nonliving beings.[33]

In the "Critique of the Teleological Power of Judgment," partner to that of "Aesthetic Judgment," Kant cautiously endorsed Blumenbach's *Bildungstrieb* as the best available name for the apparently self-generating agency of organic forms (§81; 292–93). As we saw in the previous chapter, Kantian organicism—owing, no doubt, to its privileged position within the 1790 *Critique* that Romanticists have long taken as their canon's philosophical catalyst and key—continues to monopolize critical understanding of the sphere of interaction between literary and scientific cultures, and between poetry and philosophy, in the decades "around 1800."

To review, then: Kant's "Critique of the Teleological Power of Judgment" took the "organized and self-organizing being" as its motivating problem and model object, marked by a "self-propagating formative power" unlike "any causality we know" (§65; 245–46). Set apart from the contingency and passive subservience to "blind mechanism" that characterizes inorganic, material being for Kant, the organism appears to constitute an autonomous "causal nexus," both "cause and effect of itself" (§§65, 61; 244–45, 234). Soliciting investigation under the rubric of their special, autotelic insularity, organisms authorize teleological judgment as a regulative "way of judging" not allowable otherwise (§65; 247). Moreover, as Jennifer Mensch has recently demonstrated, though Kant (unlike his predecessors and successors) grants merely heuristic, regulative validity to organicism in natural science—for the vitalist reason that life proves inexorably resistant to the kind of mechanical explanation that constitutes "determinate knowledge of nature"—the same is not true for his philosophy of mind.[34] Organicism furnishes nothing less than the authorizing logic for cognition in the critical system, which ultimately attributes the coherence of reason and experience to reason's "self-birth": an organicist "epigenesis of reason" that Kant openly champions as the logic through which ideation can be considered "transcendentally free," neither pregiven by God, nor utterly beholden to empirical experience.[35]

The whole of Goethe's changing and multigeneric output on the questions of *Bildung*, morphology, and metamorphosis has to be taken as his response to the epochal problem of living form. But it seems important simply to notice that while many contemporaries attributed formation to a power, force, or drive (*-kraft, -trieb, vim*), Goethe named it an *-osis*. *Metamorphosis* makes shape change a noun of condition, and not a power, and while it may seem trite to focus on the grammar in this way, when Goethe weighed in on the *Bildungstrieb* issue, it was to suggest (much as Saint-Hilaire would) that Blumenbach's grammar was symptomatic of a profound problem. In a short critique of the language of organic powers and materials, called simply "Bildungstrieb" and published in the second issue of the *Morphology* journal (1820), Goethe takes Kant's praise of Blumenbach as a starting point for a close reading of Blumenbach's term *Bildungstrieb*.

Goethe focuses on the way *Bildungstrieb* departs from the phrase *vim essentialem* (*wesentliche Kraft*, essential force) previously advocated by the pioneering experimental embryologist Caspar Friedrich Wolff, devastating Blumenbach's terminology with praise.[36] In replacing Wolff's overly "physical" and "mechanical"-sounding word *Kraft* (force) with the term *Trieb* (drive), Goethe argues, Blumenbach brought the expression to perfection— and thereby laid bare its deep-set anthropomorphic impulse. In substituting "drive" for "force," as Goethe put it,

> Blumenbach achieved the highest and ultimate [form] of this expression; he anthropomorphized the word of the riddle [*anthropomorphosierte das Wort des Rätsels*], and called that which was under discussion a *nisus formativus*, a drive, a vigorous activity [*einen Trieb, eine heftige Tätigkeit*], through which *Bildung* was supposed to be effected.[37]

Goethe goes on to specify how a word like *drive* effects its anthropomorphic trick: it turns the "riddle" about life into an agon between actor and object, conjuring up an agent . . . well, *verbing* . . . upon a passive, material element. He explains this way, arguing that the terminological problem reveals three intractably anthropomorphic habits of thought:

> [First], that in order to observe the matter at hand, a prior activity had to be conceded; and that, when we want to think an activity [*Tätigkeit*], we underlay it with a fit element that it can act upon [*wir derselben ein schicklich Element unterlegen, worauf sie wirken konnte*]; and lastly, that we have to think of this activity with this underlying basis from then on, as existing together and always simultaneously at hand.[38]

"Personified," Goethe continues with proto-Nietzschean (or neo-Lucretian) iconoclasm, this actor-object "monster confronts us as a god, as creator and sustainer, which we are summoned to honor and to praise in every way." In this hidden idolatry, such strains of epigenesis strangely resemble the overtly theological preformism they so noisily oppose, such that the controversy between "epigenesis" and "evolution" outlined in the previous chapter presents a false choice: "words with which we only hinder ourselves."[39]

Once "an organic being has emerged into appearance" (*ein organisches Wesen in die Erscheinung hervortritt*), he argues, we will need "the concept of metamorphosis" in order to grasp its activity: in order, that is, to think transformation without "anthropomorphiz[ing] the riddle" by making form the outcome of an agent or power at work upon a passive, material substrate.[40] Goethe's attempt to handle life as a *condition*, a symptomatic and context-dependent presentation, not a *power*, lays the groundwork for a perspective capable of countenancing impersonal efficacies of shape change and exchange. Here beings change their forms according to a more polyvalent and contingent set of causal relationships than the organicist notion of teleology will permit, and Goethe's writings first assume, and then insist, that apprehending the complex of causes at work on and in living bodies will require holding at bay both teleological judgment and the strict distinction between "organic" and "inorganic" things.

Fittingly, then, one of Goethe's most subtle refutations of the Kantian conception of the organic being as an autonomous "natural end" occurs in a very late essay on the *in*organic science of meteorology:

> Now here is, above all else, the principal point to be heeded: everything that is or appears, lasts or passes, is not to be thought as wholly isolated, wholly naked; one thing is always steeped, accompanied, clothed, enveloped in another [*von einem anderen durchdrungen, begleitet, umkleidet, umhüllt*]; it causes and suffers influences, and if so many beings work through one another [*durcheinander arbeiten*], where in the end is the judgment, the decision about what is the ruling and what the serving thing, what is appointed to lead and what required to follow?[41]

Goethe ultimately suggests, in a proto-ecological vein, that our tendency to view form in terms of "determinations and purposes/ends" (*Bestimmungen und Zwecke*) will fall away in favor of nonteleological consideration of "relations and connections" and the manner in which "one species, if it does not precisely arise out of, at least sustains *itself* in and through, the other."[42]

We can begin to identify here, among the bewildering welter of late-century vitalisms, vital materialisms, neo-Spinozisms, Kantian teleomechanisms, and idealist *Naturphilosophien* in which Goethe has been (justly) implicated, the as yet unnoticed through-line of a Lucretian approach to the problem of form. Its telltale, in the above passage, which begins to set up an ontology in which relation and dependency are primary, is the subtle diminution of the usual philosophical priority of being over appearing and permanence over transience. Goethe instead equitably links *sein* und *schein, dauern* and *vorübergehen*, with "or"—*was ist oder erscheint, [was] dauert oder vorübergeht*—and submits both alternatives to a more general rule: "one thing is always steeped, accompanied, clothed, enveloped, in another." The rule avers that no thing is extricable from its figurative garment, an envelope of otherness that steeps it to the bone. It exposes the designation of agent and patient, cause and consequence, as an imposition of power pertaining to a lexicon of political sovereignty rather than givenness: a "decision" about "ruling" and "serving," "leading" and "following." And it proposes, as an alternative, the complicated task of thinking how "many beings work through one another." Goethe here canonizes figuration, contingency, and otherness as "always" proper to "everything," the way things are as well as the way they seem, the way they last as well as they way they pass. To the epochal question "whence life?" in the line that extends from the multivalent pointlets in Goethe's experimental logs to this posthumously published essay (1833), one can quite rigorously answer that *life depends*, which is why it is needful to delve into the earlier modern possibility of "equivocal generation."

c. Equivocity

Goethe's infusoria experiments bear nonchalant witness to the once controversial theory of "spontaneous" or "equivocal" generation: the theory that animal life could generate, without parental seed or precedent, from mere matter, or from materials of an unrelated kind. Equivocal generation came in two technical varieties in the seventeenth, eighteenth, and nineteenth centuries: the *abiogenesis* of animalcules out of utterly inanimate materials, or their *heterogenesis* out of living or once-living, but categorically *different*, stuff.[43] The examples that seemed self-evident for centuries—for spontaneous generation was decisively refuted only with Pasteur's sterilization and fermentation experiments in the mid-nineteenth century—were the worms or flies that so frequently appeared in rotting materials. *De rerum natura* was the preeminent locus classicus for *generatio aequivoca*: "what prompts thee to believe," asked Lucretius in Creech's popular translation, "That Things, endow'd with SENSE, can ne'er derive / Their beings from INSENSIBLES, and live?" (2.836–38). Surely, he answers, mixing sticks and stones and earth together doesn't produce life. But this macroscopic scale is inadequate to the question: at stake

are miniscule elements of "convenient SHAPE and SIZE" and the differences
that arise with changes in their "POSITION, MOTION, ORDER" (2.847–48).
In a splendid example of the relay between figures and reality in his poetic
science, Lucretius moreover transforms the dead-end macroscopic analogy of
sticks and stones into a convincing *instance* of the microscopic motions it formerly failed to illustrate. Just add water. For "EARTH, or WOOD, or STONE,"
if "fermented by a timely RAIN," will indeed "Grow fruitful, and produce a
num'rous Train / Of WORMS" (2.849–52). As the lifeless bodies decompose,
their atoms "leave / their former SITE and UNION" and may "fall" into an
"ORDER, fit to make an ANIMAL" (2.853–55). The glancing way that we have
seen the adjective "lively/alive" (*lebendig*) modify the bodies under view in
Goethe's text is one hallmark of the contingent causality at issue in this illicit
type of generation. And in *De rerum natura*, it also accounts for the origin
of species: ridiculing the idea that they were "fram'd above," fixed by divine
creation, Lucretius instead insists that the same "EARTH" that still "supports"
the living creatures also "first gave Birth" to them (2.1105, 2.1109–10).

The controversy that once surrounded the now arcane-sounding doctrine
of *equivocal generation* is no longer self-evident. Put briefly, as Karl Marx still
found it worthwhile to do in his 1844 Manuscripts, "*Generatio aequivoca* is the
only practical refutation of the theory of creation." He meant that the theory
was shorthand for the feasibility of a rigorously godless universe that did not
require an immaterial source of form and meaning, as well as, more subtly, for
the possibility that such a universe did not necessarily obey an ironclad chain
of causes and effects.[44] If complex, specific, viable forms could spontaneously
generate from the fortuitous interaction of simpler materials, no transcendent
act of Creation, no formal or final causes, would be required to explain the
diversity and sophistication of existing forms. There might, moreover, be ex-
ceptions to the deep-seated natural-philosophical expectation, at least since
Aristotle, that in nature, as in art, "like produces like." Staring into infusions of
the most quotidian materials, microscopists witnessed "Substance . . . active
beyond Expression, bringing forth, and parting continually with, moving progressive Particles of various Forms, spherical, oval, oblong, and cylindrical";
"specks of mold [that] seem to become transparent and transform themselves
into Inf. animalculae"; "Globe-animalcules [that] often seem to flow together
like bubbles, and to continue on their way as a simple animal."[45] Were they
witnessing difference in kind emerge out of the matrix of the familiar, mere
matter propel itself across the category and species thresholds thought to
separate the Creation?

The fact that Marx would find a moment in the 1844 Manuscripts, and again in *The German Ideology*, to avow *generatio aequivoca* suggests the doctrine's stakes not only for natural but also for *historical* materialism, in the fugitive Romantic complicity of these two discourses that my final chapters, especially, elaborate.[46] Early writings that find Marx arguing strenuously that "History itself is a *real* part of *natural history*" reveal a key way in which historical materialism situated itself in solidarity with old materialist natural philosophy: here, the possibility of a thoroughly natural *and* a thoroughly social history occupy the same, mundane side—the side Marx called "real," "natural, corporeal, sensual, objective"—of an opposition structured instead against theological and immaterialist logics of origination, (re)production, and transformation.[47] As for Blake, naturalizing humans serves, rather than undermines, the project of historicizing them. Thus, perhaps, Marx dignifies the audibly scholastic term *generatio aequivoca* with his adjective of highest praise: "practical." It names the stand on causation that risks millennia of philosophical ridicule in order to reject otherworldly determination of the forms and purposes of biological and social life. But the point for the moment is that this conception of a nature *inclusive* of human, social history requires that *contingency* be inscribed in nature's core.

Thus, as Joseph Priestley representatively fumed in response to Erasmus Darwin in 1803, equivocal generation presents the basic scandal of "*events without a cause*":

> [O]rganized bodies, of specific kinds, are maintained to be produced from substances that could not have any natural connexion with them, or particular relation to them. And this I assert is nothing less than *the production of an effect without any adequate cause.*[48]

In the visible system of nature, Priestley maintains, "causes have a regular connexion": we everywhere observe that like parents produce like offspring via an "organized germ" particular to their species, and the same must hold true for microscopic life. "*Events without a cause*" belong to that department natural philosophy ought to leave to theology—miracles—which is why Priestley declares any doctrine that attributes uncaused variation to the natural world "unquestionably atheism."[49] It is telling to read the important early experimental biologist and Buffon collaborator John Turberville Needham boast of witnessing "exuberating ductile Matter" produce "what I may call a microscopical Island, [of] Plants and Animals" out of a single plant clipping, and then laboriously explain away this world of heterogenous life as a set of licit permutations *within* the "certain and determinate," God-given bounds of a single species, "because the specific *Semen* of one Animal can never be moulded into another." Naturalists took pains to guard against "any Danger upon [these] Suppositions of falling into equivocal Generation," as Needham put it; at the same time, we can hear the way it covertly prepared the way for the species flexibility requisite to evolutionary theory.[50]

And notice that, beyond its specific relevance to early biological theory, equivocal generation was already synecdoche for the basic causal subversion that we would now call, following Althusser's own late work on Epicurus and Lucretius, the "aleatory materialism of the encounter."[51] "Equivocal generation" named the life-scientific permutation of the broader Epicurean materialist (non)principle of contingent causation. *De rerum natura* taught that all the apparent regularities and necessities of this world derive from the fact that loose atoms "knocking together by chance . . . heedless, without aim, without intention," sometimes hit upon a combination that "could become in each case the beginnings of mighty things, of earth and sea and sky and the generation of living creatures" (2.1058–63). Althusser chimes with Priestley and his contemporaries in concluding that the basic scandal of the thesis was to put sheer chance before causation at the origin of the nature of things:

> Epicurus tells us that, before the formation of the world, an infinity of atoms were falling parallel to each other in the void. They still are. . . . [This] implies that, before the formation of the world, there was no Meaning, neither Cause nor End nor Reason nor Unreason. . . . Then the clinamen supervenes. . . . induc[ing] *an encounter* with the atom next to it, and, from encounter to encounter, a pile-up and the birth of a world. . . . The idea that the origin of every world, and therefore of all reality and all meaning, is due to a swerve, and that Swerve, not Reason or Cause, is the origin of the world, gives some sense of the audacity of Epicurus's thesis.[52]

Within *De rerum natura*, the *clinamen* is given without reason, left suitably unexplained. But Lucretius is eloquent and explicit about the ethical imperative of thus holding open a loophole or lapse in Necessity, lest "cause . . . follow cause from infinity," lest "motion [be] always one long chain, and new motion [arise] out of the old in order invariable" (2.251–55). The *clinamen* keeps nature, in its human and inhuman varieties, from "having necessity within it in all actions," allowing it sometimes instead—"at no fixed place and no fixed time" (*nec regione loci certa nec tempore certo*) (2.293)—to "break the

bonds of fate" (*fati foedera rumpat*) (2.254). "Freedom," as Jonathan Kramnick perfectly puts it, "is not so much a condition of the agent as a quality
inherent in the universe."[53]

Equivocal generation thus designated the exhilarating possibility of breaking precedent and necessity within the spheres of ontogeny and phylogeny.
And though common wisdom and traditional scientific historiography declare spontaneous generation defeated with Pasteur's germ theory, modern
theories of life's origins have to grant the aleatory, material abiogenesis of the
first organic compounds, and it is arguably impossible to finally prove the
negative.[54] It is no accident (or an accident in that very special, Lucretian
sense) that the doctrine's "defeat" was a model case for Bruno Latour's postpositivist reorientation of the history of science to accommodate the "relative
existence" of things.[55]

Like most contemporary publications, the one that inspired Goethe's infusoria experiments in the 1780s was guarded on the subject of equivocal
generation. Wilhelm Friedrich von Gleichen's *Abhandlung über die Saamen-
und Infusions-Thierchen* (Treatise on Seed and Infusion Animalcules) (1778),
whose microscopic trials Goethe set out to imitate, asked coyly whether the
fact that his own observation "happens to transgress the bounds of our current concept of generation" *must* make it, "without exception, the theory of
Generatio aequivoca?" But a reader could be forgiven for inquiring, since von
Gleichen next asks if it is not "far more sublime and noble" to envision a God
who "impressed the elements with the power to effect the formation of bodies
through the coming-together of atoms determined for organization" than one
who "pieced them together himself, like a workman of men":

> For we see every day that it is given to the elements to compound and to
> disintegrate according to general laws, and that they are in this way the re-
> pository of the principles—genesis, preservation, annihilation—of all pos-
> sible beings of our planet.[56]

If we put aside the obligatory overlay of divine determination, von Gleichen's
stress on the "coming-together of atoms" that "compound and . . . disinte-
grate" in ways that might produce "all possible beings" is quite technically
true to *De rerum natura*, which, crucially, does not invest its material elements (atoms), or a certain class of them, with life (or virtually any other feature), instead casting "all possible beings," including living ones, as outcomes
of their multimodal collocations.

In marked contrast to the absolutism that marked late eighteenth-century vitalist biological rhetoric, Lucretius gave his technically equivocal doctrine concerning matter and life in appropriately equivocal language:

> nam tibi de summa caeli ratione deumque
> disserere incipiam, et rerum primordia pandam,
> unde omnis natura creet res auctet alatque
> quove eadem rursum natura perempta resolvat,
> quae nos materiem et genitalia corpora rebus
> reddunda in ratione vocare et semina rerum
> appellare suëmus et haec eadem usurpare
> corpora prima, quod ex illis sunt omnia primis.

[For I shall begin to discourse to you upon the most high system of heaven and of the gods, and I shall disclose the first-beginnings of things, from which nature makes all things and increases and nourishes them, and into which the same nature again reduces them when dissolved—which, in discussing philosophy, we are accustomed to call matter, and bodies that generate things, and seeds of things, and to entitle the same first bodies, because from them as first elements all things are.] (1.54–61)

That is, instead of importing Epicurus's Greek *atom* into Latin, *De rerum natura* gives a lush list of alternative translations: matter *and* seeds *and* life-giving bodies *and* first bodies; "*Grundstoff* [raw material] / *Allerzeugende Körper* [all-generating bodies], *die Samen und Stoffe der Dinge* [the seeds and stuff of things], / *Auch ursprüngliche Körper* [also original bodies]," as Knebel's list has it (1.54–56); as Creech brusquely condenses the situation, each "Name[s] / We use promiscuously; the Thing's the same" (1.75–76). In so doing, *De rerum natura* calmly equalizes terms that must have sounded, in late eighteenth-century ears, like loaded buzzwords of the contemporary life-scientific debate—the preformists' "seeds," the epigenesists' organic particles, the equivocalists "mere" matter—retroactively resonating with Goethe's rare sense that such terms were more interchangeable than antithetical. As we have seen, the infusoria papers distribute the adjective "lively" with similar nonchalance. For as Lucretius makes clear by stressing such terminological alternatives as customs of philosophic speech, these are post hoc designations that work backward from the diverse kinds of *res* that have come into (plural, empirically perceptible) being.[57] Here sentience and significance are second-order effects of the configurations between individually inanimate and insensate parts.

As Goethe puts it in a related meditation on the purported ontological difference between botanical and animal life, the "first beginnings" of an animal or plant "are determinable in either way [*nach beiden Seiten determinabel*]": a "life-point [*Lebenspunkt*] fixed, mobile, or half-mobile is that which is scarcely noticeable to our mind."[58] The word *determinable* is noteworthy here: at stake, just at the threshold where human sense strains to catch life in the act, is neither an innate and category-specific power, nor the kind of pure *indeterminacy* that has become critically fashionable since, but a contingent susceptibility to context. The syntax, moreover, suggests that such collateral susceptibility is exactly what Goethe thinks contemporary human thinkers are most likely to miss: equivocity, the state of being "determinable in either way," constitutes "that which is [*das was ist*] scarcely noticeable to our mind," especially given a cultural formation keen to see living forms as "self-organizing," teleological "natural ends."

Inanimate and virtually featureless in themselves, for Lucretius, first elements generate things—vital, material, and textual—by collocating in the plural. When it comes to giving an account of their emergence and destruction, then, both Goethe and Lucretius give much greater weight to the *relations between* elements than to their independent endowments. Once a plural thing, what Goethe calls a "being-complex" (*Komplex des Daseins*) and Lucretius a "contexture" (*contextum*), has come into appearance, the labor of empirical observation and representation is to attempt to adequate its changes of shape: "for," to follow Lucretius, "nothing remains like to itself; everything moves about; Nature alters everything and forces everything to change"; or, in Goethe's words, "nowhere do we find a stable, resting, or closed thing; rather everything fluctuates in constant motion."[59]

Thus Goethe and von Gleichen are particularly interested in the ways minimal bodies compound and disperse, collaborating to form and unform living and nonliving entities. Goethe's notes are at their most expansive and enthusiastic in describing points in relationships of dance, play, collaboration, and urgency, such as the "endlessly small but very beautiful point-animalcules that seemed to linger in a bunch and play with one another." The animalcules of the potato infusion evince a particularly "sociable being," "gliding much more gently up to one another, around one another, returning again, and seeming to sniff one another with their forward, pointy tips," and all in all, "behaving toward one another in a way that would have befitted much more organized animals." (Xenophobes, however: when Goethe adds a drop of peppercorn animalcules from the neighboring infusion, these sociable potato beings curl up and play dead!).[60]

The notion that a living being, when it "obtrudes into appearance," is already a collective of others, their aggregate effect, seems to stem from the microscopic communities observed during the 1780s and to intensify as the *Morphology* journal project takes its notably composite shape. In an introductory essay, "The Intention Introduced," dated 1807 and published in *On Morphology*'s first issue (1817), Goethe explains his notion of a "Being-complex" this way:

> Each living thing is not singular, but rather a plurality; even in so far as it appears to us as an individual, nevertheless it remains an assembly of living, independent beings. . . . Some of these beings are already bound together from the start, some find and conjoin with one another. They divide, seek each other again, and in this way effect an endless production in every way and in every direction.

> [Jedes Lebendige ist kein Einzelnes, sondern eine Mehrheit; selbst insofern es uns als Individuum erscheint, bleibt es doch eine Versammlung von lebendigen selbständigen Wesen. . . . Diese Wesen sind teils ursprünglich schon verbunden, teils finden und vereinigen sie sich. Sie entzweien sich und suchen sich wieder und bewirken so eine unendliche Produktion auf alle Weise und nach allen Seiten.][61]

Bodies, "particularly" though not exclusively living ones, are fractious, unfinished, and heterogeneous forms, shaped by departures, incorporations, and reunions, and subject to a not always sensible politics of parts. *Morphology*, it turns out, is the logic of this kind of plural, *dis*organizing form, and the poetic science suited to its extravagant transformations. Goethe writes that the science requires "keeping ourselves just as mobile and plastic [*beweglich und bildsam*]" as the natures under view.[62]

If the *On Morphology* periodical is any indication, this means summoning all possible genres to match the incessant formal mutations of the objects of study. Less intuitively, and in contradistinction to the early infusoria experiments that have been the focus so far, the *On Morphology* journal offers a set of exceptionally "historical" empirical protocols as suited to this task, a practice of experimentation whose results depend upon periods of latency, revision, and reception.[63] At a moment when natural-scientific journals were proliferating as punctual means of establishing priority of discovery, the *On Morphology* journal curiously foregrounds the place of obsolescence and lapsed time in the "fates" of biological texts, authors, and specimens.[64] In places, the belated and genre-bending *On Morphology* practices, between biology and autobiography, a form of late life writing that casts senescence not only as its inevitable necessity but also as its desirable method. It is to this biopoetical innovation that we now turn.

2. OBSOLESCENT LIFE

d. Going to Dust, Vapor, Droplets

Goethe's late science and poetics of decadent self-dispersion is most graphically at work in an essay called "Dissipation, Evaporation, Exudation" that appeared in the *Morphology* journal's third issue (1820).[65] The apparently minor, disorderly "Dissipation" essay has received little critical attention and been presented dismissively by recent editors as a "loose grouping" of observations that modern science can now identify as pertaining to "asexual propagation in mushrooms."[66] But published on the heels of the *r*eprinting, in the *Morphology* journal's first issue, of Goethe's celebrated *Attempt to Explain the Metamorphosis of Plants* (*Versuch die Metamorphose der Pflanzen zu erklären*) (1790), the "Dissipation" essay, in its very aimlessness, needs to be read as an experiment in displacing the telos of Goethe's earlier metamorphic theory.

The 1790 *Metamorphosis of Plants* treatise had expressly named and excluded certain metamorphoses from consideration—"accidental" or "fortuitous" metamorphoses, metamorphoses effected "from without"—on the precise grounds that taking them up would "displace our aim [*Zweck*]."[67] That aim was heterosexual reproduction, the "apex of Nature" toward which healthy metamorphosis, Goethe then argued, teleologically "ascends."[68] Perhaps, he speculated at the time, "an occasion will arise elsewhere to speak of these . . . excrescences," to revisit, that is, what Theresa Kelley calls "the prohibition against something like queer plant sex."[69] Tracking Lucretian materialism in Goethe makes the "Dissipation" essay recognizable as that occasion, at a distance of thirty years. Here Goethe faces up to the role of contingency in plant morphology, revisiting botanical form without regard for its purportedly procreative aim, and permitting a set of decadent, accidental, and fruitless processes to enter and decenter the category of life.

The essay's German title, "Verstäubung, Verdunstung, Vertropfung," plays on the work of the common verb prefix *ver-*, which indicates an object's transition into the state named by the stem, but can also indicate that the stem action has gone wrong: while *laufen* is "to go" in German, *verlaufen* is "to get lost." This being a (largely) botanical essay, the first term, *Verstäubung*, plays darkly on the central issue of pollination, *Bestäubung*, letting that traditional apex of plant procreation chime instead with a verb for passively gathering dust (*verstauben*), and bringing into relief the place of mere dust (*Staub*) in the word for pollen (*Blütenstaub*).[70] Grouping *Verstäubung*'s solid particles (*Staub*, "dust"), with *Verdunstung*'s gaseous and *Vertropfung*'s liquid ones (*Dunst*, "vapor"; *Tropf*, "drop"), the title follows matter through all three states, bringing botanical effluvia into touch with their inorganic correlates. Indeed, the nonvital inflections of the essay's title justly gloss an essay that, on my reading, experiments in altogether sidelining procreation as the defining issue of botanical life. The phrase "Going to Dust, Vapor, Droplets" better translates the essay's errancy into the material, for here morphology turns its attention to particles emitted in old age, after death, under duress, to excess. Untimely when it comes to sexual reproduction, these effluvia may instead constitute media of perception, communication, and even botanic artistry.

e. Trying Not to Think about Sex

Goethe in fact borrows the term *Verstäubung* from the botanist Franz Joseph Schelver (1778–1832), with whom he collaborated at Jena between 1803 and 1806, and appreciating Goethe's provocation means grasping his essay's sly ventriloquism of Schelver's heterodox botany. Goethe opens by reflecting on the dismal reception history of Schelver's doctrine of *Verstäubung*, which amounted to a full-scale assault on the sexual paradigm that had structured botanical classification since Linnaeus. For Schelver refused the basic assumption that plants reproduced sexually at pollination, when pollen grains produced by "male" anthers were said to fertilize "female" ovaries. Reflecting on sixteen years of resistance (including his own) to Schelver's ideas, the Goethe of the 1820 "Verstäubung" essay is troubled by botanists' stubborn refusal to follow Schelver's invitation to think outside the sexual system. Even in 1820, at the late date of Goethe's writing, a recent review ended by confessing nothing less than "dread" at the implications of Schelverian botany. Sexuality, the anonymous reviewer concluded, "is a thread that runs meaningfully and pleasantly through the whole of nature. Tear it in the Plant Kingdom, and understanding loses footing in nature and natural history."[71]

In the "Verstäubung" essay, Goethe fully confesses his prior complicity in this "*dogma* of sexuality," even admitting to having urged Schelver to keep silent back in 1804, lest his "heretical" idea jeopardize Goethe's own academic reputation by association (212). In fact, the institutional and ideological contingency of scientific progress is thick in Goethe's revisionary text, which repeatedly presents botanical consensus in the language of "dogma," "heresy," "faith," and "conversion," and uses the fate of Schelver's *Verstäubungslehre*—which we might gloss as the "doctrine of miss- or missed pollination"—to exemplify the sterility of untimely scientific publication. As Goethe now acknowledges, his own *Metamorphosis of Plants* adhered "religiously" to the sexual system.[72] Yet in revisiting Schelver's idea, Goethe proves himself capable of dispensing more thoroughly with the analogy to animal hetero-sex than Schelver ever intended to do.

For Schelver's controversial *Critique of the Doctrine of the Sexes in Plants* [*Kritik der Lehre von den Geschlechtern der Pflanze*] (1812) in fact aimed not to discard the "doctrine of the sexes" altogether, but rather to reserve it for the highest—i.e., animal—order of life. The notion of sexually reproductive plants amounted to a category violation within the three-tiered hierarchy of being that Schelver advanced in his *Naturphilosophie*: at bottom was inert, mineral matter, "chained in exteriority and mass"; next came plants, "striv[ing] to free themselves from the earth"; and finally, ensouled and freely moving animal life (74–75).[73] Only such animal life, "turning in on itself," Schelver argued, had internalized, as sexual difference, "the antithesis between male and female" that manifested botanically as a struggle between environment and plant (75, 67). Schelver saw plant life as "the fertile, receptive wife of Nature, whose husband is still the general, external goad to development"—the water, warmth, and light that excite seeds to life in the earth—rather than "an actual male" (66–67).

Reading Schelver, that is, it becomes painfully clear that his botany never got far from the gendered paradigm he proposed to "critique." Instead, in order to elevate procreative masculinity into the unrivaled apex of the kingdom of living nature, Schelver wrote a vast, assiduous botany against plant sexuality because it presumed to find, among mere vegetables, something akin to male sperm (65). Rejecting pollination outright, Schelver argued that the moment of plant conception takes place not in flowers, but in the ground, where general Nature excites his wifely seeds into development (73). The release of anther pollen is not *Bestäubung*, procreative pollination, but *Verstäubung*, a climax of disintegration: the "explosion" through which the material dregs of the "old" plant are cast off to hasten the arrival of the "youthful" seed in the ground (78).

As Goethe notes in his own "Verstäubung" essay, by denying gender difference among plants, Schelver purified Goethean botany of the taint of "something external" (*ein Äußeres*) at the scene of metamorphosis, something "interacting" within, "beside," or "even apart from" the individual plant. Schelver preferred to depict that plant as "raising itself upwards . . . by its own force and power" (214). But there is a gentle parody of Schelver's purism in Goethe's revisitation of both their botanies. Just as there are "ultra"-liberals and "ultra"-monarchists, Goethe remarks, Schelver was an "ultra" when it came to metamorphosis (214): "He assumes the most proper concept of healthy and regulated metamorphosis . . . which progresses, ennobling itself, such that everything material, low, common is little by little left behind," permitting what is "higher, better, spiritual to emerge in great freedom" (214–15).

Goethe's new inflections of *Verstäubung* increasingly avoid this rhetoric of purification and transcendent interiority, following not the upward trajectory of spirit, but the lateral movements of the material "left behind." For as we have seen, Goethean morphology tends to relax, rather than enforce, the bounds between mineral, vegetable, and animal life and their associated sciences, and also to stress that organisms, even the "highest" ones, need to be conceived "in and through" their elemental environments. Looking ahead to the next chapter, on Goethe's "tender empirical" method, we can add to this list that Goethe does not accept the total inactivity of inanimate beings; and, moreover, that he revalues that characteristic Schelver ascribes to feminine vegetables—"receptivity"—as central to an empiricism sophisticated enough to surpass the self-actualizing rhetoric that marked the conjoined discourses of organism and aesthetics in the Kantian style.[74]

But what Goethe seems to have found worthy of imitation in Schelver's thinking is its daring perspectival inversion: with quasi-Copernican bravado, Schelver designated a *different* point as the origin and end of vegetable life (germination in the earth, rather than pollination in the flower). To cast a pollen grain as a particle of death rather than of prolific life—"the moment of maturity, of death"—was to grant significant structure, function, and time to *dying* in the ostentatious "life" of plants. And now Goethe recognizes the gambit as an interesting and important, if unintentional, challenge to his original theory of metamorphosis, which gave little room to death, none to decay, and pride of place to heterosexual reproduction. Belatedly rising to the challenge, Goethe takes Schelver's provocation as a cue to reexamine natural life under the sign of dissipation rather than procreation. In his own "Verstäubung" essay, pollination comes to constitute just one item in a rich catalogue of *non*reproductive scatterings and exhalations—moisture, powders, scents, contagions, and invisible influences—that the morphologist observes among aged, diseased, atypical, and dying plants and animals.

Here plants produce pollen-like dust or nectar-like droplets on parts other than their purported sexual organs and at times other than blossoming: *after* flowering, for instance, reproductively useless "pollen/dust points" (*Staubpunkte*) appear on the *stem*-leaves, not the flowers, of the *Berberis* shrub (215). Corn inflicted with necrosis, Goethe notes in a tone of cautious admiration, is capable of "belatedly" scattering a seemingly "endless quantity of black dust" (216). This kind of emission is permitted not just to reinforce but to redefine, in cheekily political rhetoric, the taxonomic "realms" of nature by extending equal "citizenship" from "regular" to "abnormal" botanical "points":

> So we see that one might well indeed rank anther pollen, to which one would not deny a certain organization, within the kingdom of mushrooms and fungi. Indeed, one has already accommodated abnormal *Verstäubung* there; now one grants the same citizenship to the regular kind. (216)

For anther-pollen grains, this honorary citizenship would be a dubious re-assignment. Calling anther pollen "fungi" strips them of their ennobling analogy to male, animal sperm, relocates them among asexually reproductive spores, and places them within Linnaeus's notorious twenty-fourth and last class of plants, the *Cryptogamia*, a catchall for sexual misfits with "very peculiar, hidden, or unrecognizable fertilizing parts." In this category, Kelley argues, Linnaean botany "hides and harbors the very class of plants that undermines his claim to have created a global systematic based on visible criteria."[75] Dispatching a plant's seed into this suspect space while declining to reclassify the rest of its body, the passage also satirizes the exaggerated importance of sexual organs in orthodox taxonomy. And it renders pollen flexible and cryptic, capable of occupying several positions on a spectrum that stretches from deathly "black dust," to fungal spores, to angiosperm pollen, to, as we shall see, animal "life-points."

In the lengthy case studies that fill out the bulk of his "Verstäubung" essay, Goethe investigates these emissions "that appear against the law" by tracing a complex network of circumstances external to the organism and to organic science. In the case of an old linden tree's prolifically fruitless "exudation," an untimely sweat of nectar that would otherwise have nourished the coming fruit, for instance, the morphologist correlates meteorological observations, the effect of temperature on the plant's capacity to retain fluid, the activity of insects, and chemical analyses of the nectar.[76] Here externality and accident reenter the scene of metamorphosis, formerly a dynamic play of internal tendencies, and lives change form according to an oblique and polyvalent set of causal relationships inconceivable through the paradigm of an autotelic *power*.

Schelver had questioned the assumption that plant reproduction ought to be dignified with the name "sexual." Goethe's lush repertoire of dissipations sidelines sexual reproduction altogether, and with it, the focus on embryogenesis and the organization of *new* life that had been foundational to the epistemological and institutional legitimacy of the young life sciences. What, the morphologist asks instead, might life look like from the perspective of senescence, pathology, and the particulate losses that mediate *between* beings? What formal de-compositions could communicate such processes?

f. Natural Simulacrum

Goethe's most striking instance of communicative decay is the following experiment in fungal self-portraiture:

> One lays a not yet open white mushroom, with a cut stem, on a piece of white paper, and it will shortly unfold itself, and so regularly pollinate the pure surface, that the entire structure of its inner and under folds will be drawn most conspicuously; which illuminates that the *Verstäubung* does not occur here and there, but rather that every fold yields its portion in its native direction. (216)

The case gathers together many of the improprieties evidenced in the essay's other examples: the mushroom performs an unwilled and untimely emission that will produce no progeny; it does so not from a dedicated sexual organ, but from "every fold" of its body; the emission is occasioned by external circumstances and concurrent with decomposition. But what emerges as powerfully here as the reproductive futility of this act is its graphic success: dying, this mushroom makes an image of itself—"draw" (*zeichnen*) is the verb Goethe uses—so that "the entire structure of its inner and under folds" can be seen "conspicuously" on the white paper.

Goethe's matter-of-fact treatment of this act of nonhuman representation ought to occasion some twenty-first-century surprise. Though metaphors of Nature as artist or artisan abound in *natura naturans* philosophies, in such rhetoric a particular fungal shape is itself the artwork: a mode of Nature's unitary, plastic artistry. (There is plenty of this in Goethe.) But this experiment catches a single white mushroom, quite cut off, projecting its likeness on paper. Nor can it be said that a human observer is needed to do the cutting and catching: in a pun never lost in these botanical writings, pieces of paper are called, as in English, leaves (*Blätter*), so that the various powders, vapors, and droplets that accumulate on (plant) leaves as the essay progresses only reinforce the suggestion that there are forms of drawing, painting, and printing at work that are indifferent to the morphologist's looking. And while there are undeniable elements of organicist rhetoric here—one could point to the emphasis on *self*-reproduction and to the wholeness of the image—it is important that in this accidental likeness, the spores behave neither like individually epigenerating points nor like unfolding preformist seeds. Instead, by chance and by aggregate, they make a reproduction in another medium: a granular, atomized image in dust (like charcoal, pencil, or chalk), rather than a progeny in flesh. The "Verstäubung" essay thus renders a body's "points" circumstantially "determinable" in yet another direction: toward signification. The incident exemplifies the credibility, from the perspective of Goethean science, of nonhuman acts of representation: the reality of material icons and indexes that emanate, in the Lucretian style, not just from persons, but also from things (which will also be to say, from persons in and not despite their reality).[77]

In fact, once the mushroom has graphically expressed the possibility, a pattern emerges throughout the loose catalogue that is the "Verstäubung" essay: often, in "acts of scattering dust" (or vapor, or droplets), things leave images on each other's surfaces. Goethe's attention was attracted, for instance, to the linden tree's excessive honeydew because the tree so "regularly spritzed" (*regelmäßig gespritzt*) the stones beneath with its "glossy" and "gummy points" that it produced a circle of "lacquered" (*lackiert*) gravel mirroring the circumference of its branches (219–20). After the mushroom's spore-print comes an example of animal *Verstäubung* in dying flies: "by and by [they] spray a white dust from themselves," with "ever increasing elasticity . . . so that the fine dust shows its traces over an increasing distance, until the resulting nimbus measures an inch across" (217, 217n243). The aging morphologist sounds increasingly impressed by arts of botanical and animal mortality.

The mushroom episode has another notable effect: though initially given as an example of *Verstäubung*, the middle of the passage temporarily drops the *Ver-* prefix that connoted "irregularity" to report that the mushroom will "regularly pollinate" (*regelmäßig bestäuben*) the paper leaf on which it lies. In the next sentence, the term *Verstäubung* returns, but in depathologized form: it now describes a "regular" and thorough full-body activity that just happens to land on paper instead of soil, thus producing a picture instead of progeny. The easy vacillation between "pollination" and "dissipation," polemical antitheses in Schelver's botany, reveals the extent to which the question of the pollen's male sexual potency has diminished in importance. This communication is indexed no longer to genital sex but to the essay's diverse catalogue of particulate attritions; the stark antinomy between organic and inorganic being has given way to a situational dependency by which a particle's status as germ, dust, or grapheme, again, simply *depends*—upon reception, configuration, and timing—rather than an inherent power or impotence.

The mushroom's spore print is a particularly "conspicuous" and artful item on a continuum of particulate effluences that are characteristic of botanical bodies subject to time, weather, influence, and interaction. Goethe chronicles the "effluvia" that "take shape as oil, gum, and sap on leaves, twigs, stems, and trunks"; the "delicate exhalations" upon which insects feed, and the ones that make a plum "appear blue to our eye" (222–23). The catalogue concludes by attesting to the insensible but consequential "atmospheric element[s]" that must communicate "reciprocal activity from plant to plant," since certain plants appear to be able to promote or inhibit each other's growth over short distances (223). Passing from the graphic clarity of the mushroom's sporeprint, to the cloud form ("nimbus") emitted by the dying fly, to these insensible influences, Goethe's essay begins a life science of the material atmosphere *between* transient beings, morphologists included. Contributing to such an atmosphere, laden with everything from odors to images, beings would also seem to contribute to the medium through which they are perceived by the naturalist attuned to their nonhuman communications.[78] It is a startling anticipation of what biosemioticians call the *semiosphere*.[79]

g. Writing Decadent Life

Channeling the "ultra"-organicist Schelver, Goethe's opening sentence positioned decadent activity as a fleeting, self-negating instant in the service of Life's ceaseless forward march: perhaps one could legitimately view the three titular decompositions, *Verstäubung, Verdunstung, Vertropfung,* "as symptoms of an Organization progressing inexorably forward, hurrying from life to life, yes, through annihilation [*Vernichtung*] to life" (212). But Goethe cannot, like Schelver, challenge the heterosexuality of plant generation while leaving the life sciences' founding fixation on generation ("from life to life") intact. Instead, his essay delays the vitalist juggernaut that would hurtle "through annihilation to life" by dilating, indeed, expending its whole length, on the specific instances of "annihilation" that such a notion of supra-individual Organization would subsume. As this happens, a range of activities come into view that are neither heterosexually reproductive, nor definitively self-destructive. They are, as Goethe puts it, "productive in an abnormal way" (215)—not of offspring, but of signs (representations, communications, influences, and perceptions). In *dis*organizing acts of going to dust, vapor, or droplets, things leave their shapes on neighboring surfaces and scatter their influence into the air. By the end of the "Verstäubung, Verdunstung, Vertropfung" essay, "from life to life" has come to connote not the frighteningly inexorable forward march of the species through particular, disposable embodiments, but the lateral transmissions between beings that share the same time.

One notices in Goethe's rhetoric a careful admiration for this casting off of material, an "elastic" ability that actually seems to increase with the body's age.[80] I want to suggest, in closing, that *Verstäubung* constitutes not just a late scientific interest for Goethe, but also a late poetics—an "art of losing," in Elizabeth Bishop's words—that ceases to intend creative generation of whole, self-sufficient works in favor of acts of serial self-dispersion.[81] Just as the late morphology forgoes the teleological version of "life" advocated by organicist biology, so the late style forgoes the analogy between aesthetic and organic forms, situating itself, instead, among the events of communicative, material dispersion that define embodied transience.[82]

In fact, the *Morphology* project begins by characterizing itself as the serial issue of obsolescent life-pieces, "papers [leaves?] . . . in which we were earlier moved to set down a piece of our being." Like many of the most dramatic *Verstäubungsakte* (acts of going-to-dust) of the plants and animals in the essay, the journal's untimely release was instigated by circumstances of duress. Goethe would never have cast his views "into the ocean of opinions," he explains in *Morphology*'s opening "apology,"

> if we had not, in the hours of danger just passed, so vividly felt the value to us of those papers in which we were earlier moved to set down a piece of our being.
>
> So let that which, in a youthful mood, I often dreamt of as one work, come forward as a draft, even a fragmentary collection, and work and be of use as it is . . .
>
> JENA, 1807.[83]

The dateline, "Jena, 1807," has a peculiar effect. Specifying the "hours of danger" to which the passage alludes, the date ties the project's inception to the battles of Jena-Auerstädt, Napoleon's unexpectedly swift and embarrassing defeat of the Prussians of the Fourth Coalition. Weimar was plundered after-ward, and the events both laid bare German civic and military belatedness relative to French modernity and marked the local beginning of the end of the *ancien régime*.[84] But since the prefatory text appeared for the first time in 1817, in *On Morphology*'s first issue, the date also antiquated by a decade the pas-sages a current reader had just absorbed. For Goethe, these "hours of danger" seem to have occasioned a newly provisional and dispersive approach to the science and writing of "life," one keen to keep mortality and senescence at its center, and sensitive to the way a punctual address to the historical present might be shot through with old and distant things.

CHAPTER 3

Tender Semiosis: Reading Goethe with Lucretius and Paul de Man

Once more you near me, wavering apparitions
That early showed before the turbid gaze.
Will now I seek to grant you definition,
My heart essay again the former daze?
You press me! Well, I yield to your petition
As all around, you rise from mist and haze.

[Ihr naht euch wieder, schwankenden Gestalten,
Die früh sich einst dem trüben Blick gezeigt.
Versuch ich wohl, euch diesmal festzuhalten?
Fühl ich mein Herz noch jenem Wahn geneigt?
Ihr drängt euch zu! Nun gut, so mögt ihr walten,
Wie ihr aus Dunst und Nebel um mich steigt.]

GOETHE, *Faust*

The previous chapter examined Goethean morphology as a multimodal refutation of the ideal of autotelic organization in biology. Relinquishing that ideal, I closed suggesting, enabled Goethe to set aside the discipline's engrossing theme of procreation and to notice instead how signification and communication contribute to the processes of shape change and exchange he called "metamorphosis." This chapter pursues this poetic dimension of Goethean morphology, examining the way its protocols of perception and representation resuscitate the lost semiotics that supported Lucretius's classical exercise of materialist poetry as science. In Goethe's neo-Lucretian poem "Dauer im Wechsel" (Permanence in Change) and a series of related morphological essays and epigrams, beings transpire into sensible signs of their constitutive shape and movement. In the first part of this chapter, I unravel this unfamiliar account of sensation and semiosis in Goethe and Lucretius, examining how it differently answers questions Paul de Man made central to rhetorical readings of Romanticism: questions of the relation between linguistic and natural materiality, between symbol and allegory, and between literary criticism and other disciplines. Goethe's poetry and morphology, I argue, follow through on an extraordinary implication of Lucretian poetic science: that the objects of empirical observation, as well as of poetic address, can materially contribute to the figures through which they are known—given, that is, a practice of poetic empiricism open and alert to their participation, in sensation and in words.[1]

In the chapter's second part, I turn to the experimentalist and poetic practices Goethe called "tender empiricism" and "objectively active" thinking: epistemic attitudes geared, I argue, to register such transfigurative activity among beings at the hinge of a long disciplinary transformation that would strip objects of this activity and poetry of its objective claims.[2] The "tenderness" in Goethe's "tender empiricism" is a double-sided virtue that will recall the sweet science of Blake's Antamon: a mode of "beholding" an object in "beautiful flexible hands" that averts the violence of experimentalist practice not only by handling objects gently but also by acknowledging the experimenter's own tractable softness and vulnerability to their touch. Indeed, I argue in what follows that Goethe's "objectively active" poetic science slyly exposes the Kantian ethics of distanced "impartiality" in aesthetic and scientific observation as built to guard against tenderness in the second sense.

Throughout, I hope to demonstrate that the strategic anachronism in Goethe's revival of Lucretian semiotics operates not to restore a prelapsarian system of divinely underwritten correspondences between things and names or a redemptive unity between mind and world, but in anticipation of logics of life and signification whose needfulness we would name only *after* the modern disciplinary separations that his early nineteenth-century writings were intent to resist: biosemiotics, systems theory, and informatics. These discourses are beyond the scope of this chapter, but their spaces and motives are limned, I think, in Goethe's counterdisciplinary poetic empiricism.

1. PHENOMENALITY AND MATERIALITY IN GOETHE AND LUCRETIUS

a. The Skins or Signs of Things

Morph.
Formation, Transformation, Gestalt
Moment of skin-shedding.

[Morph.
Bildung, Umbildung, Gestalt
Moment der Häutung]

GOETHE, *manuscript note*, LA 2.9b

De rerum natura and Goethe's *On Morphology* journals correspond profoundly in their fixation on the skins and surfaces of things. Both texts link them literally and materially with the obtrusion of bodies into perceptibility, to the way, as Knebel's Lucretius has it, a being "comes forward into the realm of light" (*tritt hervor in dem Lichtraum*), or, in Goethe's echo, "comes forward into appearance" (*in die Erscheinung hervortritt*).[3] That is, precisely at the moment when biological beings, as Foucault and Jacob told it, defied the "lines, surfaces, forms, reliefs" that were the self-evident data of taxonomic natural history to "turn in upon themselves" and their own inscrutably self-organizing interiors, Goethe makes a curious case for the scientific pertinence and epistemic reliability of the visible. A morphological axiom insists that everything that "is" appears or "shows," and everything that appears *is*, in radical validation of phenomenal appearance:

Morphology
Rests on the conviction that everything that is must also indicate and show itself. From the first physical and chemical elements to the intellectual expression of humans, we affirm this basic principle.

We refer equitably to all that has *Gestalt* [form, shape, figure]. The inorganic, the vegetative, the animal[,] the human[:] all indicates itself, it appears as that which it is to our outer and to our inner sense.

Form is something moving, becoming, passing away. The doctrine of form is the doctrine of transformation. The doctrine of metamorphosis is the key to all the signs of nature.

[Ruht auf der Überzeugung daß alles was sei sich auch andeuten und zeigen müsse. Von den ersten physischen und chemischen Elementen an, bis zur geistigen Äußerung des Menschen lassen wir diesen Grundsatz gelten.

Wir wenden uns gleich zu dem was Gestalt hat. Das unorganische, das vegetative, das animale das menschliche deutet sich alles selbst an, es erscheint als das was es ist unserm äußern unserm inneren Sinn.

Die Gestalt ist ein bewegliches, ein werdendes, ein vergehendes. Gestaltenlehre ist Verwandlungslehre. Die Lehre der Metamorphose ist der Schlüssel zu allen Zeichen der Natur.][4]

Again leveling the tenacious metaphysical priority of being (*Sein*) over appearance (*Schein*), as well as that of "sense" (*Sinn*) as "inner," intellective meaning over "sense" as "outer," sensory perception, the passage's last two lines position transience—"moving, becoming, passing away"—as the conduit between these terms, whose difference, in the object as in the subject, attests to a lapse of time rather than of reality or veracity.[5] Moreover, the temporal difference manifest in changes of form or shape (*morphé, Gestalt*) is also the difference of signification: in *metamorphosis*, Goethe insists, something "indicates," "signifies," adumbrates or gestures toward itself (*sich andeuten*); it "demonstrates," "points," shows, or displays (*zeigen*). The science of metamorphosis Goethe calls *Morphology* concerns "signs of nature," but these "marks" or "figures" (*Zeichen*) are handled not as divine signatures but as indexes of shape change—as transient and transitive disclosures incident upon "all that has *Gestalt* [figure, form, shape]," rendering signification a feature of physical metamorphosis.

In his ecumenical extension of signification "from physical and chemical elements" to plants, animals, and "the intellectual expression of humans," Goethe anticipates what since the 1980s has been called "biosemiotics," as well as the potential cooperation between poetic figuration and not-exclusively-human processes of shape change and exchange.[6] Biosemiotics begins by relocating peculiarly human communication ("anthroposemiosis") within a more general study of the interdependence of life and signs. It diminishes the privilege of verbal language as the model type of signification (linguistics as the paradigm for semiotics), recontextualizing words as species-specific forms of "zoosemiotics," alongside the "phytosemiotics" of flora and the "endosemiotics" by which bodies internally communicate among various somatic systems.[7] Goethe's best-studied writings on botanical metamorphosis are particularly sensitive to what one late lyric calls "Der Grammatik, der versteckten,/Deklinierend Mohn und Rosen" (The grammar of hidden,/Declining poppies and roses).[8] Yet *Morphology*'s sensitivity to nonhuman signs will be more capacious still, extending past biota to encompass "inorganic" semiosis in the formation of colors, minerals, and clouds. In many late instances, Goethe proves less concerned than the key theorists of biosemiotics with biological exceptionalism, and less concerned than his sharpest biosemiotic interpreter with "the heightened semiotic freedom that is specific to human language" and the dynamic roll of human consciousness in an earth conceived as an integrated, living totality.[9] For the stranger Goethean possibility of a general *hylo-* or *physio*semiosis— shot through with lapses, losses, and gaps that require a thinking of void and incompletion, rather than totality—Lucretius may still be our best theorist.

De rerum natura is in fact a tremendous help for understanding morphology's basic conviction that each thing "must also indicate and show itself." A sheaf of Goethe's notes related to Knebel's translation shows intense work around a fragment of a line from book 2: "[uti docui,] res ipsaque per se / vociferatur" ([as I have taught], and as the thing itself proclaims) (2.1050). Knebel settled on the more emphatic "es spricht die Sache sich selbst durch sich laut aus" (the thing itself speaks itself aloud), true to Lucretius's redundant *ipsa . . . per se*.[10] The way this happens, however, is more like what Goethe calls the morphological "moment of skin-shedding" (*Moment der Häutung*) than like articulate speech. Lucretius explains that all bodies are constantly shedding atomic skins, evanescing into films that hold their shape, for a while, as they copiously disperse into the air. In Knebel's nineteenth-century German voice:

> Also sag' ich, es senden die Oberflächen der Körper
> Dünne Figuren von sich, die Ebenbilder der Dinge;
> Häutchen möcht' ich sie nennen, und gleichsam die Hülsen von diesen;
> Denn sie gleichen an Form und Gestalt dem nämlichen Körper,
> Dem entflossen umher sie die freien Lüfte durchschwärmen.

> [So I say, the surfaces of bodies send
> Thin figures from themselves, the likenesses of things;
> Little skins I'd like to call them, and the sheaths of these;
> For they resemble in form and figure that very body
> From which they've flowed, swarming about through the open skies]
> (4.48–52)

De rerum natura goes on to explain that these airborne exfoliations furnish the data of sensation, transporting visible, audible, palpable, odorous semblances of bodies across the distances between them.

In Knebel's deft digest, "The images contain the reason for the becoming-visible of bodies"; or, to borrow from Goethe's summary of the case in *Color Theory*, "sight occurs because images detach themselves from objects and enter the eye."[11] It is, then, through these literally extravagant signs that bodies appear at all, "indicating themselves," as Goethe put it, to one another. Exfoliating into images, beings decay into each other's senses, perceiving one another by means of an unwilled traffic in "tenuous figures" (*dünne Figuren*) that has metamorphic consequences for all involved. Being a body in time means sending likenesses "swarming through the open skies," figuratively representing oneself in a process that produces the stuff and the possibility of others' temporal and phenomenal experience.

Calling these slight but real "little skins" *imagines, simulacra,* and *figurae,* Lucretius enlists them at once for *De rerum natura*'s physics and for its poetry, a doubly provocative gesture that, like Goethe's linking of metamorphic and semiotic change, affords physical reality to signs and semiotic activity to physical bodies.[12] Not restricted to strategies of consciousness or linguistic effects, figuration is cast as an inevitable aspect of corporeality, the gradual coming undone of particulate bodies that is the necessary corollary of their atomic com-position. Like the "wavering shapes" (*schwankenden Gestalten*) that initiate Goethe's *Faust* by "arising," "approaching," and "pressing in" on the speaker in this chapter's epigraph, in *De rerum natura,* things "ostend" themselves in figures and evanesce (2.1024–25).[13] Lucretius's favorite macroscopic analogies for such simulacra—the cast-off skin of snakes, the bark of trees, the coats of insects (4.51, 4.58, 60–61)—echo through another "important axiom of organization" by which Goethe introduces the *On Morphology* project:

[T]he whole activity of life [*Lebenstätigkeit*] requires a covering [*Hülle*] that shelters it from the external raw element. . . . Whether this envelope appears as bark, skin, or shell, everything that emerges into life [*alles was zum Leben hervortreten*], everything that acts vitally, must be enveloped. And so, in time, everything that is turned outward belongs, bit by bit, to death and decay. The bark of trees, the shells of insects, the hair and feathers of animals, even the epidermis of humans, are eternally self-detached, cast-off envelopes, given over to nonlife [*sich absondernde, abgestoßene, dem Unleben hingegebene Hüllen*], behind which new coverings are always forming, and beneath these, then, superficially or deeper, life brings forth its creative web [*Gewebe*].[14]

As Knebel's Lucretius has it, "there is always something outermost welling up prolifically from things," "ceaselessly flowing and slipping off from bodies" (4.146, 4.145).[15] Goethe suggested earlier in this piece that in contrast to morbid anatomy, which works by dissection, morphology would attend to the "outer, visible, graspable parts, in their connection." But this final passage makes clear that the morphologist's commitment to the visible amounts not to an evasion of death, but rather, as chapter 2 suggested, an intensive engagement with senescence and actions and passions of appearing. Morphology's interest in the "moment of skin-shedding" and the object's capacity to "show itself" has less than expected to do with acts of self-realizing, metamorphic exuberance, of which becoming-butterfly (from physis to psyche) would be the ready example. This "axiom of organization" instead enjoins us to grasp *dis*organization as part of the "activity of life"; to notice the way the newest thing that shows is already aged, "given over to nonlife"; to consider the way beautiful features of living bodies—their hair, feathers, shells—are not strictly alive at all. (As a telling counter-point, Kant had struggled to fit such material "concretions [*Concretionen*]" as the "skin, hair, and bones" into his account of how each part of an animal body must comprise a teleologically integrated organ.)[16] If vitalist physiology preferred to define "Life," in Xavier Bichat's words, as "a permanent principle of reaction" against the death threat leveled by the influence of "inorganic bodies," Goethe's mortalist life science here construes transience as a life's defining motion and even its protective integument, refiguring the skin as a site where organic and inorganic, living and dying, lend themselves to one another.[17]

Not only does the passage, lingering with decaying surfaces, elevate them as worthwhile objects of life-scientific attention; not only does it insist, with the "living creatures" in Blake's later epics, that the "whole activity of life" requires a "covering"; it also suggests that "behind" and "beneath" them are only more "coverings"—and indeed that "life" consists in issuing this figural "web." For if Goethe initially sets up a life/nonlife opposition allied to that age-old, metaphysical (and new, biological) preference for depth over surface, he soon empties the valuation in the blithe equation "superficially or deeper." Here animation and decay belong to the same phase of activity;[18] living is a form of decadent disclosure, a "turning outward" toward "nonlife" that issues, at once, in time and the order of the visible. The morphologist's sense perceptions concern—*are*, in the Lucretian sense—belated vestiges of their living objects, "self-detached, cast-off envelopes," "turned outward" and "given over to nonlife."

As Goethean morphology stresses in a very technical, Lucretian sense when it emphasizes, together, the moment of an entity's *obtrusion into appearance* and its status as a plural collocation or assembly (*Versammlung*) of parts, this broad-based semiotic possibility belongs to the necessary connection between appearing, plurality, and transience in *De rerum natura*. ("Each living thing is not singular," you will recall Goethe arguing, "but rather a plurality [*Mehrheit*]; even in so far as it appears to us as an individual, nevertheless it remains an assembly [*Versammlung*].")[19] According to *De rerum natura*, only atoms are strictly "individual," etymologically, physically, and conceptually "in-divisible," which is also why they do not appear, and why they do not die.

From the Lucretian point of view, to *appear* is possible only in the plural: it requires having parts to lose into the figural act of apparition. This is also to say that things appear on the same condition that renders them subject to disintegration, and that anything that has a figure, a body compounded of parts, emits figures, *simulacra*, as those parts fall away. *Schein*, by this logic, is the decomposition of things into signs. Offering no separate origin story for the figures that distinguish poetic language, those that make up the data of sensation, and the bodily shapes of things, *De rerum natura* seems to understand its poetic substance as an instance and effect—an index—of the world's decadent self-disclosure, its broad rhetoric of evanescent bodies. Lucretius's names for this kind of "self-detaching" integument and sign, *figurae*, *simulacra*, *imagines*, and Knebel's complex of translations, *Bilder*, *Gebilde*, *Gestalten*, are worth retaining because they keep the figural aspect of atomist sensation close at hand, holding open a continuum between bodily shape and trope that invites and advances poetry as the fittest verbal register for the empirical apprehension of realities in their transience.

b. Another "Rhetoric of Temporality"

Yet these terms also make audible the way Lucretian materialist poetics trouble the rhetoric of Romanticism Romanticists have learned from Paul de Man. For by de Man's lights, the collaboration between physics and phenomenal appearances just outlined in Lucretian sensory figuration would be ideology *pure*: "What we call ideology is precisely the confusion of linguistic with natural reality, of reference with phenomenalism."[20] The same can be said for the material, if attenuated, indexical connection between the Lucretian signifier and referent: "It would be unfortunate," writes de Man, "to confuse the materiality of the signifier with the materiality of what it signifies" (11). I want to consider, in this section, whether such confusions in fact take place in Lucretian and neo-Lucretian semiotics, and whether their effects are necessarily ideological.

In de Man's classic essay "The Resistance to Theory" (1982), the above "confusion" is designated ideological in relation to specific premises about where the pretended connection between "linguistic" and "natural" reality resides: among them, that it is the *logical* dimension of language—grammar, rather than rhetoric—that purports to connect the linguistic field to the world in general. (This is because, as de Man glosses the classical history of the disciplines, only the verbal science of grammar was thought to match the nonverbal science of mathematics in its logical rigor, thereby affording language "passage to the knowledge of the world" [13, 14]).[21] Hence the disruptive and counterideological force of "literariness, the use of language that foregrounds the rhetorical over the grammatical and logical function," which destabilizes the logical juncture between words and world with the prospect that language may not be "a reliable source of information about anything but its own language" (11).

A rhetorical trope, de Man argues, "operates on the level of the signifier and contains no responsible pronouncement on the nature of the world—despite its powerful potential to create the opposite illusion" (10). That illusion is named "aesthetic," in that it feigns a motivated fit between the sensuous-"phenomenal" aspects of a signifier and those of the signified; and it is called "powerful" in that, for de Man, "rhetoric" technically names the performative use of language as *force* (9–10).[22] Dovetailing with Jakobson's teaching that the poetic function of language "deepens the dichotomy of signs and objects" through "focus on the message for its own sake," and with Saussure's emphasis on meaning as a product of intralinguistic differences between signs, rather than of their reference to "extra-semiotic" entities in "the world," de Man's mode of deconstructing the illusory relation between words and world gave Romanticists a lasting method and motive for rhetorical reading.[23] Tropes are deconstructed for the self-referential, linguistic materiality that punctures their pretended continuity with "natural reality," temporarily exposing their status as arbitrary impositions of performative power. The inexorable return of the figural illusion in signification constitutes the ideological resistance that permanently motivates theory (19–20).

Yet there is a moment when a different, neo-Lucretian kind of sign may be said to have *almost* entered de Man's incredibly influential treatment of Romantic figures. In "The Rhetoric of Temporality" (1969), de Man makes passing mention of Goethe's neo-Lucretian poem "Dauer im Wechsel," in a near-miss that offers a telling glimpse of the two traditions' differences on the question of natural and signifying materials. Once brought back into view, Goethe's neo-Lucretian trope illuminates some of the unseen premises operative in the demystifying materiality of the signifier as de Man taught Romanticists to wield it, while nonetheless carrying forward his commitment to the contrapuntal labor of rhetorical reading.

In "The Rhetoric of Temporality," the contest outlined above between the perspective that conflates linguistic sign and phenomenal world and the perspective that debunks that illusion plays out *within* rhetoric, between the tropes of symbol and allegory.[24] The essay takes contemporary literary criticism to task for accepting as Romantic literature's signal linguistic achievement the synthesis between subject and object, consciousness and nature, epitomized in symbolic representation. In symbolic diction, de Man summarizes, "image coincide[s] with substance," purporting seamless participation in natural totality and brooking no "disjunction between the way in which the world appears in reality and the way it appears in language" (207, 191). In what has proved a lasting characterization of symbol as the trope of "self-mystification" and of allegory as its historically responsible "unveiling," de Man argued that when early Romantic literature finds its "true voice," it renounces the moment of symbolic unification between "material perception" and verbal sign in favor of a manifestly disjunctive and linguistically self-referential allegorical mode. "If there is to be allegory," de Man argues, "the allegorical sign [must] refer to another sign that precedes it," instantiating a temporal difference between signifiers that is the inalienable condition of the mode (207). Thus Rousseau, as de Man argues in his splendid, signal example, "does not even pretend to be observing," as he constructs Julie's emphatically allegorical garden in *La nouvelle Héloïse*: this is a garden assembled almost exclusively from borrowed elements of *Le roman de la rose*. Allegorical language is thus "purely figural, not based on perception": in fact, it "submits the outside world entirely to its own purposes" (203). The phrases illuminate not only the fundamental antithesis between perception and figuration in de Man's thinking, but also an existential agon between linguistically bound human subjects and the "outside world."

Indeed, between these two tropological uses of language, de Man discerns nothing less than an "ethical conflict," in which allegory "always corresponds to the unveiling of an authentically temporal destiny":

> Whereas the symbol postulates the possibility of an identity or identification, allegory designates primarily a distance in relation to its own origin, and renouncing the nostalgia and the desire to coincide, it establishes its language in the void of this temporal difference. In so doing, it prevents the self from an illusory identification with the non-self, which is now fully, though painfully, recognized as non-self. It is this painful knowledge that we perceive at the moments when early romantic literature finds its true voice. (206–207)

In a movement that becomes protocol for rhetorical Romantic criticism, the possibility of coincidence between self and nonself is the illusion punctured by Romantic literature's "true voice."[25] And we soon learn what kind of "nonself" de Man has in mind. Allegorical disenchantment takes place "in a subject that has sought refuge against the impact of time in a natural world to which, in truth, it bears no resemblance" (206). Though routed through careful and convincing readings of Wordsworth and Rousseau, in this culminating phrase the argument assumes categorical form: in truth, a subject bears no resemblance to a natural world, from which it is riven by language and time. And this metaphysical dictum, I think, undergirds ethical and theoretical mandates throughout de Man's oeuvre: from here to "Resistance to Theory" to the late writings on the special materiality of the letter that "disarticulates" nature and the body, the cardinal ideological illusion is for subjects of conventional language to presume to resemble or partake in the extrinsic "natural" and "phenomenal" world to which they refer.[26] While the next section shows this prohibition aligning felicitously with Kant's sublime extraction of the subject from "nature within us" and "nature outside us," it also rules out of court the rival realist, naturalist, or materialist ontologies that might enjoin a subject to understand diverse entities and processes as constituents of a common universe. (In the Lucretian case, a commonality of chance rather than design: "Nature," as Deleuze has justly written on the subject, as "an infinite sum" that is not a whole, "which does not totalize its own elements.")[27]

A Lucretian *simulacrum* or *figura* of the perceptual-rhetorical kind examined in the previous section fails to sort into de Man's schematic contest between symbols and allegories in several interesting ways. Quite literally a husk of its referent, the atomist sign would seem to carry out the mystifying fusion of material perception and representation that de Man ascribes to the symbolic ideology. But having necessarily traversed a distance of space and time between perceiver and perceived, such a sign also asserts a temporal discrepancy akin to the one de Man reserves for allegorical demystification. Though fragments of the referent arrive, really, the illusion of complete presence is unavailable, not least because the referent alters through each particulate production of an image. And as we have seen, this mode of figuration in *De rerum natura* (as well as in Goethean morphology) is so thoroughly constitutive of phenomenal appearance and temporal transience at once that its very sensuality reveals, rather than veils, what de Man calls the self's "temporal predicament." And yet, according to his schema, must not this very rhetoric of temporality amount to some form of ideological-symbolic recuperation, given the commonality it tolerates between selves produced and bound in conventional language and "natural" nonselves? Are we even talking about language in Lucretius, or merely about sensation?

De Man's argument against the symbolizing self who attempts to borrow, from nature, a permanence improper to him attaches lightly to Goethe in "The Rhetoric of Temporality." Specifically, de Man invokes the title of the Goethe poem "Dauer im Wechsel," which Goethe eventually integrated into the late cosmological cycle that was one product of his long-professed intention to write neo-Lucretian didactic epic.[28] De Man's first description of the synthetic "temptation" of the symbolic view proceeds as follows:

> The movements of nature are for him [Wordsworth] instances of what Goethe calls *Dauer im Wechsel*, endurance within a pattern of change, the assertion of a metatemporal, stationary state beyond the apparent decay of a mutability that attacks certain outward aspects of nature but leaves the core intact. . . . Such paradoxical assertions of eternity in motion can be applied to nature but not to a self caught up entirely within mutability. The temptation exists, then, for the self to borrow, so to speak, the temporal stability that it lacks from nature, and to devise strategies by means of which nature is brought down to a human level[.] (196–97)

What is so mysterious about this assessment is that the Lucretian materialist tradition Goethe summons in "Dauer im Wechsel," the tradition de Man designates as a "temptation," in fact instructs the self in precisely the lesson de Man wishes the self to learn: that there is permanence in nature, but not for us. As *De rerum natura* repeatedly insists, single atoms are indeed indestructible, but the products of their intricate configurations—such as self, sentience, consciousness, and memory—are irretrievably lost at death, when the contextures cease to hold and the composite pieces, as Goethe puts it in "Dauer im Wechsel," "hasten likewise to element."[29] Acceptance of absolute personal mortality is the foremost ethical imperative of Epicurean philosophy, which traces every form of violence and exploitation to (futile) attempts to defend against it.

True to its model, the poem "Dauer im Wechsel" works not to anchor "a self caught up entirely within mutability" in nature's metatemporal core, but as an emphatic instruction in the self's temporal *in*stability. The speaker actually enjoins his "Du" to accelerate her or his passing, to surpass other objects in swiftness of accession to transience: "Schneller als die Gegenstände / Selber dich vorüberfliehn!" (In David Luke's prose translation, "Let yourself speed by even more swiftly than these objects!")[30] After two rather unsurprising stanzas of blooming and fading blossoms and other *carpe diem* topoi, the speaker confronts the addressee directly. "Now you yourself!" (*Du nun selbst!*), as if to say, "Now, as for you . . ."

Du nun selbst! Was felsenfeste
Sich vor dir hervorgetan,
Mauern siehst du, siehst Paläste
Stets mit andern Augen an.
Weggeschwunden ist die Lippe,
Die im Kusse sonst genas,
Jener Fuß, der an der Klippe
Sich mit Gemsenfreche maß.

Jene Hand, die gern und milde,
Sich bewegte wohlzutun,
Das gegliederte Gebilde,
Alles ist ein andres nun.
Und was sich an jener Stelle
Nun mit deinem Namen nennt,
Kamm herbei wie eine Welle,
Und so eilts zum Element.

[Now you, yourself! What, firm as rock
Rose up before you,
Walls you see, you see palaces
With constantly different eyes.
Vanished is the lip
Once restored in kisses,
That foot that scaled the cliff
With mountain goat bravura.

That hand that gladly and gently
Stirred to do good,
The articulated structure,
All is different [another] now.
And what in that place
Now calls itself by your name
Came here like a wave
And hastens likewise to element.][31]

Most relevant and astonishing here is the second quoted stanza's sudden redescription of the beloved's hand (or is it her whole limbed body, or the

whole jointed poem?) as "Das gegliederte Gebilde": a "segmented," "mem-
bered," "compounded," or "articulated" structure—or "shape," "figure,"
"formation"—at once composite, vanishing, and impersonal. Despite the earlier blazon, which rather clinically separates eyes, lip, foot, and hand for serial, third-person description, nothing quite prepares us for an address to the
beloved as "articulated structure" or "compounded figure." The stark, stand-
alone line may appositively redescribe "that hand," or may represent a new
item in the list of emblazoned and vanishing body parts; it can also take in the
whole limbed figure of the beloved, or his/her anatomized image in the poem.
And, of course, the phrase "compound figure" implicates the structure of the
poem itself, an intricate arrangement of linked units on many orders of magnitude and in many sensory registers: a five-part stanzaic image composed of
quadratic stanzas, each comprising eight octosyllabic lines; each line's four
trochees themselves linking letter/word and phoneme/morpheme combinations, and culminating in end-rhymes that make audible another jointed
structure (*abab-cdcd*). Maximally alliterative, the phrase *gegliederte Gebilde* is
itself sensibly compounded of repeating and transposable visual and auditory
units (ge, de, te), aurally and visually mimetic of its modular meaning.[32]

We can affirm, with de Man, that both selves and non-selves partake in a
"paradoxical . . . eternity in motion" here, but an eternity whose exact cost
is any pretense at self-consistency in time. Instead, in this stanza of "Dauer
im Wechsel," a flurry of indexicals—*That, that, all, now, what, that, now,
itself, your, here*—studiously evacuate the beloved's name and person, leaving
a bare set of relational coordinates as a place and time from which to speak.
Through this hollowed frame a substance of dubious identity—"*what* in *that*
place / *Now* calls *itself* by *your* name"—rolls calling toward the speaker, one in a
presumptive series of disintegrating figures that "Came here like a wave / And
[hasten] likewise to element." The compound figure (*Gebilde*) in question
establishes the speaker's present place and time, demarcating a "here" and
"Now" between actions given in the poem's past and vocative future tense.
But the figure does nothing if not impress the speaker with the belatedness of
this perceived present: flowing into images, the beloved changes in the interval of those images' arrival; s/he is "*another* now." And touching the speaker
out of the past, as we think starlight does, the images shadow his perceived
present with temporal difference: "another *now*."

At issue, then, is neither the bad-faith simultaneity of subject-object, mind-
nature, word-world that marks symbolic diction for de Man, nor their salutary
renunciation in favor of temporally conscious, intralinguistic allegorical refer-

ence. Goethe has instead staged the strange and estranging arrival of an atom-ist indexical sign, in its joint capacity as verbal figure and datum of percep-tion. *Gebilde*, as we saw, is one way Knebel rendered the airborne figures of Lucretius's theory of sensation: "Die leichten Gebilde der Dinge / Schnell ab-fliegen, und sich in dem Augenblicke verbreiten" (The light figures of things fly off swiftly and disperse themselves in the blink of an eye) (4.165). Like the word *figure* in English, *Gebilde* ran the gamut of material, mental, and artificial usages in Goethe's period, from concrete, physical forms, to the tenuous ag-gregate shapes of clouds, to delicate drawings, images, and ghostly or dreamt "figments."[33] In this the term rather perfectly captures the range and the work of the Latin words with which Lucretius renders Epicurus's Greek theory of the sensuous *eidola*. *Wallen*, "to flow or undulate," and *[ent]fließen*, "to flow [away]," are among the verbs with which Knebel translates Lucretius's verbs for simulacra-diffusion (*fluo,-ere*, *[dif]fundo,-ere*); they echo in Goethe's phrase, "like a wave."

This collapse of what de Man construes as ontologically distinct "linguis-tic and natural reality," which ought to allow the self "to borrow, so to speak, the temporal stability that it lacks from nature," instead has the reverse ef-fect. It would seem that there are figures, neither symbolic nor allegorical in de Man's sense, for which this indistinction draws out, rather than masks, the temporal discrepancy between signs and referents and the saturation of rhetoric by time. The neo-Lucretian figure positions the poem as one among numerous jointed structures that can be said to last only at the expense of their self-consistency, by way of their susceptibility to comprising "an *other*, now," or "another now."

Such a sign arrives out of a philosophical formation for which physical and poetic, natural and linguistic structures are understood to be aspects of the same natural universe—where "natural" is understood not in opposition to "conventional," but in opposition to divine, and where "universe" capa-ciously denotes the unfinishable sum of extant things and processes, *inclusive* of human culture and convention. And so the physics and poetics of *De rerum natura* each presume and constantly rediscover material gradations between poetic and physical shapes, which shade into each other like the figure of the beloved in "Dauer im Wechsel," whose hand decays into images of its own compound structure. It is possible that the figures of poems—both the "im-ages" their words effect and their structured composition at various orders of magnitude—are not different in kind from the other compound bodies (*geglie-derte Gebilde*) and the appearances they make. But this suggestion requires closer scrutiny.

c. Atoms, Letters, Figures

What might it mean to call Goethe's *gegliederte Gebilde* an instance of "atomist indexicality"? Goethe's poem itself provides a splendid exposition of the notion of indexical signification as Charles Sanders Peirce would come to theorize it toward the end of the century. In contrast to *icons*, which mean by resemblance, and *symbols*, which mean by convention (here fall the arbitrary systems of verbal signs that would be the focus of structuralist linguistics and literary theory's twentieth-century uptake of semiotics), the *index*, Peirce argued, means "by virtue of being really affected by [its] Object."[34] Not necessarily verbal, an index "arises as a physical manifestation of a cause," and even in an artwork, the preponderance of meaning seems to stem from the sign's participation in the continuum of the physical world, rather than its differential relation to other signs, or the artist's creative invention.[35] "The index is physically connected with its object," Peirce writes, and "the interpreting mind has nothing to do with this connection, except remarking it, after it is established."[36] In *De rerum natura* such acts of "remarking," of noticing that which is already "present before our eyes" as itself a notice of unseen causes, are the principal interpretive habit of mind that the poet-empiricist teacher seeks to impart to the pupil: *nonne vides*, "do you not see," Lucretius asks no fewer than thirteen times in the poem, training the student-reader to attend to the ubiquitous unremarked signs that offer tracks or traces for knowledge to follow, *vestigia notitiai* (2.113, 2.124, 2.196, e.g.).

Perhaps the greatest difficulty surrounding Peirce's notion of an "index" has been the heterogeneity of his examples. These are mostly nonverbal: a barometer as a sign of approaching rain, a weathervane as a sign of the force and direction of the wind . . . a sundial, a footprint, a knock on the door, a pointing finger. But Peirce also includes *verbal* deictics, those demonstrative and personal pronouns ("this," "that," "here," "there," "now," "then," "I," "you," "he," "she") that depend on context for their meaning. *De rerum natura*'s whole speculative-empirical project most obviously hinges upon indexes of the first type, relying upon what Peirce would call the "dynamical connection" between the seen and the unseen to uncover realities of subliminal magnitude.[37] For Lucretius and for Peirce alike, the invisible yet undeniable actions of wind and temperature, detectable only in their effects on other bodies, are key examples of relays between perceptible and imperceptible realities and the semiotic force they carry.[38] Yet the logic behind classing such nonverbal indices with verbal deixis is obscure: Mary Anne Doane comments that while in both cases "the index is defined by a physical, material connection to its object," other features make the two varieties hard to think together. While the nonverbal set entails a "disconcerting closeness" to the object, a contiguity or "fullness" that "raises doubts as to whether it is indeed a sign," verbal indexicality would seem to exemplify signification at its emptiest, "a hollowness" contingently awaiting occupation.[39]

It is precisely at this juncture that Goethe's "Dauer im Wechsel," which capitalizes on both varieties of index, illuminates their complicity. As we have seen, the poem's prolific deictics dramatically relinquish the speaker's power to specify and to name, reducing language to the minimal provision of an outward-opening, relational frame. In the poem, an elemental wave pours through this aperture, such that the deictic use of language is shown to enable something profoundly alien—"what"—to protrude through the intimately known and presumptively human body of the beloved; the lips of the poem's addressee, the "thou" to its lyrical "I," begin to ventriloquize an "it" who "now calls itself by your name." The astonishing sequence redefines the quintessentially lyric, vocative power to call, the Adamic prerogative to name (*nennen*), and the subjectifying privilege of saying one's name, in terms of the racing, cresting, breaking expression of broad and powerful physical forces: "like a wave."

Verbal and nonverbal indexicality, the poem teaches, are interdependent: deixis would be the mode, within language, of ceding place to nonverbal indications. Put differently, it facilitates the subsumption or reintegration of language into its purportedly extrinsic context, support, or elemental ground, which surges into constitutive relief, in Goethe's poem, as the speaker's vocative address dwindles into third-person deixis.[40] The indexical artwork as a whole, Rosalind Krauss observes in her classic essay on the subject, "functions to point to the natural continuum, the way the word *this* accompanied by a pointing gesture isolates a piece of the real world and fills itself with a meaning by becoming, for that moment, the transitory label of a natural event."[41] Krauss's comment on American art in the 1970s provides an unexpectedly apt gloss on the Goethean stanza that turns the beloved's name into a "label" precariously affixed to a wave.

It is important that the stanza's first overtly rhetorical figure (in the narrow sense of verbal trope) occurs here, in the simile "wie eine Welle." Two more effects become remarkable at roughly the same time: the poem's interpellation of its reader as the "you" in question ("What in that place / Now calls itself by *your* name"), and a certain special, neo-Lucretian pun on the word *Element*. We'll take the last first, the pun Goethe intones through a preparatory rhyme: St*elle* / W*elle* . . . *Ele*-ment. Famously, the word *elementa* in Lucretius's Latin can mean both the alphabetical letters that compose words and the material minims (atoms) that compose bodies.[42] But the way *De rerum natura* in fact tightly controls this double meaning helps to answer de Man's important challenge regarding the relationship between "the prosaic materiality of the letter" and "the materiality of what it signifies." In sensation, as we have seen, a figural trace of the referent touches the observers' senses. Is the same true for reading?

It is not strictly *im*possible for Lucretius that the "materiality of the signifier"—the atomic components of the ink-printed letter on the page or dark pixels on a screen—formerly belonged to the body of the referent. This would be by chance, rare and not to be relied upon. But the materiality of a letter might nonetheless contain, to return to de Man's formulation, a "responsible pronouncement on the nature of the world," albeit through a different form of resemblance and responsibility. The letter's capacity, in conjunction with other letters (*elementa*), to produce a signified that it neither possesses nor resembles in itself *is like* the atom's capacity, in conjunction with other atoms (*elementa*), to produce phenomena that it neither possesses nor resembles in itself: secondary, tertiary, quaternary "accidents" and "events" of matter in combination. Here is how Lucretius puts it in one of five places in the poem where this likeness between the two kinds of *elementa* is drawn, lest his auditor assume that the atoms themselves "permanently possess that which we see floating upon the surfaces of things [*quod in summis fluitare videmus/rebus*]":[43]

> Moreover, it is important in my own verses with what and in what order the various elements are placed. For the same letters denote sky, sea, earth, rivers, sun, the same denote crops, trees, animals. . . . position marks the difference in what results. So also when we turn to real things: when the combinations of matter, when its motions, order, position, shapes are changed, the thing also must be changed.

> > [quin etiam refert nostris in versibus ipsis
> > cum quibus et quali sint ordine quaeque locata;
> > namque eadem caelum mare terras flumina solem
> > significant, eadem fruges arbusta animantis;
> >   . . . verum positura discrepitant res.
> > sic ipsis in rebus item iam materiai
> > concursus motus ordo positura figurae
> > cum permutantur, mutari res quoque debent.]
> > (2.1013–22)

Whereas for de Man the radical discrepancy between the arbitrary and conventional "materiality of the letter" and "the materiality of what it signifies" debunks the illusory connection between linguistic and natural reality, reference and phenomenalism, Lucretius takes *this very discrepancy* in signification as exemplary of the differential dynamic of reality and phenomenality in general—of the way elements (verbal or nonverbal) in accidental combination can occasion effects of a qualitatively different order.

In this respect, the phenomenon of meaning effected by individually insignificant alphabetical letters is like the phenomenon of life from individually inanimate atomic elements, and the phenomenon of time from individually eternal ones: "So much can elements do, when nothing is changed but order" (*tantum elementa queunt permutato ordine solo*) (1.827). Words and other things are similar in the *dissimilarity* they evince between their simple, elemental and their complex, composite being: a difference made by way of collocations arbitrary in origin, but replicable and routinized through continuity of context and custom, rather than internal necessity. As in Blakean epigeneses, "things which have been accustomed to be born will be born under the same conditions" (*quae consuerint gigni gignentur eadem/condicione*) (2.300–301), without ascribing or inscribing that fate to or in their materials.

In "Resistance to Theory," de Man jokes that "no one in his right mind will try to grow grapes by the luminosity of the word 'day'" (11). But for Lucretius, the same absurdity lodges in the very physics of day, for how *does* luminosity come into being out of particles that singly and technically possess no heat and no light? In *De rerum natura*, too, forgetting the qualitative difference between sensible and insensible, phenomenal and atomic realities produces a cosmic kind of laughter: "if you think that whatever you see among visible things cannot come into being without supposing that the elements of matter are endowed with a like nature . . . it will follow that they guffaw shaken with quivering laughter, and bedew their cheeks and faces with salty tears" (1.915–20). Atoms will laugh at you for thinking they can laugh, the joking poet points out, which laughter would bring the whole idea of "first beginnings" to "an end." To endow atomic primordia with attributes that they in fact obtain only in concert with others and intermingled with void would be to afflict them with a mortal softness: could a "Grin," as Creech has it, "distort their little Face," it would express not indestructible singleness but the permeable plurality that subjects a body to dissolution (1.925).

"Things and words," explains Michel Serres in a suitably poetic *and* fluid dynamical exposition of *De rerum natura*, are "negentropic": under conditions of universal entropy, verbal and nonverbal compounds alike express the possibility of local and temporary retention, much the way a spiral eddy in a river reinstates "its own conditions" within the general flux. Serres is particularly true to Lucretius in advancing, through multiple metaphors, an understanding of signification that refuses to imagine meaning, form, law as inscribed on or upon the body of a word or thing by an incorporeal source. Instead, in the first fact of their accidental coherence, a body constitutes "from end to end" and "directly" the "tablet of its own law":

> Things and words are negentropic tablets and by declination they escape, for the time of their existence . . . the irreversible flux of dissolution. And all this is necessary, it is perhaps the most advanced discovery of ancient atomism: the code avoids entropy for the time of memory. For we have to-day no other conceptual means to draw order from noise or a system from disorder.[44]

"Language," Serres comments, is "too strong or too weak a word" for what Lucretius has in mind, and the provisional alternatives Serres offers— "coding," "memory," "writing"—lessen the risk only slightly.[45] They tempt us to overgeneralize the specificities of human communication, or overestimate its originary power (too strong), which is the same as narrowly defining the set of processes to which languages belong (too weak). Language is a subset of composition in general as the significant coherence of arbitrary difference: a case, maybe, of equivocal generation.

Shining a light on this dimension of corporeality would be the way that, to return to de Man's formulation, language, in *De rerum natura*, constitutes a source of information about something other than itself. But the similarity should not be taken too far, and certainly does not license the conclusion that the universe is made of or by words. Words exemplify the discrepancy between elemental and phenomenal reality, and the consequential retention of arbitrary conjunctions; but they are not the world's exclusive or exhaustive representatives. Dancing dust motes in a sunbeam exemplify atomic turbulence, a snake's shed skin exemplifies the atomic media of sensation, a thinning ring on a finger exemplifies atomic attrition. Each partially images reality because it partially instances reality, as a case of broader necessities and chances. Each will probably (in the strict sense of probability) have analogues, precedents, and consequences among other kinds of corporeality. To make words the exclusive paradigm for things, alphabets the origin of atoms, would be to mistake effect for cause, case for law, species for genus—and to go deaf to the instruction of the numerous other kinds of signs "striving ardently to fall on your ears" (2.1024-25).

Such are some of the instructive potentials and limits of the "prosaic materiality of the letter," to borrow de Man's phrase, and of literal language—language defined as "of the letter" and free from figuration—as an instance of materiality in general in the tradition to which Goethe's poem points us. Part of what makes this tradition exceptional is that it affords figurative language a broader instructive and exemplary scope, as when the marked trope "wie eine Welle" (like a wave) enters Goethe's stanza and reorients the discourse toward material force. One way of putting this is that while not all atoms, joining up, compose letters in *De rerum natura*, all compose figures. In Erich Auerbach's estimation, moreover, Lucretius's "most ingenious" use of the word *figura* was to apply it to atoms themselves—"though small, the atoms are material and formed; they have infinitely diverse shapes . . . he often calls them 'forms,' *figurae*."[46] This is no small thing, in the economy of *De rerum natura*, for although the group activities of atoms produce the whole gorgeous panoply of phenomenal attributes, the atoms themselves are kept quite bare of intrinsic properties.

This austerity is an ethical and political program in Lucretius: it allows the preponderance of materialist explanation to tip from necessity and volition toward contingency and context, to redescribe most of what regularly happens by nature, as well as by subjective action, as "historicity-effects and body-effects," in Jacques Lezra's apt terms—*events* of matter in its "various connections, weights, blows, concurrences."[47] This is not to deny that things, as they are, are extraordinarily tenacious, but to render them, in principle, tractable: products of chance whose endurance is attributable to momentum, repetition, and cultural and contextual support, rather than inexorable necessity. Lucretius's immediate examples of such second-order qualities/accidents (*eventa*) "that may come and go while the nature of things remains intact" are telling: though he might have chosen color, texture, or aroma, he instead lists "slavery," "poverty and riches, freedom, war, concord," as if to refute, in a phrase, the charge that a thoroughgoing naturalism must *naturalize* the history of power (1.455-58).

But "figure" is among the atoms' inalienable properties, a property being that which "can never be separated and disjoined" from a thing without its "destructive dissolution." Which makes of *De rerum natura* the rare philosophy for which there is no getting under figure, no literality or logic behind or beneath or before it, no reality (no thingfulness) without it, and no deconstruction required to make it say so. Yet rather than conclude that everything is therefore language, we are instead invited to admit the astonishingly simple possibility that facts aren't plain. Lucretius, who speculatively strips the first bodies of every dissociable quality, leaves behind the one historically easiest to dismiss as inessential "ornament." He then proceeds as though there is nothing epistemologically diminishing, embarrassing, or disqualifying, nothing "mere"—merely "figurative," "merely metaphorical"—about the figural heart of physics. To assume that figures betray anthropomorphic distortion and linguistic imposition is to assume that *anthropos* alone has the capacity to make them. And if giving atoms figure in this way works against the derogation of figuration as inessential, the more famous doctrine of the *clinamen* works against the derogation of figuration as "improper," since it affirms, as the best available account of the evidence of existing and appearing things, a deviant tendency to turn, to trope, among their parts. Tropic impropriety is proper to matter: to banish poetry from physics or from natural philosophy would sorely delimit their fidelity to theme.

The materialist, Lucretius seems to say, the one who wishes to understand the world from among and not beyond the bodies that make it up, cannot, even speculatively, dispense with the shape in which the particularity of the world's items inheres. Thus, the atoms, he argues, will defy the philosophical temptation to render them pure and interchangeable: "since they exist by nature and are not made by hand after the fixed model of one single atom, [they] must necessarily have . . . different shapes as they fly about [*dissimili inter se quaedam volitare figura*]" (2.378–80). This passage of *De rerum natura* seems quickly to digress. Soon we are talking about animals that seem, to us, to belong to a single species but, among themselves, "are known clearly to each other no less than men are" because "each differs from each in shape" (*inter se differre figuris*) (2.351, 2.348). We follow, for more than twenty lines, the distraught mother of a sacrificed calf, blind to every pleasure or distraction as she "bereaved wanders through the green glens, and seeks on the ground the prints marked by the cloven hooves [*quarit humi pedibus vestigia pressa bisulcis*]," searching inconsolably "for something of her own that she knows well" (2. 352–67). The episode is setting up a macroscopic analogy for the irreducible figural variety of atoms against a fundamental modality of philosophical abstraction, the establishment of kinds (*genera*), whether through Aristotelian taxonomy or Platonic ideal forms. Yet as usual with analogies in *De rerum natura*, and as especially befits an analogy about the limits of formal abstraction, the figural connection exceeds the initial formal or logical similarity between atomic and bovine genera.[48] For it also reveals the bereft

animal mother to be engaged in the same semiotic operation as the poetic empiricist who is learning and teaching from her case: both are following figural
traces (*quarit . . . vestigia*) in search of something unseen, engaged in a form
of hermeneutics that is not limited to the *genus humanum* and that pressures
the stability of kinds from another side, that of transspecies commonality in
the production and interpretation of signs.

"What is so very wonderful in this business," Lucretius asks regarding the
(surely multiple and repeated) origins of human communication, "if the *genus humanum*, having active voices and tongues, could distinguish things by
varying sounds to suit varying feelings [*cui vox et lingua vigeret,/pro vario
sensu varia res voce notaret*]?" (5.1056–58). So many other animals do. And
as with the mother of the slaughtered calf, it is animal parenthood that presses
people toward this kind of verbal semiosis: in *De rerum natura*'s conjectural
natural history, various families and neighbors struggle toward gestural and
verbal signification in order to urge nonviolence (*nec laedere nec violari*) for
the sake of such small, vulnerable persons, "signifying by voice and gesture
with stammering tongue that it was right for all to pity the weak" (5.1020–24).
In terms of this chapter's indexical focus, it is notable that Lucretius posits
the deictic gesture of pointing as the precondition of articulate speech in ontogeny, as well: infants are driven to "point with the finger at things that are
before them" in a way that conditions and intimates their developing capacity
of language (5.1031–32), much as the calf uses the pointy beginnings of horns
on his forehead to butt and push in anger even while these are scarcely grown.
The origin of verbal languages is a question of how groups come to signify
with their roughly similar bodies, a question of convenience, utility, and convention among beings with "vigorous tongues."[49] As will come as no surprise
by this point, this conception of the history of verbal languages does not serve
to adjudicate between nature and culture.

To bring to a close the long excursus that Goethe's "Dauer im Wechsel"
opened out of de Man's "The Rhetoric of Temporality," it must be conceded
that Goethe's poem indeed ends in a sudden recuperation of immortality. But
this compensation is not borrowed from nature but gained by sudden dis-
identification with it:

> Laß den Anfang mit dem Ende
> Sich in Eins zusammenziehn!
> Schneller als die Gegenstände
> Selber dich vorüberfliehen!
> Danke, daß die Gunst der Musen

Unvergängliches verheißt,
Den Gehalt in deinem Busen
Und die Form in deinem Geist.

[Let the beginning and the end
Draw themselves together in one!
Faster than the objects,
Speed yourself away!
Give thanks that the favor of the Muses
Promises an imperishable thing,
The content in your breast
And the form in your mind.]
(33–40)

Goethe's abrupt, Muses-ex-machina guarantee rings hollow against a "Du" whose proper contents and second-person form were systematically evacuated in the prior stanzas, and it introduces form-content and body-spirit dichotomies that the prior stanzas' notion of composite figure did not require. (Is there then a note of sarcasm, then, in the gratitude owed the Muses? *Thanks a lot, Muses.*) The poem's publication in Goethe's final authorized edition of his works, the *Vollständige Ausgabe Letzter Hand* (literally, "Complete Edition of the Last Hand"), heightens the irony of its final assurances. Goethe secured virtually unprecedented, exclusive and (some thought) megalomaniacal legal powers to oversee the edition and publication of his definitive collected works.[50] His attempt to secure his own posthumous literary monument folds the wish for "Permanence in Change" of the consolatory fifth stanza back into the hard-boiled fourth: "That hand . . . is another now," mobilizing the English and German pun in the edition's title that makes "last," as "ultimate" or "final," indistinguishable from "last" as "latest" or "most recent." The text and its pronouns pertain to whatever hand they are in, their address "imperishable" only by way of a deictic and figural contingency that rigorously submits both "form" and "content" to extradiegetic determination.

The poem's contribution to the rhetoric of temporality, then, occurs not in the last stanza's suddenly sentimental self-description, "imperishable thing," but in the rival formation, *gegliederte Gebilde*, wherein poetic, natural, and bibliographic corporealities coincide. Here the morphologist, by way of Lucretius and his own frustration with contemporary accounts of shape formation, limns the kind of structure that would become vital to Marx and Althusser: "the structure, is immanent in its effects . . . the structure, which is merely a specific combination of its peculiar elements, is nothing outside its effects."[51] In the next chapter, Shelley's poetry will press the potential of Lucretian composite figuration toward historical thinking in that sense. But first, Goethe has more to say about its implications for an ethics of experiment and a historicist mode of natural-scientific writing hospitable to the poiesis of other beings.

2. TENDER EMPIRICISM

In a profound essay that both credits and questions the restraint-based model of ethical action that extends from Kant to Levinas, Barbara Johnson asks whether ethics ought to aspire to something past "mere defense of the Other against the potential violence of the Subject":

> [I]f ethics is defined in relation to the potentially violent excesses of the subject's power, then that power is in reality being presupposed and reinforced in the very attempt to undercut it. What is being denied from the outset is the subject's *lack* of power, its vulnerability and dependence. Respect and distance are certainly better than violence and appropriation, but is ethics only a form of restraint?[52]

Taking up the fundamental feminist critique of the objectification of women, Johnson goes on to ask whether the problem with "objectification" might stem from asymmetries of power rather than "from something inherently unhealthy about *willingly* playing the role of thing." "What if," she asks, testing the Kantian stricture against using people as "means" against the idea of "transitional objects" in Donald Winnicott's developmental psychology, "the capacity to become a subject were something that could best be *learned* from an object?"[53] In this scheme, the durability of objects, their capacity to withstand a certain degree of subjective rage or misprision, instructs a person in the limits of his or her power.

The scientific method Goethe called "tender empiricism" (*zarte Empirie*) hinges on a willingness to play the object in something like Johnson's sense. *On Morphology*'s methodological essays advocate an "objectively active" disposition that cultivates "the subject's *lack* of power, its vulnerability and dependence" (Johnson) as an ethical and epistemic virtue for an empiricism fit to register the subtle impact of nonhuman figurations and to sophisticate the rhetoric of autonomous power marking idealist organicism and aesthetics at the turn of the nineteenth century. Against the mounting valorization of scientific "objectivity" as "knowledge that bears no trace of the knower,"

Goethe, I argue here, hijacks the term to advocate the observer's susceptibility to transformation by the objects under view, casting that transformation as an indispensible register of those objects' efficacy and activity, even though it requires abandoning every claim to universality for the knowledge produced in their cooperation.[54] "The moral is simple," as Donna Haraway would conclude much later: "only partial perspective promises objective vision," which "turns out to be about particular and specific embodiment and definitely not about the false vision promising transcendence of all limits and responsibilities."[55] Goethe's surprising concurrence with what feminist science studies would come to name "situated knowledge" is no accident, but the effect of a concerted attempt to allow the hyperfeminized virtue and liability of *Zartheit*, "tenderness" or "delicacy," reconstruct the empirical method.

In their monumental history of scientific objectivity, Lorraine Daston and Peter Galison argue that objectivity as we still know it needs to be understood as a strategic, mid-nineteenth-century negation of the overbearing new form of subjectivity documented in Kantian critical epistemology. In contrast to their impressionable, sensationalist Enlightenment predecessors, Daston and Galison argue, post-Kantians regarded the self as a "coiled spring" whose projections threatened the possibility of empirical observation.[56] This new sense of subjectivity helped polarize the cultural roles of "Poet" and "Man of Science" (in Wordsworth's terms) over the course of the first half of the nineteenth century, as experimentalists developed self-effacement protocols to mitigate the subjectivity of their representations and artists gained cultural prestige by heightening the subjectivity of theirs.[57]

The broad-brush thesis has the virtue of making clear the extent to which post-Kantian scientific objectivity complements the disinterested form of aesthetic subjectivity that literary critics know better. For as if the perfect counterpart to scientific judgment that would bear no trace of the subject, aesthetic judgment, as we know, aspires to a form of auto-affection that bears no trace of the object: as the first paragraph of the "First Moment" of the "First Book" of the "First Section" of the "Critique of the Aesthetic Power of Judgment" informs readers, "aesthetic judgment" concerns "the feeling of pleasure and displeasure, *by means of which nothing at all in the object is designated*, but in which the subject feels itself as it is affected by the representation" (§1; 89, my emphasis). At stake, of course, will be a form of "taste" devoid of appetite or desire whose self-reflexive feelings paradoxically propel the subject toward intersubjective community. Daston and Galison's *Objectivity*, however, is also symptomatic, of something else: of the extent to which Kantian critical philosophy retroactively controls the way historians of literature *and* science reconstruct the period's emergent, disciplinary modernity. For if the third *Critique*'s division into aesthetic and natural-philosophical halves helped authorize and helps explain the nineteenth-century institutionalization of opposed human and natural sciences, this outcome was by no means a foregone conclusion.

Against this broad backdrop, Goethe's resignification of the term "objectivity" in the 1820s emerges in all its subversive mischief. For Daston and Galison, among others, Goethe exemplifies a pre-Kantian notion of scientific representation, one in which the experienced observer deduces the ideal type (*Urtypus*) beneath the myriad phenomenal forms of nature.[58] But I want to touch on the much-belabored crux of Goethe's Kant engagement in order to show that Goethe's late forms of poetic and scientific objectivity are not relics of pre-Kantian epistemology and natural philosophy, but subtle—even "tender"—refutations of Kant's premises about the ethics of observing and knowing organisms and artworks. The method Goethe calls "objectively active thinking" responds to a conviction, like Johnson's, that the Kantian ethos of subjective restraint harbors covert forms of self-aggrandizement. It is an ethos so concerned to curtail the subject's passive vulnerability to extrinsic determination that it risks foreclosing not only the sensuous and epistemic fruits of passivity, but also the activity of anything or anyone other than the self.

Goethe's late, counterdisciplinary essays on method break from the Kantian ethics of self-containment to suggest how a knower might entangle passionately with things and yet fruitfully fail to consume, exhaust, or appropriate them because of the passivity such knowing also entails. *On Morphology*'s methodological essays make clear that the empiricist observer needs to understand that "at bottom he is *nature*" (as Karl Marx would put it when he took up the subject of "objective activity"), one among morphology's living objects and inexorably constituted, like them, in dependency and equivocal determination.[59] But these essays also revalue such "tenderness" and such ineluctable partiality as fortuitous, as well as necessary, for knowledge: they open onto a disavowed dimension of experience subject to shaping by the poiesis of other beings.

d. Kant's Immodesty

The Goethe of the *On Morphology* periodicals gently mocks Kant's celebrated epistemological modesty—his abstention from philosophizing about things-in-themselves—as "roguish" and "ironic." Kant's ostentatious concern, in Goethe's paraphrase, to prevent the subject from "presumptuously . . . tack[ing] a whim that runs through his brain to the objects" is undermined by "asides" that beckon the chastened subject "over the bounds that [Kant] himself has drawn." Thus Goethe depicts the famously austere philosopher as a kind of "delicious" tease ("der köstliche Mann"):

> He seems concerned to constrain our capacity for knowledge to the ut-most. . . . But then, after he has sufficiently driven us into the corner, even into despair, he settles on the most liberal expressions, leaving to us what use to make of the freedom to which he has, more or less, entitled us.[60]

The tiny essay from which this passage is drawn furnishes, as an example, Kant's tantalizing gesture toward "anschauende Urteilskraft" in the "Critique of Teleological Judgment": an "intuitive power of judgment" that Kant calls possible, but not for us. Such a mode of knowing, Kant hypothesized, could go from the intuition of a whole to the parts and would therefore know "no contingency" in the connection between parts and whole (§77; 276). Since Eckart Förster's influential scholarship on the subject, Goethe's nod toward the Kantian passage has been taken as a key, perhaps *the* key, to Goethean philosophy of science, which is said to seek an experimental method capable of realizing the holistic and contingency-free power of contemplation that Kant carefully withheld from human access.[61] But much that we have already seen from Goethe cautions against reading his epistemology as the enthusiastic fulfillment of any such Kantian intimation: his skepticism about teleology as the fittest mode of judging form; his deliberate revision of his own metamorphic theory in order "to know," rather than eliminate, as Kant has it, "contingency"; and the *Morphology* journal's biological and aesthetic commitment to punctual and partial forms—to serialization, as Eva Geulen puts it, "not for the sake of connections but, on the contrary, to stave them off."[62] Indeed, it is worth pausing over the first task taken up in Goethe's "Anschauende Urteilskraft" essay, which is to lay bare the subjective license performatively garnered in the quintessentially Kantian gesture of epistemological restraint.

For the Goethe of the *Morphology* periodicals, Kant's concern to prevent the subject from "presumptuously and arrogantly" projecting a "whim that runs through his brain" onto the objects screens a much more significant hubris: the presumption that a person would be capable of producing whims in utter independence of objects; and, more importantly, that what is important about a subject is the way in which he is *not* an object.[63] The counterexample of Goethe's morphology writings brings into relief just how much the simultaneous codification of aesthetics and the organism in Kant's critical philosophy relied upon discounting the activity of "natural" nonselves, within and without.[64] In the aesthetic part of the *Critique*, cleaving the subject from the effects of objects is not tragic, but aspirational: what the text celebrates in the experience of both beauty and sublimity is the way, in taking habitual cognition by surprise, such experiences make the inner workings of one's own faculties of mind "palpable" as a discrete, "pure" feeling—an auto-affective "feeling of the free play of the powers of representation"—distinct from the occasioning object (§9; 102). To carry through a single aesthetic judgment, or to cultivate aesthetic judgment in general, means to sharpen this distinction to the point where a subject is indifferent to the very *survival* of the object: "One must not be in the least biased in favor of the existence of the thing" and "what matters is what I make of [its] representation in myself" (§2; 91).

In the case of sublime experience, this indifference tips toward a feeling of marked "dominion" over objective nature inside and out: sublimity exists "only in our mind, in so far as we can become conscious of being superior to nature within us and thus also to nature outside us (in so far as it influences us)" (§28; 147). The resulting consciousness of superiority unleashes nothing less than the extraordinary, democratic movement of discursive, intersubjective "humanity" in Kant, articulated more directly in the case of the beautiful: aesthetic experience throws a welcome wrench in the smooth operation of individual cognition-as-usual, propelling a person to seek consensus with other formally like-minded subjects. In the sublime remnant that exceeds natural determination, "we" are made to feel that "the *humanity* in our person remains undemeaned even though the *human being* must submit to that dominion" (§28; 145, emphasis added).

The plot of the *Critique*'s biological side mirrors this high point of sublime experience: contemplating the organism as self-organizing "natural end" culminates, for Kant, in the conclusion that the human being "contains the highest end itself, to which, as far as he is capable, he can subject the whole of nature, or . . . at least he need not hold himself to be subjected by any influence from nature" (§84; 302). One sees here just how much the regulative principle of teleological judgment that Kant attaches to "organized and self-organizing" natures supports his aspirational vision of "humanity" as that which exceeds natural determination by containing its own ends.[65] The possibility of autotelic causation on display in organic forms corroborates the possibility of a free humanity in the third *Critique*; in fact, it even corroborates the possibility of "humanity" *tout court*, where the very concept means individual and collective freedom from extrinsic determination. Without calling into question the nobility of these aims, it is worth asking, with Goethe, how such weighty stakes bias Kant's description of aesthetic and living forms, and whether a polemically anti-empirical philosophy, wrought against "any influence from nature," ought to give the rule to natural-scientific looking—and even to aesthetics in its historical sense as a science of sensation.

The very first sentence of the very first issue of Goethe's *On Morphology* describes an attitude of subjective dominion over objective nature that is strikingly familiar from Kantian sublimity. But Goethe represents such an attitude not as the apex of cultivated judgment (aesthetic or biological), but as a naïve and aggressive first attempt at scientific looking:

> When the person summoned to lively observation begins to struggle with Nature, he feels at first an immense drive to subordinate the objects. It is not long, however, before they press in on him so forcefully, that he well feels how much cause he has to recognize their power and to honor their influence.[66]

These first two sentences have exactly reversed the two-step progress of Kantian sublime experience, which seeks to move from a perception of Nature's indomitable power to a subjective experience of human dominion; the next sentences subject the Kantian stress on impartial judgment to an unending "training" in receptivity. The observer who acknowledges this "reciprocal influence" (*wechselseitigen Einfluß*), Goethe continues, accesses "the possibility of an infinite training, in so far as he renders his receptivity [*Empfänglichkeit*], just as much as his judgment [*Urteil*], adroit at constantly new forms of accommodation and reaction [*Formen des Aufnehmens und Gegenwirkens*]."[67] Welcomed, the pressure of external objects (for they "press in," like those of Faust and Lucretius) initiate a rival, epigenetic logic of compounding, collaborative receptivity, rather than self-activity: here each interaction enlivens the observer through new forms of response.[68] To document the complex, recursive temporality of such "reciprocal influence," Goethe explains, morphological experiments and observations will have to be told "historically," rather than once and for all.[69] Fittingly, he unfolds this ethos of receptivity and its concomitant historicist method in an essay that itself responds to another's "witty word."

e. Active like an Object

The anthropologist Johann Christian August Heinroth sent Goethe a copy of his *Lehrbuch der Anthropologie* (1822), with a note directing Goethe to a page where he would find himself praised as "the creator of the true scientific method."[70] There, Goethe finds his thinking described as *gegenständlich*, or objective, a word rich enough to motivate a new *Morphology* essay, "Meaningful Progress by Way of a Single Witty Word" (*Bedeutende Fördernis durch ein einziges geistreiches Wort*) (1823).[71] Part of Heinroth's (unintentional) wit, from Goethe's point of view, seems to have stemmed from the way his usage of "objective" departed from what Goethe (like Daston and Galison) characterized as the increasingly "sharp sundering of the objective from the subjective" characteristic of his age.[72] Incorporating Heinroth's praise virtually verbatim, Goethe's own essay elaborates "objectively active" (*gegenständlich tätig*) thinking this way:

> [Heinroth] means to express: that my thinking does not sunder itself from the objects, that the elements of the object, the intuitions, enter into my thinking and are permeated to the inmost by it, that my intuiting is itself a thinking, my thinking an intuiting . . .

> [(Heinroth) aussprechen will: dass mein Denken sich von der Gegenständen nicht sondere, daß die Elemente der Gegenstände, die Anschauungen in dasselbe eingehen und von ihm auf das innigste durchdrungen werden, daß mein Anschauen selbst ein Denken, mein Denken ein Anschauen sei . . .][73]

This version of objectivity precisely deflates the possibility and desirability of disinterested observation toward which Kantian judgment in both its aesthetic (maximally objectless) and scientific (maximally subjectless) varieties aspires, as well as the critical philosophy's emphatic refusal to presume a real relation between things in themselves and their phenomenal representations to consciousness. (Coleridge, who also comes across Heinroth's praise of Goethe, is predictably unimpressed.)[74] Placing "intuitions" (*Anschauungen*) and "elements of the object" in synonymous apposition, stressing that those "elements of the object" enter the observer, and vice versa, unapologetically identifying thought with sensuous intuition, and modeling reciprocal action down to its chiastic syntax, Goethe's account multiply refutes impartiality as a basis for knowledge. Here the event of thinking entails a literal exchange of parts (elements, points), such that knowing is irremediably "partial" (*parteilich*) in the double sense of biased and incomplete. The sensations that ground empirical inquiry entail a penetrating form of contact understood to reconfigure observer and observed alike.

"Every new object, well observed," Goethe concludes in the essay's most trenchant resignification of objectivity, "opens up a new organ in us" (*Jeder neue Gegenstand, wohl beschaut, schließt ein neues Organ in uns auf*).[75] Indeed, Goethean revisionary empiricism is abbreviated in this astonishing phrase, which shuttles "us" from observer of metamorphosis to metamorphic object, and back again. To observe well, Goethe maintains, is to permit oneself to metamorphose in the act of observation, as the morphologist experiences the opening of an "organ" adjusted to the metamorphic object that occasioned this development.[76] To do so is to become the object of an object's action, to assume the passive position in a process (and in a sentence) whose outcome is to produce an observing subject once again, retrofitted with a "new organ" fit for further actions and passions of observation, with further morphological consequences.

In fact, such transpositions of morphology's subject and object, its observer and observed, make a fine definition of "experiment" for Goethe, who entitled an earlier essay "Experiment as Mediator between Subject and Object" ("Der Versuch als Vermittler zwischen Subjekt und Objekt"). And such will be the meaning of the tenderness in Goethe's "*tender* empiricism" (*zarte Empirie*): the method sets out to confront, and ultimately to cultivate, the observer's vulnerability to the object under view, recasting the scene of experiment as one of bilateral metamorphosis that puts the subjects and objects of knowledge into a spiral of "accommodation and reaction."[77] A key motive and asset of Goethe's revisionary empirical method is to attend to the forms not of unity or identity, but of interactivity and transfiguration, that disappear into the Kantian cleft between human representations and "things in themselves."

Just as Kant's stress on impartial viewing undergirds both the aesthetic and the biological wings of the third *Critique*, Goethe's subversion of this principle quickly extends from scientific experiment and representation toward the poetic kind, as he spreads Heinroth's descriptor "objective" from a "method in the observation of nature" (*Verfahrungsart im Naturbetrachungen*) to an artistic method: "What has now been said of my *objective thinking* [*gegenständlichen Denken*]," Goethe writes, "I may well symmetrically relate to an *objective poetry* [*gegenständliche Dichtung*]" ("Meaningful Progress," 307). Each is redescribed as an art of being *acted upon*—opened, impressed, populated, impregnated—by one's objects. Of his poetic productions, Goethe notes that certain "motifs and legends . . . pressed themselves so deeply into my mind that I sustained them, active and alive inside, for forty or fifty years," and the same language recurs to describe his process of arriving at a scientific theory: "all the objects that I had inspected and studied for fifty years" collaborate to "activate a representation and conviction in me from which I cannot now desist" (307, 309). To the open-textured living bodies we examined in the previous chapter, we can add that of an equally permeable artist, scientist, and historical subject, gestating "seeds" of his objects—whether series of osteological and plant specimens, legends, or even the French Revolution—which are liable to issue unexpectedly as literary and scientific "fruits" (307–9). To document this "reciprocal influence," then, the "Meaningful Progress" essay concludes, morphology will have to be told "historically": "I will therefore take the liberty of representing my former experiences and observations [*Erfahrungen und Bemerkungen*], and the ways of thinking that sprung from them, historically in these pages [*in diesen Blättern geschichtlich darzulegen*]" (309).

Notice how cleverly the then- and still-prevailing understanding of "objectivity" has been derailed: "objective" here means neither a subject's capacity to impartially assess an object nor his ability to make a claim that would hold true independent of his looking. In "objective activity," the subject, above all, consents to *be* an object.[78] Pressing Heinroth's phrase to the point of this apparent contradiction is the wit of Goethe's own essay, which exposes how another opposition—between activity and passivity—so underwrites the "sharp separation" of subjects from objects that to speak of the "activity" of objects sounds anthropomorphic, oxymoronic, or magical. And Goethe proves equally willing both to speak in this way and to act out the more scandalous, implicit correlate to Heinroth's witty word, modeling a passivity of the subject that contemporary subjective idealisms were structured to preclude. Or, more precisely, a passivity they were structured to *effeminize*: figuring himself as mentally opened, penetrated, and impregnated by his objects, Goethe keeps *empfangen* in the sense of becoming pregnant awkwardly audible in his revaluation of epistemological *Empfänglichkeit* (receptivity) (307).

In this sense, as we saw him do with Schelver's notion of *Verstäubung*, Goethe has subverted the source of the "witty word" that so deftly captures his own method: Heinroth could be expected to squirm at Goethe's characterization of objective thinking as a variety of unpredictable pregnancy. In his chapter "Sexual Difference in Humans" in the *Lehrbuch der Anthropologie*, Heinroth elaborates that polarity between "masculine" and "feminine" that we have heard the botanists contend "runs through the whole of nature."

The "positive, ruling, conceiving" male organism is "firmer, stronger," while the female is "negative, serving, bearing" and characterized, above all, by *tenderness*. Her organism is "lower, *zarter*, weaker," her skin "softer, *zarter*"; her mind defined by "receptivity" (*Empfänglichkeit*) in contrast to his "self-activity" (*Selbstthätigkeit*).[79] This is why, according to Heinroth, the great achievements of human art and science "have for all time been primarily the property of men."[80] But in Goethe's appropriation of Heinroth's anthropology, as in his appropriation of Schelver's botany, a revaluation of the feminine-gendered qualities of "tenderness" and "receptivity" is central to his revisionary, poetic empiricism, which attempts to let the ways the subject is (also) an object enter, or reenter, natural philosophy and its representation.[81]

The peculiar centrality of the word *tender* in Goethe's oeuvre has been noticed before: in a conceptual history devoted exclusively to enumerating and classifying the senses of the word *zart* in Goethe's lexicon, John Hennig concludes that Goethe's investment in the term stems precisely from its "objective/subjective double-meaning," the way it "lacks a sharp division between passive and active, substantive and functional, material and personal . . . psychic and physical."[82] He wagers that Goethe used the term more often than any other German writer, documents that this usage increased with age, and affirms that Goethe mostly applied the adjective to objects that (from "noodle-dish" to "Galvonometer") had never before received it.[83] Yet even in 1980, Hennig seems unnerved by Goethe's tendency to praise not just female but *male* artists and scientists (Bacon, August Batsch, Herder, Schelver, Correggio, Laurence Sterne) for their "tenderness." Hennig is quick to reassure readers—albeit without his essay's former abundance of textual evidence—that "Goethe was fully cognizant of the danger of tenderness [*Zartheit*] slipping into sickly or frail softness . . . [his] orientation leads rather to inner cultivation and rigor."[84] On the contrary, it seems that permissive softness and plasticity are just what Goethe is after, in natural-scientific as well as aesthetic looking, and in a Kantian context in which rigor was all the rage. The objectively active thinker, he elaborates in a further response to Heinroth, is "pervaded by the poet or, better, by the artist and art connoisseur"—precisely in so far as he observes a natural object like those specialists observe great paintings—he "lets it work upon him" (*er läßt es auf sich einwirken*).[85]

Indeed, the categorical uniqueness of "the aesthetic" as we inherit it from the *Critique of the Power of Judgment* diminishes with the rift it was supposed to bridge once representation making is a tendency human consciousness shares with other things. At issue in the Goethean alternative is a practice of observation that is neither particularly concerned to defend, nor particularly well comprehended by, the distinction between aesthetic and natural-scientific looking.[86] Goethe's concern is, instead, to restrain the self's "tremendous drive to subordinate the objects" and to question that self's corollary tendency to define its freedom in terms of its dominion over all external causes of its being. Frederick Amrine has aptly described Goethe's empiricist methodology as the "controlled development of new ways of seeing as such," "as many 'modes of representation' as possible, or better, to cultivate the mode of representation that the phenomena themselves demand."[87] But though initially, nominally sensitive to the plurality of "modes of representation" demanded here, he and many other expositors of Goethean philosophy of science do not permit the journals' own disjunctive morphology to unsettle their conviction that Goethean looking ultimately aspires toward a transcendent, organic unity between subject and object, idea and experience.[88] Echoing Coleridge's famous account of the "translucence" of symbolic representation, Amrine concludes that in Goethean observation "universal and particular, idea and experience . . . become reciprocally determinative, and, in that sense, a unified organic whole . . . the universal shines through the particular while inseparable from it, as in a symbol."[89] But this chapter and the previous one have tracked a different kind of sign in Goethe, one that renders observation instead inseparable from obsolescent particularity and phenomenal disclosure as a casting-off of skins (and dust, vapor, and droplets): subtle, figural, and particulate effluences through which beings contribute to the perceptual and semiotic processes by which others come to know them.

After all, it was Goethe who, in somewhat notorious opposition to Newton, polemically refused to abstract vision from the murky, material medium in which things appear.[90] "Turbidity," *Trübe*, was Goethe's name for the inalienable medium of sensuous perception, which he explained in language that draws together phenomenality and materiality under the aegis of a final, explicitly neo-Lucretian form of tenderness:

> The first diminution of the transparent, that is, the first, gentlest filler of space, the first approach to something bodily, nontransparent, as it were, is the turbid [*die Trübe*]. It is thus the most tender material [*die zarteste Materie*], the first lamella of corporeality.[91]

Goethe here explains that Lucretius's term *rarus*—loose and tenuous, opentextured because shot through with void—provides the best description for this "first lamella of corporeality," upon which his whole theory of perception rests. It denotes a "net-like film" of "dissimilars," in which "disparate pieces, though sundered, still attach to or float near one another."[92] And Goethe also here claims the Latin *turbo-* as the etymon of *Trübe*, insinuating into his media theory a sense of "unity, stillness, and coherence *disturbed*" that derives from the Latin verb through which no lesser a later neo-Lucretian than Serres would choose to unfold his entire exposition of *De rerum natura*.[93] The aging, underappreciated Goethe of the morphology writings was finding, by way of Lucretius, a version of corporeality at once heterogeneous, disjunctive, and "most tender": a materialism generative enough to allow him to depart from the integrity and totality of organic form and symbolic representation at the level of biology, epistemology, and semiotics.

In addition to its unwieldy title, *On Natural Science in General, Particularly on Morphology*, Goethe's double-journal project has a genre-bending *sub*title: "Experience, Observation, Consequence, Connected by Life-Events" (*Erfahrung, Betrachtung, Folgerung, durch Lebensereignisse verbunden*). The more one reads *Zur Morphologie*, the less clear it becomes that "connected through life-events" indicates the individual life of the journals' human editor at all. For a naturalist who was coming to insist that no being could be accurately "thought as wholly isolated" from its constitutive "infusion" or "envelope" of otherness, the experimental "life-events" that bind might be those that disclose experimentation as a joint life.[94] The "event" or advent of this kind of joint life would be marked by a temporary loss or alternation of subject and object places; its procedure would render experimentation less replicable than irreversibly "historical"; its representation would have to register the turbid, evanescent zones of transcorporeal contact and participation—maybe through a mode of classical figuration whose tenuous apparitions Knebel kept calling *zart*.

CHAPTER 4

Growing Old Together:
Lucretian Materialism in
Shelley's *The Triumph of Life*

Pinocchio becomes a real boy when his body is entirely smooth. Organic form is thus, among other things, an erasure of articulation. This may be why Western cultures are intolerant of any lines on the body—any wrinkles or signs of experience—especially in a love object.

BARBARA JOHNSON, *Persons and Things*

a. Prologue: Montaigne's Face

What, in time, makes up a face? Nearly two-thirds of Michel de Montaigne's late essay "Of Physiognomy" (1585–88) passes without mention of the correspondence between a facial feature and the personal, let alone racial or national, character of the one who wears it. The very physical heading instead fastens together matters we might label sociohistorical, medical, moral, and rhetorical: local outbreaks of plague and of the ongoing religious civil wars, petty cruelties among neighbors, the virtues of Socratic "simplicity" of expression, and a series of luxuriantly self-reflexive accounts of piecing together the text at hand as a garment of borrowed flowers. In fact, it is here, describing his "treatise on physiognomy" as "a bundle of others' flowers" (*un amas de fleurs estrangers*), that Montaigne first uses the word "physiognomy" at all.[1] Contrary to later modern understandings of corporeality, the essay's relationship to its title suggests that the task of interpreting the features of a physical body or face (*physio+gnomen*) discloses not psychological character or national type, but a local and contingent history: the complex sequence of incidents, affects, and influences that shape a personal face over the course of a shared historical interval. The essay makes one wonder not only what bundle of strangers' words the "I" animates in this form of life writing but also whether the physical body we now call "biological" is not similarly animated, whether its living morphology might be conceived as a more fortuitous bundle of others' "flowers" and "troubles."

The first overt aim of Montaigne's essay—as of its echoes in a certain familiar strain of Romantic aesthetics—is to extol the virtue of "unadorned," "natural" expression, a "childlike" purity of self-presentation that Montaigne (dubiously) attributes to Socrates. Yet from Socrates's incongruously unpleasant face onward, the problem of interpreting faces in the essay, the problem "Of Physiognomy," instead propels lush digressions on local minutiae and links them frankly with the problem of growing old. For that is what Montaigne is doing in these late essays: gently mocking—and clearly, copiously relishing— the "folly" of "putting one's decrepitude in print."[2] The essay allies senescence with "yielding" oneself, prolifically, to the impress of "the age" and the burden of "borrowed ornaments": "I load myself [*je m'en charge*] with them more and more heavily every day beyond my intention and my original form, following the fancy of the age and the exhortation of others."[3] The collective experience of the civil wars is described similarly, as a "load": "a mighty load [*charge*] of our troubles [*nos troubles*] settled down for several months with all its weight right on me."[4]

For Montaigne, then, physiognomy concerns the transfigurative burden (*charge*) "of the age" upon the small figure of an embodied self: over time, he writes, a highly specific series of impacts, influences, and incorporations transform what is mine—"my intention and my original form"—into a decadent record of diverse transactions and borrowings. In this sense, an aging person is increasingly "of the age." And yet to age in this way implicates a stream of minutiae so particular in their sequence and so local in their place—"A thousand different kinds of troubles assailed me in single file," writes Montaigne— that to be so shaped is to become not typical, but irreproducible, to acquire a unique face.[5] That is, the corporeality of old age takes extremely particular and collective form in Montaigne; senescing into an increasingly "borrowed" self, the essayist claims to write from a position of generous and nontraumatic alienation: "opulently," and in "ignorance."[6]

b. Morphology and Shelley's "Shapes"

This chapter is not, directly, about Montaigne's late life writing, but about Percy Shelley's "poetry of life."[7] Still, Montaigne's weathered and composite physiognomy serves as its emblem because those earlier modern wrinkles knit together biological and historical ages much as the wrinkled faces in Shelley's *The Triumph of Life* (1822) can be seen to do. In Shelley's last, long, fragmentary poem, faces age under the pressure of a markedly historical atmosphere, one in which others' affects condense into erosive "drops of sorrow" and a long-distance image of Napoleon in chains physically transfigures a person's face: "I felt my cheek / Alter to see the great form pass away" (lines 224–25). The poem ends picturing senescence as the unintended work of multitudes, each wrinkle attesting, in a way not legible like a text and not evident in the texts of monumental history, to the attenuated impacts of a shared historical present.

The Triumph of Life was made famous, in late twentieth-century criticism, for the way its "disfigured" faces allegorized the linguistic and material violence inherent in figuration as a function of reparative reading. In this chapter, however, I attempt to show how *The Triumph*'s final, neglected "new Vision" pointedly ceases to construe figuration as a principally verbal or cognitive process at all.[8] Shelley ultimately deposits readers in the midst of a passage from *De rerum natura* that makes figural face-exchange a mundane transaction among *all* bodies, not just a covert violence of words, or of "two subjects involved in the process of reading."[9] In this atomist scene, "Vision" itself is a figural effect of material dissolution, conditioning a mode of empirical poetic seeing that revises the vital subject of *The Triumph of Life* under changed philosophical and poetic premises about the relation between life, matter, and trope.[10] For against the life-destructive materiality Romanticists have learned to see in this poem, we are in fact directed toward a materiality defined neither by life nor by antagonism toward it; one marked instead by a tendency to trope.

While Shelley's debts to a broadly Lucretian philosophy of aleatory flux have been explored, the ramifications of *The Triumph*'s specific allusion have gone largely unattended.[11] What Shelley's poem expressly retrieves is the same astonishing poetics of sensation that so interested Goethe: those passages in which beings transpire into tropes, collectively producing the entwined realities of temporal and perceptual experience. This Lucretian possibility enables *The Triumph*'s weathered faces to confront us with an extraordinary concurrence of vital, rhetorical, and historical eventfulness that needs to be understood for the challenge it posed to then-triumphal means of representing biological, poetic, and historical form. "I define life as *the principle of individuation*," wrote Coleridge in 1816, "or the power which unites a given *all* into a *whole* that is presupposed by all its parts."[12] Against such individuating, holistic, and teleological investments, Shelley's fragmentary poem about "Life" ends by scrutinizing the "living storm" between bodies, as though life's science and its history concerned less their self-developing progress than the transfigurative touch of an atmosphere they both generate and endure.

In a study that takes an extraordinarily long view of Western figurations of nature, Pierre Hadot remarks that from the end of the eighteenth through the beginning of the nineteenth century, "nostalgia for a Universe as Poem" reached its two-thousand-year zenith: "People dreamed of a new Lucretius."[13] Far from triggering what Noel Jackson worried was "the extinction of a poetic species," we now know that, in the English context, Erasmus Darwin's neo-Lucretian epics produced a veritable vogue—a major "Lucretian moment" linked to the intellectual circles Noah Heringman has studied under the sign of antiquarianism, and Martin Priestman of atheism and free-thought.[14] This chapter and the next focus on Shelley's versions of neo-Lucretianism because they most frankly demonstrate what I take to be a crucial and unexamined motive for the revival of *De rerum natura* in Romantic poetry and life science. As I began to intimate through Goethe, the materialist poetics of transience enabled some Romantic-era thinkers to open the biopoetic ideal of "organic form" to historical figuration, without thereby alleging the unilateral shaping, by language, of the matter of natural life.[15] In this way, as I hope this chapter and the next illustrate, *De rerum natura* becomes a resource for unfamiliar versions of Romantic historicism, providing an alternative lexicon, logic, genre, and imaginary—a poetics—for connecting the epochal epistemic problem of biological life to the period's pressing new sense of its own historicity.

Shelley's oeuvre in fact puts Lucretian figural exfoliations to use in bio-historicizing efforts that range from the poetics of quickening leaves in the "Ode to the West Wind" to the mental films hovering in "Mont Blanc," the "mutual atmosphere" that sustains botanical communication in "The Sensitive Plant," and the cast-off "ugly human shapes and visages" that "Past floating through the air" after *Prometheus Unbound*'s bloodless revolution.[16] Such neo-Lucretian moments are valuable precisely because, at least after *Queen Mab*, they produce no new universal epic, but instead stage the dramatic differences that ensue when an overdetermined issue (e.g., "Life") is remediated through Lucretian figures that do not assume and perpetuate a rift between matter and meaning. This chapter and the next focus on what I take to be the two most interesting, in biological and political terms: first, *The Triumph*'s interface between senescent morphology and the pressure of "the age"; then, *The Mask of Anarchy*'s composition of an alternative political allegory whose didactic decomposition reveals the systematicity of an apparently extraordinary event of state violence. Like Goethe's "Dauer im Wechsel," *The Triumph of Life* and *The Mask of Anarchy* stage the embarrassment of a certain set of poetic and epistemological conventions by Lucretian kinds of signs. In Shelley's hands, however, this switch or upset heightens the poems' dexterity in rendering the social, historical, and political dimensions of what Shelley called "the apprehension of life."[17]

And if Goethe thought that life's apprehension required a *morphology*, a not exclusively vital science of "shape" as the meaningful coincidence of rhetoric and reality, Shelley offers a perfect counterpart from the poetic point of view: for deeply related reasons, he makes "Shape" the protagonist of his "poetry of life" and styles his lyrical "I" after an exfoliated leaf, a recombinant cloud, or Mutability tout court. It is from the complementary angles of poetic morphology and morphological poetics, then, that each poetic empiricist turns to Lucretian materialism to ask after what was becoming unrepresentable as the study of life defined itself absolutely against the inorganic sciences, and as disciplinary expectations consigned figurative language to the immaterial experience of subjects and scientific prose to the material being of things.[18] What emerges in each case is an alternative logic that discovers susceptibility to impress and influence, to sympathy, senescence, and chance, at the core of what constitutes viable form.

I hope the resonances between this chapter and the previous one sound out an alternative terrain for comparative Romanticism in an Epicurean tradition against which Coleridge's gorgeous and better-known reception of German idealism was wrought. The similarity between Goethe and Shelley is not principally a matter of direct influence, but instead passes through what Paul Hamilton calls *The Triumph*'s "French Connection," attesting to the persistent interest in Terror-tinged, radical materialism in British and German cultural contexts officially squared against what Coleridge called "Demiurgic Atomy."[19] And although Goethe and Shelley were born more than four decades apart (1749 and 1792, respectively), it is no accident that the writings under examination stem alike from the late eighteen-teens and early twenties. This is a moment sociologists mark as a perceptible watershed in scientific professionalization and power, during which the untroubled, high Enlightenment polymathy that Goethe so well exemplified becomes, as his methodological essays of the period show, an attitude that can be maintained only through considerable self-conscious scrutiny and justification.[20] Such an attitude, in other words, must be *theorized* and *defended*, as Shelley tries to do in 1820 by extolling Poetry as "that which comprehends all science, and that to which all science must be referred."[21]

This chapter draws out the significance of Shelley's neo-Lucretian gesture in *The Triumph of Life* across the several contemporary domains with which it communicated. First, that of the medical sciences and the philosophical vitalism that controlled both sides of the early-century debate in London over life's principle and cause. Second, that of Romantic historicism and what Kevis Goodman and James Chandler have diagnosed as the period-specific problem of how to represent "contemporaneity" as heterogeneous and unsettled "history in solution."[22] And third, that of the disciplinary status of poetry, which Shelley thought needed a "Defence" in 1820 from a sentiment expressed in print, not quite jestingly enough, by his own friend Thomas Love Peacock: poetry is a "frivolous and unconducive" pursuit from which "the progress of useful art and science, and of moral and political knowledge, will continue more and more to withdraw attention."[23] Shelley worries, on the contrary, that the consensus that these are "unpoetical times" to which only correspondingly unpoetical forms of "art and science" can answer stands to turn modern specialization into so many expert ways of not asking whom progress serves.[24] In these contexts, *De rerum natura* offers the powerful counterexample of an unapologetically poetical claim to social, material, and natural realities. *The Triumph* demonstrates how Lucretius's poetic science— this figural materialism that does not know, need, or support the deepening division of labor between literary and scientific cultures in general, or the categorical uniqueness of "biology" in particular—is of Romantic and ongoing interest for its countercultural capacity to reattach biological to historical and rhetorical materials and to think through the way apparently individual persons collectively produce and integrate passages of historical time. Over and against vitalist valorization of *new* life in its self-generating power, this vantage attends to wrinkles as life signs of the collective age.

c. Shelley, Wrinkled

The posthumously published *Triumph of Life* consists of a series of nested visions narrated in recursive terza rima by a dreaming lyric speaker and his interlocutor, a very wrinkled person who self-identifies as "what was once Rousseau" (line 204). The visions unfold various perspectives on a "shape" called "Life," driving a triumphal chariot at breakneck speed and towing a "captive multitude" of historical personages in chains. These include "Kant," "Voltaire," "Frederic" (the Great, of Prussia), "Catherine" (the Great, of Russia), "Leopold" (II, of the Holy Roman Empire), "Napoleon," "Caesar," "Constantine," "Plato," and "Bacon," among others, in a cast weighted slightly toward the recent history of "the times that were / And scarce have ceased to be" (lines 180, 178, 233–36, 224, 284, 254, 269). As M. H. Abrams once wittily summarized, "Rousseau" interprets this procession of "great" and "unforgotten" (line 209) names and faces as ordered according "to the degree of their failure": the degree to which they succumbed to "those material and sensual conditions of everyday existence which solicit, depress, and corrupt the aspiring human spirit."[25] Indeed, much of the first part of the speakers' dialogue, and many critical interpretations of the poem, concern whether it is possible to "forbear / To join the dance" of mortal life unfolding before them, as a "sacred few" spirits seem to have done: whether it is possible to shun the sensuous life of the decaying body, which life "Rousseau" regrets as a "Stain" that his spirit "still disdains to wear" (lines 188–89, 128, 205). The interlocutors repeatedly and famously ask after the problem of bodily life in terms shared by the two newly developed genres of life writing, autobiography and biology: "Shew whence I came, and where I am, and why," and "what is Life?" (lines 398, 544). The poem, I will argue, experiments in answering these demands with the history of what stains and wrinkles their mortal bodies, as if life's logic depended upon it.

Though Romanticists understand Shelley as an acute theorist of history, even epitomizing "the age of 'the Spirit of the Age,'" this has rarely been by way of the mortal, wrinkled, terrestrial corporeality that preoccupies *The Triumph of Life*.[26] Yet Paul de Man's "Shelley Disfigured" (1979) made *The Triumph of Life* a privileged test case for the *rhetorical* dimension of face change and exchange—the trope of prosopopoeia—in Romantic writing and its critical reconstructions. The poem is practically "synecdoche," Tilottama Rajan remarks, "for the self-effacement of language and of romanticism as a cultural project that continues to mobilize the economy of criticism."[27] De Man took the fragmentary, "mutilated" *Triumph*—and the "disfigured" faces of its personages—as an allegory for the "sheer," "unrelated," "violent," "power" at work in every act of reading.[28] Readings establish historical relation, cognitive meaning, and aesthetic pleasure by way of an "endless prosopopoeia," de Man argued, a giving of face (French: *figure*) to the dead that cannot but disfigure and efface even as it animates and monumentalizes.[29] For each reparative act of "cognition and figuration" must violently instantiate the originary "madness of words"—the "positing power" of "language considered by and in itself," a "random" power indiscriminately destructive of all relation, sense, and sequence.[30]

De Man's lasting provocation was to suggest that in registering, through disfigurement, a power radically alien to life, relation, and sense, *The Triumph* comes closest to history, to the materiality of "actual events, called 'Life' in Shelley's poem."[31] De Man mimes the gesture by invoking Shelley's "mutilated" corpse to exemplify the matter of historical eventfulness, pointing out that Shelley's death at sea in 1822 left the poem and the poet correspond-

ingly "defaced."[32] Since "Shelley, Disfigured," the relation of Shelley's poetics to material history has elicited Romanticists' finest elaborations of a negative aesthetic radicalism that takes de Man, Kant, and Theodor Adorno as
its touchstones and casts "the materiality of actual history" as a paradoxical
precipitate of the lyric negation of contemporary life. On this view, lyric registers "previously obscured aspects of the social"—the writer's and the reader's
respective presents—precisely in so far as it "eschews the relation of self to
society as explicit theme," severing its contacts with the age.[33] Sheer negation
thus defines this version of historicopoetic "material": Forest Pyle's exemplary reading, for instance, takes cinders and ash as Shelley's quintessential
materials—charred minima that mark the site where a remarkable array of illusions (of historical insight, ethical choice, sensuous facticity, verbal intelligibility, revolutionary efficacy, theological redemption) go up in smoke, on
contact with the "vacating radical or the aesthetic itself."[34]

There is ample evidence for such negative poetic materialism in *The Triumph of Life*: interpreters have focused on the "shape all light," de Man's
"model of figuration in general," whose (metrical) feet ruthlessly trample "the
gazer's mind" to "dust."[35] Yet the famously "disfigured" figure of the speaker's
dream guide, "Rousseau," whose prosopopoeia has seemed to emblematize
the damage, is amenable to a different kind of reading. In a way perhaps too
obvious for notice, "Rousseau" is not a figure animated and cut down, faced
and effaced, all at once in Shelley's poem. The "grim Feature" (line 190) with
whom *The Triumph*'s dreaming poet-speaker converses might be less "mutilated" by the hermeneutic rush "to reconstruct, to identify, to complete"[36]
than simply *decomposed*: "what was once Rousseau" speaks while becoming
landscape, by way of a prosopopoeia so slight, so incomplete that it amounts
at first only to "strange distortion" and leaves the face it draws mistakable for
"root," "grass," and "hill side" (lines 204, 183, 182, 185, 183). "What was once
Rousseau" is left mistakable, that is, for "what was once Rousseau" in a more
literal sense at the time of Shelley's writing: a body more than forty years bur-
ied. Against the (admittedly wayward) norms of lyric apostrophe, the speaking "I" does not address but is *addressed by* this feature of the landscape, surprised when "what I thought was an old root" begins to speak to him. This
might be a typical ruse of lyric self-production, were it not a touch funny:
taken aback, the interrupted speaker expends his own apostrophic powers in
a little parenthetical gasp of surprise—"(O Heaven have mercy . . . !)"—and
begins to discern a "grim Feature" still eminently continuous with the surrounding landscape (line 181).[37]

This face is presented less as the work of an animating, lyric "I" upon an inert object than as a figure capable of obtruding itself into the subject's reverie, indeed, of startling that dreaming subject into noticing the action of his surround. To frame the argument in terms of genre, a strategy I pursue at length in the next chapter, we might say that here lyric poiesis is startled by a kind of poetry "strange" to it, a didactic kind that wishes to share tropological power and credit with represented things. The speaker collaborates in this low-level animation by calling his unexpected interlocutor a "grim Feature," a mere part of a face, leaving intact his inhuman and inanimate resonances: textually, the "grim Feature" is a Milton citation; minerally, it is a feature of a landscape.[38] Historically, as Orrin Wang astutely points out, the question of "what was once Rousseau" was synecdoche for the recent history of Europe, from radical enlightenment to Napoleon's fall, a shorthand question "people asked in order to interrogate the era they inhabited."[39] Here this historical question cannot fully extricate itself from a "hill side," or from *Paradise Lost*, or from the aging body, participating instead in a subtle, partial, multilateral prosopopoeia that eventually leads the interlocutors into a lengthy exposition of Lucretian figural materialism.

d. Life, Triumphant

The Triumph of Life opens with a scene that studiously conflates biological and imperial senses of "power," a vision in which Life's process is pictured as the triumphal procession celebrating "some conqueror's advance / . . . the true similitude / Of a triumphal pageant" (lines 112, 117–18). But this image is not, as has frequently been assumed, a condemnation of life as such: life as such was a matter of tense public debate in early nineteenth-century England, where both medical and general publics viewed rival physiological accounts of "life" as supporting competing systems of moral, political, and theological order.[40] The Shelleys' own physician, William Lawrence, had recently played a principal role in a heated public debate on the nature of physiological "vitality" that was fought over the legacy of Blake's "Jack Tearguts" (John Hunter). Sparring with his former mentor John Abernethy over whether life ought to be considered immanent to animal organization or divinely "superadded" to it, Lawrence's lectures in the late teens transfixed and scandalized broad segments of the English public and elicited retaliation from state authorities, not least a blasphemy ruling from Lord Chancellor Eldon (who appears as "Fraud" personified in Shelley's *The Mask of Anarchy*).[41]

"Life," as Blake and Goethe have shown, was a topic of ideologically freighted dispute, and Shelley's *Triumph* needs to be understood as dissenting from a particular, presently triumphant, and rhetorically triumphalist way of researching and speaking about it: the vitalist presumption that the question worth asking about living beings concerned the "power(s)" that categorically distinguished them from other (frequently "material") beings. The "triumphal pageant" of life in the poem's first scene is a visual pun on the medical and imperial senses of "power," inviting readers to consider how vitalism might collaborate with power in its more overtly historical and political modalities. Taking place at the conjunction between rhetoric and biopolitics that Sara Guyer has termed "biopoetics," that is, *The Triumph*, like Blake's *Book of Urizen*, caricatures one biopoetical mode and offers up another.[42]

In fact, in a way almost too obvious for notice, the very title of Shelley's poem intimates the thesis we saw Foucault, Canguilhem, and Jacob make axiomatic in the twentieth century, pointing to the pancultural "triumph" of a newly sovereign concept, "Life," in reconfiguring the early nineteenth-century epistemic landscape. If, for some (disputed) period, living beings had resided inconspicuously on the continuum of natural beings in general, by the late eighteenth century the difference between organic and inorganic, living and nonliving, had polarized into a definitive, ontological distinction for much medical and natural-philosophical thinking. Blumenbach's representative statement from *On the Formative Drive* (1781) might be recalled in this context: "one cannot be more inwardly convinced of something than I am of the powerful gulf that nature has fixed between the organized and inorganic creatures."[43] As Shelley shows more forcefully in *The Mask of Anarchy* (chapter 5), the epistemic triumph of "life" also opens new domains for the exercise of governmental power.

Shelley's titular provocation to life, triumphally construed, challenges the unquestioned importance of vitalism at both ends of the political spectrum. As a fellow of the Royal Society and professor of anatomy and surgery of the Royal College of Surgeons, the Shelleys' physician, William Lawrence, was an important conduit and advocate for continental life science, even likely the first to import the word "biology" into English.[44] He translated Blumenbach's *Handbuch der vergleichenden Anatomie* (1805) into English as *A Short System of Comparative Anatomy* (1807) and dedicated to him his incendiary *Lectures on Physiology, Zoology, and the Natural History of Man* (1819).[45] Lawrence's lectures were broadly, provocatively "materialist" from the perspective of the clerical, medical, and governmental authorities they outraged. Defending the right of physiological research to "sanctuary from the incursions of extra-physical or metaphysical chimeras, and from the intrusion of immaterial agencies," Lawrence insisted that from within this sanctuary, "life" appears "immediately dependent on organization"; vital powers seem to be inseparable from animal structures with which they are invariably found and to be capable of proceeding without supernatural interference.[46]

Yet what has been underestimated here is Lawrence's staunch avowal of the fundamental vitalist conviction that the "laws of life" constitute an absolute "exception" to those of physics, chemistry, mechanics, and mathematics and "cannot be explained on these grounds."[47] As a strong advocate of the autonomy of biology, he resolutely rejects the possibility that "the vital processes can be explained on the same principles as the other phenomena of matter."[48] When the "vital principle" came to engrossing public debate between Lawrence and John Abernethy in London in the teens, that is, both sides of this dispute—including what the press designated as its "materialist" left, in the person of Lawrence—shared a vitalist commitment to uncovering the principles that categorically distinguished the living from other material bodies and processes.[49]

New studies are brilliantly variegating the spectrum of Romantic vitalisms, but it is worth asking whether all approaches to the problem transpire within vitalist limits, however openly and experimentally defined.[50] Shelley's poem expressed an intuition that this rhetoric of vital power—whether that power was cast in Abernethy's terms, as a transcendent "superadded principle," or in Lawrence's, as a power immanent in organization—risked collaboration with power of the imperial and colonizing kind. This was not an abstract risk in the early 1820s. Lawrence's 1819 *Lectures* graphically clarified the moral and political eventualities of vitalist organicism: ranking men in "fixed varieties" according to their degree of "organization," Lawrence discerned a hierarchical array of "National Characters" even in fetal specimens.[51] In light of the "retreating forehead and depressed vertex of the dark varieties of man," he remarked, abolitionists ought to temper their expectations to match the "less perfect" organization of enslaved peoples. The *Quarterly Review* dryly editorialized that a "slave-driver in the West Indies" might use Lawrence's book to soothe his "qualms of conscience."[52]

More subtle, and more subtly contested in Shelley's *Triumph*, is Lawrence's equation of degree of life with degree of organization in the *Lectures*: "Just in the same proportion as organization is reduced, life is reduced; exactly as the organic parts are diminished in number and simplified, the vital phenomena become fewer and more simple."[53] As the *Quarterly*'s reviewer implied, when medical science judges specific forms of life to be less alive, it helps to render their deaths less reprehensible—a logic familiar to twenty-first-century readers from its realization in fascist thanatopolitics.[54] Without defaming at a stroke the whole diverse history of philosophical and medical vitalisms, there is ethical value to noticing that we are discovering, within Romantic-era writing, an underrecognized pattern of dissent from the vitalist, organicist, and autonomy-of-biology views, which Goethe, Jean-Baptiste Lamarck, Etienne Geoffroy Saint-Hilaire, John Thelwall, and Shelley, for instance, each criticize for "absolutism."[55] At issue is the possibility of thinking "life" in ways that are less prone to cast those presently deficient in power as deficient in life; of keeping open the possibility that the logic of life (biology) bears important relationships to "inorganic" physical, textual, and historical processes; and of ceasing to "disdain" the vulnerable, transient body as a "Stain" upon transcendent "spirit" (the position "Rousseau" expresses in Shelley's poem [lines 205, 201]).[56] Indeed, vitalist rhetoric has a way of celebrating "Life" as dissociable from, or triumphing over, particular mortal "lives." To speak of a power "hurrying from life to life, yes, even through annihilation to life," as Goethe put it in pointed caricature of Schelver, is to abstract the issue to the point that it offers little protection against the annihilation of any one.[57]

Despite the personal connections Shelley and Lawrence shared, and despite their like tarnishing, by mutual adversaries, as "materialists," *The Triumph of Life* takes a hard view of the version of life exemplified in Lawrence's *Lectures*. Indeed, the poem depicts triumphal life, the vision of vitalism, as another damaging episode in a long history of triumphal processions. Like "some conqueror's advance," life's "triumphal pageant" consists of vaunting "Conqueror," renowned "mighty captives," a "ribald crowd," and anonymous prisoners in chains (lines 112, 118, 129, 135, 136). That "Life" here triumphs over various and sundry prior triumphalists, reducing them to "spoilers spoiled," is the hollowest kind of victory (line 235). As Wang shows, what the poem laments is the disturbing persistence of triumphalist forms in Revolutionary and imperial rhetoric alike.[58]

The poem's speaker and "what was once Rousseau," who speaks with shame of his time-ravaged body, return repeatedly to the question of whether the speaker can "forbear / To join" the "sad pageantry," or whether one must change "from spectator" to "Actor or victim in this wretchedness" (lines 188–89, 176, 305–6). But "forbear" can mean to tolerate just as much as to abstain, and thanks to the particular logic of parades that makes the "spectator" party to the event, the speaker's dilemma around whether to "join" is a manifestly empty choice. By the light of the poem's first visions, then, Shelley countenances the grim possibility that to represent life and historical process at present—a Regency and Restoration present in which monarchies continued to absorb the most promising democratic shocks—might mean a species of involuntary complicity in "The progress of the pageant" (line 193) that Walter Benjamin would later call the "triumphal procession" (*Triumphzug*) of victors' history.[59] The most disconcerting feature of post-Waterloo historical experience for radicals, Rei Terada suggests, may have been the growing perception that it was no longer "even possible to tell revolution and restoration apart," especially amid the "lunatic spectacle" of Regency pageantry designed, as Joel Faflak puts it, "to manufacture subjects' consent through the illusion of their agency."[60] Is there a less triumphal manner of viewing and representing the epochal and, Shelley suggests, interlinked problems "of Life" and of "the times that were / And scarce have ceased to be" (lines 233–34)?

To start to read *The Triumph of Life* as contravening vitalism for the sake of life, or better, for the sake of "those who live," begins to redress a symptomatic asymmetry in the poem's critical reception.[61] On the one hand, Shelley's poem is a privileged test case for deconstructive readings; on the other, the only one of Shelley's poems to have "Life" in its title has received curiously little attention within the recent outpouring of scholarship on the Romantic life sciences.[62] Nor is this because "science," then as now, as Ross Wilson argues, is ill-suited to Shelley's sense of "life," fixing and objectifying an issue that the poet keeps rigorously vague and resistant to "apprehension."[63] For, in fact, as we saw in chapter 1, the early experimental life sciences and natural philosophies were marked from the outset by just such imprecision: vitalist varieties, in particular, pioneered Wilson's conviction that their object was ontologically resistant to conceptualization; "this essential element," writes Lawrence, "is, by its very nature, fluctuating and indeterminate," and "effectually precludes all useful application of mathematics" and "calculation."[64]

The problem is rather that, viewed from within the frame of vitalist organicism that these studies have reconstructed, *The Triumph of Life* seems not to pertain to life at all. As Denise Gigante observes from this perspective, "Rousseau" in the poem "may speak the word *Life*, but . . . he appears not to know what it means."[65] Not quite: Shelley's characters are dissenting from what much of mainstream literary-scientific culture says it means. *The Triumph of Life* is countering the vitalist celebration of new, powerful, autonomous, self-organizing life with a vision of life as a flagrantly disorganized and disorganizing effect: an obsolescent and transitive expression between bodies that multiplies their relations and exacerbates their contingency. Here "women foully disarrayed / Shake their grey hair in the insulting wind," parading their aging bodies against the insult of their exclusion (lines 165–66). The poem is polemically preoccupied with vitalism's constitutive exclusions, which are not incidentally deconstruction's watchwords: the inorganic, the material, the mechanical, and the dead.[66] But Shelley shows that these terms are not necessarily anathema to "life," nonvitalistically conceived.

It is therefore through, not despite, its formal and thematic interest in death and decay and its special strain of anachronistic materialism that Shelley's poem participates in the cultural debate about living form. *The Triumph* does so by drawing us thoroughly outside the vitalist frame, summoning readers to an ancient materialist philosophy that accounted for life without implementing an ontological distinction between the living and other beings, without investing the matter of their bodies with an immaterial "superadded" principle, and even without understanding that matter as distinguished from other kinds by an immanent power or property. Atoms and void, *De rerum natura* insists, are the universe's sole components, and they are not alive. Nor is nature in its totality. Life, self, and sentience issue as second-order effects of the atoms' intricate, accidental concrescences—and they are lost completely at death, when the component pieces disperse and persons, as Shelley's poem puts it, "die in rain" (line 157).[67]

Turning to Lucretius to roll back the agon between organic and inorganic being, Shelley positions morphological susceptibility to influence, to decay, and to rhetorical transfiguration as central, rather than inimical, to life. The next part of *The Triumph of Life* turns the organicist focus on "unity . . . produced *ab intra*" inside out, focusing instead on the "living storm" that passes between beings.[68] In its untimely ignorance of the polemical uniqueness of modern "life," as of modern "literature," Shelley shows, *De rerum natura* has the force of a cultural counterweight, of keeping the science of the living in contact with other kinds of composed and decomposing forms. It does so by affording to matter neither sentience nor spirit, but poetry—a tendency to trope.

e. A Thousand Unimagined Shapes

A "new Vision" follows on the destructive heels of the "shape all light" in *The Triumph of Life*, atomizing its aesthetic absolute into "a thousand unimagined shapes" (lines 411, 491). These notably plural shapes, "like atomies that dance / Within a sunbeam," need to be understood as an alternative "model of figuration in general," in de Man's terms, with dramatically different consequences for the relation between the materials of mortal life, of history, and of trope (lines 446–47).[69] Dancing dust motes in a sunbeam are a famous Lucretian topos, a privileged figure in *De rerum natura* for the chance encounters and entanglements among atoms that, according to the poem's physics, first produced the structures of the universe and continue to propel its incessant change.[70] Lucretius explains that the little ricocheting dust motes are "an image and similitude" (*simulacrum et imago*) of "the first-beginnings of things . . . ever tossed about in the great void" (2.112, 2.121–22).

But this "simulacrum" leads the poet to ask his auditor to contemplate the sunbeam's turmoil for a further reason. The motes are a telling *instance* of the ongoing atomic commotion that they image: subsensible atomic interactions are compounding to produce these dusty dances just big enough to see. The motes' double status as image and instance of their own subperceptible constitution goes to the heart of a Lucretian physics of figuration that is of sustained importance to *The Triumph of Life*.[71] Not only a figure for atomic motion, the dancing motes are a figure for the reality of figures, here credited as conduits between sensuous perception and realities of otherwise inaccessible scale.

In *De rerum natura*, poetic images are just as much found as made: the dusty *simulacrum* at issue, says Lucretius, is "always moving and present before our eyes" (*ante oculos semper nobis versatur et instat*) (2.113). Things offer themselves up to perception in tropes, ostending their sensible and subsensible substance toward the attentive observer. Yet often "like dust clinging to the body," like "the impact of a mist by night, or a spider's gossamer threads," such forms of nonhuman disclosure "are so exceedingly light that they usually find it a heavy task to fall" (3.381–83, 3.387–88). The poet's task is less to impress the reader with the creations of his expressive consciousness than to register and amplify what is already "striving ardently to fall upon your ears," the "new aspect of things to show itself" (*tibi vementer nova res molitur ad auris/accidere et nova se species ostendere rerum*) (2.1024–25). *De rerum natura* is the poetry that works to receive and amplify the subtle impact of what often passes unfelt: it "arrests," in Shelleyan terms, "the vanishing apparitions which haunt the interlunations of life."[72]

In *The Triumph*, the change to a Lucretian optic obscures the grim procession of History writ large and makes visible, instead, a surrounding storm of minute and nonlinear relations: a nontriumphal form of liveliness at work in the interstices of imperial pageantry. Turning away from the famous subjects of the prior visions, this revisionary "new Vision" takes as its object the turbid medium through which the spectators of "Life's" triumphal procession have been watching all along (line 411). And the air through which they move and see proves thick with spectral and equivocal being: a "living storm" rife with cast-off matter and errant particles worked into "busy phantoms" by the sunlight (lines 466, 534).[73] As Lauren Berlant observes of moments when a distinctively "*historical* present" forces itself to be felt and thought within ordinary experience, "we are directed to see not an event but an emergent historical environment that can now be sensed atmospherically, collectively"—a very apt description of what Rousseau directs us to sense.[74]

Reporting from the poem's newly immersive perspective, "plunged" within "the thickest billows of the living storm," Rousseau describes "the earth" as "grey with phantoms," and "the air" as "peopled with dim forms" (lines 467, 466, 482–83). He then explains at length his gradual realization that the whole dense atmosphere in which he is submerged is a weather system of discarded faces:

> ". . . —I became aware
>
> "Of whence those forms proceeded which thus stained
>   The track in which we moved; after brief space
> From every form the beauty slowly waned,
>
> "From every firmest limb and fairest face
> The strength and freshness fell like dust, and left
>   The action and the shape without the grace
>
> "Of life; the marble brow of youth was cleft
>   With care, and in the eyes where once hope shone
> Desire like a lioness bereft
>
> "Of its last cub, glared ere it died; each one
> Of that great crowd sent forth incessantly
>   These shadows, numerous as the dead leaves blown
>
> "In Autumn evening from a poplar tree—
>   Each, like himself and like each other were,
> At first, but soon distorted seemed to be
>
> "Obscure clouds moulded by the casual air,
> And of this stuff the car's creative ray
>   Wrought all the busy phantoms that were there

"As the sun shapes the clouds—thus, on the way
Mask after mask fell from the countenance
And form of all . . ."
(lines 516–37)

Plunged with "Rousseau" into the "living storm," it is by now obvious that we have also been plunged yet again into book 4 of *De rerum natura* (Shelley's favorite, as he once wrote a friend).[75] As each body looses images like falling leaves—"each one / Of that great crowd sent forth incessantly / These shadows, numerous as the dead leaves blown" (lines 526–28)—Shelley remobilizes, as Goethe did, and as Marx would, Lucretius's singular theory of the *eidola*, the figural effusions through which bodies dissipate into the rhetorical medium of their perceptibility and intelligibility. "Fine, filmy floscules, from the surface fly," is how the Shelleys' John Mason Good translation attempts to capture the theory of "Images of Things," the overzealous alliteration attuned to the quick, serial emission of likenesses that subtly alter in their passing (line 531).[76] In Rousseau's retelling, the living transfer "mask after mask" from face to figure, "countenance" to simulacrum, in a process that both gradually depletes their youth and beauty ("From every form the beauty slowly waned") and nuances the "light's severe excess" (line 424). Shadowing the scene ("the grove / Grew dense with shadows" (lines 480–81)), the airborne figures seem to make inductive knowledge possible for the first time in a poem that has pilloried enlightenment as an annihilating kind of brightness. "I became aware," says Rousseau, "Of whence those forms proceeded which thus stained / The track in which we moved," attesting to a kind of knowledge that proceeds from contact with figures turbid rather than bright, the perception of which is also an intimation of their causes.[77]

Exceedingly slight, like flakes or dust in Shelley's analogies, Lucretian "semblances and tenuous figures" (*effigias tenuisque figuras*) (4.42) are amenable to dissolution and recombination—"Each, like himself and like each other were, / At first, but soon distorted seem to be," as Shelley has it (lines 530–31). Coalescing or expanding, they join the clouds that form our atmosphere, "Obscure clouds moulded by the casual air" (line 532). Perhaps most critically for their Shelleyan adaptation, Lucretian *figurae* "wander" long distance and long term, texturing any perceived present with shades of distant and prior happenings. As Good has it, "from all things vestiges there are / Of subtlest texture hovering through the void" (4.92), the belated "phantoms" that "people" Rousseau's sight.

Like Goethe's morphology, Shelley's neo-Lucretian "shapes" point us toward materialist figuration as a mode of decadent (down-falling) and extravagant (outward-wandering) transfer: a process by which corporeal beings produce an order of sensible experience that is coextensive with their transience. This is an order no less real, material, or consequential for being figural and transitory. Here figures are fractions of the real estranged from their sources, and all bodies, not just verbal ones, are granted the capacity, indeed the necessity, of producing them: "From every firmest limb and fairest face / The strength and freshness fell like dust" (lines 520–21).[78] By Lucretian lights we can read the "thousand unimagined shapes" of Shelley's "new Vision" in a different sense. Such shapes may be "unimagined" because they are, tenuously, exogenously, real.

The possibility goes some way to account for an observation that recurs in the most sensitive assessments of Shelleyan figuration from the early nineteenth century to the present: it has a peculiar pretense at substance, at being "something more" than metaphor. Shelley, wrote Hazlitt, paints "pictures on gauze. . . . and proceeds to prove their truth by describing them in detail as matters of fact"; Walter Benjamin praised Shelley's particular "grip" (*Griff*) on allegory.[79] William Keach traced the disorienting effects of Shelleyan language to a penchant for "reversed or inverted simile," which uses "mental" vehicles to express "physical" tones; while James Chandler concludes that Shelley's figures for history never stand for "merely an affair of 'consciousness,' the immaterial internality of a material externality."[80] According to Alan Bewell, when Shelley figures imperial power as an atmospheric pestilence, he is going "beyond metaphor" to claim that "all climates *are* climates of power."[81]

As we have begun to see, for Romantic-era readers receptive to the possibility, *De rerum natura*'s model of poetic materialism offered something more than the didactic aspiration to educate and delight, despite the modest "honeyed cup" topos by which Lucretius claims to be coating harsh philosophical contents in palatably sweet poetic form. Lucretius's rare thought about *simulacra* confers a material truth to figures, and gives a figural turn to the material husks that compose the "data" of sensuous, empirical perception. *De rerum natura* exemplified the mutual implication of poetry and empirical knowledge in multilateral processes of figuration that must be the object, means, and medium of any investigation of nature. To summon it was, increasingly, to resist the long-term social transformation that was making prose the presumptive medium for communicating social and natural realities and securing poetry its enduring place at the prestigious or "unacknowledged" margins of cultural life.[82]

If, as Peacock remarked, "the empire of facts" was increasingly "withdrawn from poetry," the facts, from the neo-Lucretian perspective Shelley summons, were hardly to blame: in the figural acuity of poetry, facts can "clamor," as Lucretius has it, into communication, through improprieties of human language crucially assignable to neither knower nor known alone.[83] The problem, as Shelley puts it in a trenchant oversimplification, is "empire" as a model and effect of knowledge as a collective enterprise: "The cultivation of those sciences which have enlarged the limits of the empire of man over the external world, has," Shelley argues in *A Defence of Poetry*, "for want of the poetical faculty, proportionally circumscribed those of the internal world; and man, having enslaved the elements, remains himself a slave."[84] The production of what Peacock calls "methodical, unpoetical, matters-of-fact" represents a form of inquiry, as Goethe showed of Kant, so thoroughly immunized against the collaboration of its objects and others and the passivity this might demand— against the art of being "abandoned to the sudden reflux of the influences under which others habitually live," as Shelley is wont to say of a poetic habitus "not subject to the controul of the active powers of the mind"—that epistemic modesty becomes a form of domination.[85] It is in the name of such eminently modern, professional decorum that Lawrence expressly eschews the "fictitious beings of poetry" and the tempting "unnatural union" that would establish "the poetic ground of physiology."[86] As we have seen, his strict adherence to "the path of observation" issues in a racist physiognomy that bears witness less to the natural-historical facts than to the comparative anatomist's remarkable capacity to shield his criteria of observation from alteration by another kind of face.

The Triumph's last neo-Lucretian vision, of course, renders such defenses literally and figuratively futile. Here being a body in time means shedding atoms of self—involuntarily re-presenting oneself—and weathering others' particulate bombardment: "Bending within each other's atmosphere," as Shelley puts it in the poem (line 151). Beings decay into perception, experiencing each other by means of an incessant exchange of similitudes that is neither willed, nor linguistic, nor without physical cost. We can say, with de Man, that "Rousseau" is indeed telling the story of a distinctly prosopopoetic demise, and that bodies indeed are disfigured in the course of all this figuration—but gradually, slightly, even gently. If for de Man *The Triumph of Life* allegorized the inevitable, face-exchanging madness of words, Shelley is intent to revive *De rerum natura*'s incessant face exchange among *bodies*, a giving and receiving of face and figure of which linguistic activity is just one instance. The resultant changes of face mark less the special, violent interruption of figuration, cognition, and relation by an alien materiality ("actual events," the "madness of words") than the routine material condition of appearing at all.[87] In Shelley's neo-Lucretian mode, figuration and mortality nontraumatically coincide: the "new Vision" of "Rousseau" in the poem is about giving each other wrinkles, weathering the innumerable touches of each other's "busy"-ness (line 534).

"Rousseau" likens the impact of the vision's airborne simulacra to dingy, somewhat enervating snowflakes:

> "And others like discoloured flakes of snow
> On fairest bosoms and the sunniest hair
> Fell, and were melted by the youthful glow
>
> "Which they extinguished . . ."
> (lines 511–14)

In this meeting between phantom-flake and living skin a mist is produced, a rain that seems perspiration, precipitation, and passion all at once. "For like tears," the phantoms "were / A veil to those from whose faint lids they rained / In drops of sorrow" (lines 514–16). We are here confronted with the disconcerting reality of material figures: at first *like* weather (snowflakes) and *like* tears (drops of sorrow), the phantoms ultimately lose the hesitation of simile and erode the distinction between personal psychic sorrow and collective atmosphere. "They" simply "rained / In drops of sorrow" from "faint lids," lids whose owners might have been weeping, or simply rained upon.[88] It is important, though, that these sorrow drops are not strictly—or not exclusively—theirs. Well-traveled simulacra wet eyes, condition seeing, and reshape faces in this area of the poem. Nor is it all so weepy. The busy phantoms perform a remarkable variety of corrosive activities: flinging, flying, dancing, chattering, playing, nesting, thronging, and laughing, too, between the bodies that prolifically lose and loose them (lines 487–510). Where being a body means induction in an unchosen collectivity of decay and transfiguration, the weather takes on a decidedly social dimension.[89]

For, in fact, the poem's last vision in no way denies the worry over historical complicity that marks the earlier modes of viewing in *The Triumph of Life*, but rather furiously multiplies it to the point where "Rousseau's" personal selfhood and its drama of choosing come undone: "I among the multitude / Was swept," recalls "Rousseau,"

> ". . . me sweetest flowers delayed not long,
> Me not the shadow nor the solitude,

>"Me not the falling stream's Lethean song,
>   Me, not the phantom of that early form"
>   (lines 460–64)

Reciting "me . . . not," "Me not," "Me not," "Me, not," the lines rhythmically subdue this author's iconic, prolific, autobiographical subjectivity. Whereas "Rousseau" emerged in the poem's first vision to lament his decay as a "Stain" upon "that within which still disdains to wear it" (line 205), this final vision chooses to tell the history of the stain over that of the personal "Rousseau." The poem's repeated autobiographical demand, "Shew whence I came, and where I am, and why—," is answered, finally, with a history of "whence those forms proceeded which thus *stained*/The track in which we moved" (lines 398, 517-18, emphasis added). The ensuing vision of life depicts selves as steeped in a staining kind of atmosphere that, as we have seen, both animates and decimates. Persons "Kindle invisibly" and are gradually "extinguished" under the impact of the climate they collectively generate and endure (lines 152, 514).

Nearly two decades ago, Steven Goldsmith diagnosed the radical promise of negative aestheticism in Shelley and his readers as the "Demogorgon principle": a late iteration of the tradition of apocalyptic formalism whose history-ending work on words necessitates a corollary "source of pure negativity . . . free from the weight of a particular body" to do the dangerous, embodied, and ethically ambiguous work of revolution.[90] The principle is named after the (notably unwrinkled) Demogorgon, "Ungazed upon and shapeless:— neither limb/Nor form nor outline," who drags Jupiter off his throne in *Prometheus Unbound*, permitting the play's hero to make revolution by bloodless speech act.[91] Indeed, this gesture recurs plentifully in Shelley's poems, coinciding with a literally apocalyptic (off-covering) aspiration toward "the painted veil/that those who live call life." Shelley has tended to picture this, from *Queen Mab* to *Prometheus Unbound*, as a universal face-lift: "No storms deform the beaming brow of heaven," reports the Fairy Queen "in her triumph"; people, too, appear wrinkle-free "through the wide rent in Time's eternal veil": "How vigorous then the athletic form of age!/How clear its open and unwrinkled brow!"[92]

All the more notable, then, that *The Triumph of Life* reversed this trajectory, "plung[ing]" with "Rousseau" "among/The thickest billows of the living storm/ . . . /Of that cold light, whose airs too soon deform" (lines 465–68). Here the poem commits, with sustained, microscopic attention, to the production of what *Mab*'s Fairy disdained as "The taint of earth-born atmospheres" out of what *Prometheus Unbound* called "ugly human shapes and visages."[93] As Paul Hamilton puts it in a sensitive reading of *The Triumph*'s Lucretian dimension, this last vision "makes a *poetic* subject of the objective processes to which we belong and over which Rousseau imagines retaining individual proprietorial rights."[94] Eschewing autonomy as life's signal attribute, Shelley's poem instead attends to wrinkles as the cumulative signs of the collective life that makes and unmakes subjects in the first place and last instance.

f. What Shares the Air

Nor would Shelley's account of mixed and seething air have seemed "merely" metaphorical in 1822. Adam Walker, the popular scientific lecturer who taught Shelley natural philosophy at Syon House and later at Eton, described the atmosphere as "a grand receiver, in which all the attenuated and volatilized productions of terrestrial bodies are contained, mingled, agitated, combined, and separated." His lecture went on to list the sheer diversity of "attenuated and volatilized" particles in which the turn-of-the-nineteenth-century body was bathed: "mineral vapours, animal and vegetable moleculae, seeds, [and] eggs," all "dissolved in light." Such air, Walker taught, does not stay safely outside a body, but is "so subtil that it pervades the pores of all bodies and enters into [their] composition."[95] English air, moreover, as Mary Favret has shown, was increasingly understood to bear currents from farther and farther afield, as meteorology changed from a study of local exhalations to "a global system of communication."[96]

The Triumph of Life ends by challenging us to take up the position of that laden, interpersonal and impersonal air. From the perspective of this atmosphere ("living storm," "casual air," "stain"), the triumphal pageant that has been the object of each successive vision looks quite different. Powerful actors are still visible in this latest version of the procession: "pontiffs" (line 497), kings, and "lawyer, statesman, priest and theorist" (line 510)—but none of them has a recognizable face. We have here a case of "anonymous life," in Jacques Khalip's terms, geared toward "saturation *in* the world" and a "reformulation of political agency as something impersonal, asystematic, and nonintentional."[97] For the busy phantoms that make up the "living storm," the bodies of monarchs are nothing more, or less, than habitat: some "played / Within the crown which girt with empire // A baby's or an idiot's brow, and made / Their nests in it" (lines 497–500). In this way the switch to Lucretian figural materialism has enabled Shelley to depict an anarchistic form of liveliness mixed in with life's imperial show: the "old anatomies" of kings are inadvertently "hatching . . . base broods" (line 501) that will ultimately "reassume [their] delegated power" (line 503).

As the poem takes up a view, as through a microscope, that reveals how the cast-off "atomies" of subjects are transformed into new "busy phantoms" that "nest" in the bodies of sensible size, Shelley takes care to provide all the requisite topoi from the ongoing debate concerning "spontaneous" or "equivocal" generation: warmth, moisture, dead matter, and new, "busy," "unimagined shapes." To review, equivocal generation (for which *De rerum natura* was the preeminent locus classicus) held that living beings could emerge, without parents, without seeds, and without intercourse, from "mere brute matter" (*abiogenesis*) or from once living "organic particles" of a different species (*heterogenesis*). As Erasmus Darwin put it in *The Temple of Nature* (1803), "without parent by spontaneous birth / Rise the first specks of animated earth;/From Nature's womb the plant or insect swims,/And buds or breathes, with microscopic limbs."[98] As we saw in chapter 2, opponents derided the doctrine as violation of the basic laws of cause and effect, the biological version of the basic scandal of aleatory atomist causality: a "system of *causality*" Kant called "so obviously absurd that it need not detain us."[99]

But this is exactly what captivates *The Triumph of Life* as it seeks to elude the teleological insularity of organic form: the decadent exfoliations of the living are "moulded by the *casual* air," in marked (or fortuitous) distinction from *causal* (line 532, my emphasis). Equivocal generation holds open the chance that different life might emerge out of the matrix of the familiar, eluding the proper causal couplings by which like produces like. (The shed simulacra begin in univocal likeness, "Each, like himself and like each other," but "soon distorted seemed to be" [lines 530–31].) Shelley gestures toward a liveliness "wrought" from already-circulating materials—"discoloured," "grey," and "stained"—and pointedly continuous with accident and decay (lines 534, 511, 482, 517). Like the atoms of Lucretian poetic science, these materials' ontological status is not inherent, but rather dependent on the congeries into which they chance: some configurations are vital ("like small gnats"), some material ("like discoloured flakes of snow"), and some textual ("numerous as the dead leaves blown," a borrowed epic simile). They equivocate between familiar ("flies") and strange ("vampire-bats"); between dying and living ("phantoms"); between the discourses of poetry, biology, and history (lines 508, 511, 528, 484, 482).

The poem stays studiously agnostic about this vital, material, and figurative swarm, whose chancy productivity is a far cry from an optimistic faith in the future progress of "Life" or history in their madcap career. Its promise is emphatically, technically minimal: the last vision describes—without laying a hope upon—the atom-thin margin by which the present deviates from itself, offering only the hard consolation that material processes (history, life) must trope rather than repeat themselves. This tone is hard to hold, and does not, in my view, amount to a politics, whether of the affirmative kind that Hugh Roberts finds by linking Shelley's clinamen to contemporary chaos theory, or of the markedly "*un*swerving negation" that Khalip, after capturing perfectly the poem's sense of "Life" as "the word that cannot be properly assigned to any one person," conceives as an Adornian "refusal to lend itself to sanctioning things as they are."[100] Anne-Lise François, who so nuances the critical vocabulary of neutrality and givenness, may best, if incidentally, capture *The Triumph*'s ethos in this place when she writes in another context of "the suspension of the impulse to take revenge on the given and decry transience as a deprivation." "In such a mood," she continues, "one recognizes instead that the lapse between life's premise and its fulfillment constitutes no more of a cheat than the Other's difference from oneself."[101] In *The Triumph*, Shelley's neo-Lucretian mood, too, entails a cessation of hostilities against life as transiently given, which is also to say, shaped into imperfect impermanence by the impact of others' partings.

A representatively subtle effect of the troping materialist alternative at work in this part of the poem is to revise one of Shelley's lead metaphors for oppression and resistance. *The Mask of Anarchy* (1819), as we will see in chapter 5, hailed the "Men of England" as "Nurslings of one mighty Mother," enjoining them to "Rise like Lions after slumber" against systematic economic exploitation and political exclusion.[102] Under the impact of the "drops of sorrow" we have been tracking in *The Triumph*, however, the leonine image returns with a vengeance, but from a perspective of uncompensated bereavement rather than redemption:

> ". . . the marble brow of youth was cleft
>   With care, and in the eyes where once hope shone
> Desire like a lioness bereft
>
> "Of its last cub, glared ere it died . . ."
> (lines 523–26)

The lines reveal the kind of prosopopoeia of which "a great crowd" is inadvertently capable: their attenuated "thousand" touches amount to something that resembles an artist's deliberate labor, as the enjambed phrase "was cleft / With care" turns the head-splitting violence of cleft into a scene of careful craft (the poet's careful cleaving of lines, or a sculptor's work on marble); or rather, suggests that the accidental impact of others' airborne "care[s]" might amount to the same thing. Next transfiguring the blow of "cleft" into a worry line, the prosopopoeia delivers a subverted Pygmalion myth in miniature: only in creasing under the weight of care does the "marble brow" transform from stone to skin in these lines, sensible now as the mobile forehead of a living face. The conflicting senses of cleaving here—holding together or dashing apart—capture the dual aspect of Lucretian transience as the loss or waning of one's "own" materials under the sculpting pressure of others' contact.[103]

Against the passage's predictable lament for beauty lost, then, and in concert with Shelley's neo-Lucretian experiment, this kind of prosopopoetic animation makes its marble Galatea *age* into life. Moreover, here aging coincides with the increase of a different form of political desire in the passage, as something animal, maternal, and aggrieved takes over "in the eyes." In *The Triumph* we are eventually seeing by the light of this fierce maternal glare, also called "the glare / Of the tropic sun" (lines 484–85). Fusing personal senescence and colonial politics, proximate and distant, this transfigured face issues a demand the poem leaves unmet, a "desire" for redress that outlasts the "hope" of its arrival (lines 525, 524).[104]

g. Atmospheres of Sensation

Wordsworth's phrase "atmosphere of sensation" is a better name than most for Shelley's reanimation of Lucretius's likeness-laden air and the disciplinary challenge it posed. The phrase emerges in a testy section of the 1802 Preface that assesses the prospects of future collaborations between the "Poet" and the "Man of Science."[105] The Poet, argues Wordsworth in a very Epicurean vein, views "man and the objects that surround him as acting and re-acting upon each other, so as to produce an infinite complexity of pain and pleasure" (258). He would be happy to take the "remotest discoveries of the Chemist, the Botanist, or Mineralogist" as his "proper objects," on the condition that he can find among them "an atmosphere of sensation in which to move his wings" (260, 259):

> The remotest discoveries of the Chemist, the Botanist, or Mineralogist, will
> be as proper objects of the Poet's art as any upon which it can be employed,
> if the time should ever come when these things shall be familiar to us, and

the relations under which they are contemplated by the followers of these
respective Sciences shall be manifestly and palpably material to us as enjoying and suffering beings. (260)

It might be reasonable for Wordsworth to want to know about the wider import of "discoveries" that seem "remote" and specialized, but upon close inspection, this is not what the passage demands. Instead, Wordsworth wishes
to feel the *relations of contemplation* that obtain between scientific objects and
their specialists—to have those relations rendered "manifestly and palpably
material" to (the rest of) "us."

An "atmosphere of sensation" would be one name for the fulfillment of
that challenge: it suggests that the space *between*, over, and around knowers
and objects of knowledge at the scene of empirical experience, the set of affective, social, and historical relations that scaffold and mediate that encounter, must itself attain palpable form. Though it seems at first that this task
falls exclusively to the "Man of Science," in fact his assignment closely echoes
that of "the Poet" in the Preface, binding the roles together as Shelley would
seek to do. In poems, Wordsworth argues, certain *relations of contemplation*
must themselves obtrude as phenomena. To drastically abbreviate a passage
that in fact performs the tortuous "fluxes and refluxes" involved in relation-
contemplation, it is by "contemplating the relation" *between* various mental
and affective experiences that "we [Poets] discover what is really important to
men" (246–47). Wordsworth's projects for both Poetry and Science meaningfully coincide in the task of making relations "palpable and material," the same
gesture that uncovers them as "really important," as mattering, and that stands
to overcome the antithesis he momentarily poses between "Poetry and Matter
of fact, or Science" (254). The phrase "atmosphere of sensation" points us to
a role for scientific poetry or poetic science in intensifying, in making felt—
making "manifestly and palpably material"—the contemporary atmosphere,
the slight but real historical relations that pressure and support present perceptual experience, conditioning empirical induction and sensory pleasure
alike. This would be the tumultuous medium that Goethe called *Trübe*, into
which life transpires in Shelley's nontriumphal sight.[106]

Writing some two decades before Shelley's *Triumph of Life* and *Defence
of Poetry*, Wordsworth had projected his collaboration between "the Poet"
and the "Man of Science"—what Lawrence would eschew as an "unnatural
union"—in a conditional and subjunctive mood, as a hypothetical wager dependent upon conditions the science-men seemed unlikely to meet. But by
the time of Shelley's writing in the early 1820s, the high bar Wordsworth set,

of "material revolution in our condition" and Science in "flesh and blood" form, had been roundly met in an unexpected way that goes some way to explain why Shelley's *The Triumph of Life* widens the lens of biopoetic experiment to encompass what seems like the whole contemporary social field. For what happened in the intervening two decades took the science question from the dyadic "relations of contemplation" between knower and known I examined through Goethe—that intimate scene of experimental looking that Wordsworth, too, seems to have had in mind when he represented the "Man of Science" loving truth in myopic conversation "with those particular parts of nature which are the objects of his studies"—toward what Shelley's *Defence* calls the "application of analytical reasoning to the aberrations of society."[107] Shelley is marking what sociologists of science identify as an early nineteenth-century "revolution in government" driven by men of science as an increasingly powerful and professional class, who understood their vocation as providing technical solutions to the palpable Regency threat of social dissolution.[108]

h. Historical Material

Choosing "phantom" as the lead term for the figural-material simulacra that condition perception and make aging an exercise of the age in *The Triumph*, Shelley emphasizes the belatedness that adheres in Lucretius's theory of the palpable, present atmosphere. In *De rerum natura*, any sensation participates literally, particularly, in the obsolescent reality of whose expenditure it is the sign: "the spent vision of times that were / And scarce have ceased to be" (*The Triumph*, lines 233–34). And *The Triumph* stresses that what Shelley calls a "mutual atmosphere," brings a body into contact not only with figures shed from nearby persons and things but also with aged particles and long-traveled images.[109] Filling up the interstices between bodies separated in space and time, the poem's atmosphere constitutes a subtle but real medium of "attenuated and volatilized" inter-influence that exceeds conceptual clarity and linear causality. We are concerned here with what Kevis Goodman, following Raymond Williams, has theorized as "social experience 'in solution,'" an "immanent, collective perception of any moment as a seething mix of unsettled elements," and with the retroactive fitness of Lucretian poetic physics as an ontology and lexicon for describing this kind of Romantic historical experience.[110] In Shelley's poem, Lucretian materialism is helping to articulate "unapprehended relations" between events both too big to grasp and too small to notice: the medium in which the figure of a distant person's sorrow might strike impalpably, like a snowflake, and in which the diffused image of Napoleon in chains might subtly reshape a particular face—"I felt my cheek / Alter to see the great form pass away" (lines 224–25).

In his attempt, in the theses "On the Concept of History," to delineate a historical materialist practice that could resist the triumphalism of present progress narratives, Walter Benjamin also wondered if the belated business of the departed does not touch us in air:

> Doesn't a breath of the air that pervaded earlier days caress us as well? In the voices we hear, isn't there an echo of now silent ones? Don't the women we court have sisters they no longer recognize? If so, then there is a secret agreement [*Verabredung*: date, rendezvous] between past generations and the present one.

> [Streift denn nicht uns selber ein Hauch der Luft, die um die Früheren gewesen ist? ist nicht in Stimmen, denen wir unser Ohr schenken, ein Echo von nun verstummten? haben die Frauen, die wir umwerben, nicht Schwestern, die sie nicht mehr gekannt haben? Ist dem so, dann besteht eine geheime Verabredung zwischen den gewesenen Geschlechtern und unserem.][111]

Benjamin's question is one, as Shelley might put it, of "what stain[s] / The track in which we mov[e]."[112] In fact, this thesis links terrestrial happiness to the messianic-materialist "fight for the oppressed past" by way of a stain or tinge, both of which share an etymology in *tingere* (Latin): our "image of happiness" (*Bild von Glück*), he writes, "is thoroughly colored [*tingiert*, tinged] by the time to which the course of our own existence has assigned us."[113] The same is true, Benjamin continues, for the "idea of the past, which is the concern of history."[114]

At moments in these theses, the historical materialist's capacity to represent something other than "homogeneous, empty time" is described in the language of vital power: not "drained by the whore called 'Once upon a time,'" for instance, "He remains in control of his powers—man enough to blast open the continuum of history."[115] But through the second thesis's questions, which move from the thought of air that will have touched both past and present beings, to the citation of the past in present speech, to unacknowledged or unapprehended (familial) relations, Benjamin arrives at the notion of "*weak messianic power.*" Here "power," it turns out, sounds more like a slight susceptibility, a mundane being claimed or spoken for that has less to do with the perspective of angels for which the theses are famous than with the fact of having had expectant parents:

> Then our coming was expected on earth. Then, like every generation that preceded us, we have been endowed with a *weak* messianic power, a power on which the past has a claim. Such a claim cannot be settled cheaply. The historical materialist is aware of this.

> [Dann sind wir auf die Erde erwartet worden. Dann ist uns wie jedem Geschlecht, das vor uns war, eine *schwache* messianische Kraft mitgegeben, an welche die Vergangenheit Anspruch hat. Billig ist dieser Anspruch nicht abzufertigen. Der historische Materialist weiß darum.][116]

In this thesis, then, the overlapping time of terrestrial generations and their (missed) happiness are aligned with the historical materialist's antiteleological, constructivist, and constellating intuition, and against progressive historicism's triumphal procession of victors' history.[117]

In an early iteration of the argument that poetry does best to negate contact with its age, John Stuart Mill argued, closer to Shelley's own time, that poetry cannot be "tinged" with "lookings-forth into the outward and every-day," or it "ceases to be poetry, and becomes eloquence."[118] But a poetics of the weakly messianic tinge is precisely what *The Triumph* seeks as it attempts to rethink embodied life, nontriumphantly, as transience "into the outward and everyday." Indeed, life here becomes a decadently transitive figure—a touch, a tinge, a stain—between those who inhabit a particular present, and between those who are present and those who are coming.

Lucretian poetics are also, quite specifically, a poetics of the eloquent tinge and touch. Abbreviating these poetics into the topos of the "honey'd cup"— Lucretius's verse is the "sweet honey" that makes the bitter medicine of Epicurean physics palatable—we lose the fact that the poet's work, the poet's verb, is not to sweeten, but to touch together, to tinge, the action of Benjamin's weak messianic air and Mill's unlyrical eloquence. *Contingere*, says Lucretius of his work at the beginning of the book that introduces the notion of the material simulacra: *musaeo dulci contingere melle*, "I have chosen . . . to touch [my doctrine] with the Muses' sweet honey" (4.20–22). And to describe poiesis as *contingent* in this way actually fortifies its reality: after all, in *De rerum natura*, reality only ever occurs because, amid the rain of first particles falling in parallel, two come into contact, initiating, "from encounter to encounter, a pile-up and the birth of a world."[119] Shelley's poem on life seeks out this touched, tinged, contingent matter as equipped to bring the discourses of history, life, and poetry into nontriumphal and timely contact.

The poem pictures "Life" as wrinkled by these serial touches, as "a Shape /
. . . whom years deform," and who is now legible not as a negative allegory for embodied life but as a sustained experiment in the epistemology of aging, in positioning contingency and decay as constitutive of, rather than inimical to, life and its science (lines 87–88). This figure of wrinkled life in the poem has a curious "charioteer": a four-faced "Janus," Roman god of beginnings and endings, with "banded eyes" (lines 99, 94, 103). Clearly, this driver is no formative drive, no *Bildungstrieb* teleologically ensuring life's progressive development. This blind figure drives by touch instead, keeping life's course contingent upon the pressure of its historical atmosphere.

A Natural History of Violence: Allegory and Atomism in Shelley's *The Mask of Anarchy*

If the aim were to continue to establish and explore the extent of what Hugh Roberts has rightly called "Shelley's Lucretian Imagination," there would be clear places to turn to in Shelley's oeuvre, from the atomist cosmology of his "Philosophical Poem," *Queen Mab*, to the "fast influencings" of *simulacra* that the poetic mind "renders and receives" in "Mont Blanc," as well as a number of existing critical accounts on which to build.[1] But this chapter instead focuses on a fleeting Lucretian shape that appears in perhaps the least likely of Shelleyan places: in *The Mask of Anarchy, Written on the Occasion of the Massacre at Manchester*, the poet's searing verse response to a catastrophic outbreak of police brutality against nonviolent demonstrators in Manchester in August of 1819.[2] Having recourse to Lucretian poetics at an obtrusive hinge in the poem between keen-sighted political "Despair" and a kind of political "Hope" that critics have judged self-serving, deluded, and double-edged, *The Mask of Anarchy* effectively puts "sweet science" to the bitter test of political violence, summoning materialist figuration to the task of communicating an event that, in Shelley's depiction, mars the face of nature with blood.

That is, I would like to put forward *The Mask of Anarchy* as a kind of political challenge to the Romantic neo-Lucretianism I have offered in the preceding chapters, particularly to its semiotic naturalism and ethics of epistemic and corporeal vulnerability. How do tender empirical, materialist poetics accommodate the historical fact of vulnerability violated? Can such techniques register or oppose the kind of violence that, cutting lives short and breaking the skins that shelter and disclose them, would seem to sever political from natural life and historical time from natural transience, as well as to demonstrate the cost of confusing the two?

Shelley's allegorical condemnation of what was quickly dubbed the "Peterloo Massacre" does in fact hinge on a tender-empirical moment: "A sense awakening and yet tender" attributed to the poem's exploited "multitude" in their interaction with elements of earth and air (lines 136, 41). This moment—a *figura-ex-machina* moment in which a flagrantly poetic "Shape" handily finishes off the task of political struggle—has been taken for a tendentiously lyrical and aesthetic one. Consequently, *The Mask of Anarchy* elicits brilliant readings of the political risks and limits of lyrical poetic action.[3] But as an unrecognized pivot to Lucretian materialist poetics, this moment in fact reworks the old crux between aesthetics and politics, not least by altering the generic criteria by which a poem's political pretenses stand to be judged.[4] While Romanticists tend to use "aesthetic" to mean Kantian "purposiveness without a purpose" and its lyric exemplars, by harking back to *De rerum natura* Shelley raises the unapologetically purposeful specter of *didactic* epic instead. Concomitantly, the poem begins to describe the state violence of "Peterloo" as an event in natural history, obsessively fretting the question of how the material evidence of the political clash—the wrongly spilled "blood" of a "prostrate multitude" demonstrating for their rights—is absorbed and recirculated through atmospheric, geologic, and metabolic cycles (line 126).[5] Why would Shelley summon the perspective of natural history, an epistemic orientation now faulted precisely for its *lack* of historical consciousness, to produce what he considered a "wholly political" condemnation of an event of national history?[6] What does it mean that the poem's acclaimed proto-Marxist structural analyses and stirring calls to mass action are attributed not to a human speaker, but to the "indignant Earth" (line 139)?

I argue in this chapter that the poem's tender empirical turn taps into an eighteenth-century tradition of neo-Lucretian didactic poetry that moved between social history and nonhuman nature not to "naturalize" historical events but to attempt their revisionary reconstruction: to "correct," as William Cowper put it "The Clock of History . . . / . . . unrecorded facts / Recov'ring, and mis-stated setting right."[7] The positive, positively *pedantic* attitude in Cowper's lines is something I wish to hold onto in this chapter, one task of which is to argue for the historicizing intelligence and Romantic-period survival of Lucretius's didactic genre and the natural-historical vocation with which it is entwined, both of which share the dubious distinction of being declared extinct by scholars of the period.[8] In a switch in sensibility, the didactic poetic ethos I outline below begins to subject the misty historical atmospheres that have built up across the preceding pages to analysis. That is, didactic poetics not only register and amplify the sensory and affective impacts that "unrecorded facts" (Cowper) exert within the present "atmosphere of sensation" (Wordsworth), but also risk an-atomizing the feeling, de-composing Lucretian apparitions in order to narrate the particular, *particulate* histories that subtend a present scene.

Spelling out the (violent) history of particles that do not, in themselves, disclose their provenance in human language, the didactic poetic speaker pulls into view contexts and causes that make an apparently exceptional event such as "Peterloo" all too probable and all too likely to repeat. One feature of the long nineteenth-century "lyricization of poetry" that has lately come under critical scrutiny, I begin to suggest here, was to guard against this inherently partisan activity of Lucretian didactic—an activity that poems in this tradition continue to make teachers and students perform.[9] So this chapter ends by exploring the possibility that, as would befit a didactic poem, *The Mask of Anarchy*'s aesthetic politics are less about the practical (in)efficacy of poetic speech than about the politics of pedagogy: they concern the limits and possibilities of poetic instruction in radical action, historical remembrance, and emancipatory means of generating and communicating knowledge.

Overall, I see *The Mask of Anarchy* as staging a contest between two different semiotic regimes, and more particularly, between two systems of allegory and didactic allegoresis. Both deploy allegory's characteristic duplicity, its "other-speaking," to levy devastating critiques of English tyranny. But at the hinge marked by a flagrantly Lucretian material "Shape," Shelley shifts from a semiotic system in which the allegorical second-sense of the violence at Peterloo is itself violently imposed upon the bodies of "the multitude" from above, to one in which such allegorical meaning is, to crib from *The German Ideology*, the "direct efflux of their material behavior" at the scene, reflected back for further interpretation. And while the poem's first form of allegorical double-speak encodes a *moral* condemnation of the massacre, this second kind speaks social theory instead, deriving from the "Shape" a systematic critique of ordinary immiseration, wage slavery, and the money form. In moving from one mode of allegory to another, that is, Shelley's poem has much to say about the kind of semiotics conducive to expressing materialist conceptions of history and biopolitical analytics of power—as well as about the combination of feeling and teaching that critical, materialist pedagogy requires.

a. "The Occasion of the Massacre at Manchester"

As *The Mask of Anarchy*'s subtitle professes, the poem was "Written on the Occasion of the Massacre at Manchester," and constitutes Shelley's fast and furious verse response—a long-distance response, from voluntary exile in Italy—to news of the slaughter of unarmed demonstrators by English militiamen at Manchester on 16 August 1819. Writing from newspaper reports of the massacre quickly dubbed "Peter-loo," Shelley first submitted his poem to Leigh Hunt for immediate publication in the *Examiner*. By spring of the following year, Shelley characterized *The Mask of Anarchy* as the first in a series of some ten post-Peterloo poems he proposed would constitute "a little volume of *popular songs*, wholly political, & destined to awaken & direct the imagination of the reformers."[10] But in the repressive British political climate after Peterloo—fearing revolution, Parliament levied "Six Acts" against "radical reform," thoroughly constraining freedoms of press, speech, and assembly—Hunt declined to publish *The Mask* for more than a decade. It emerged, finally, in 1832, to celebrate the Reform Bill's realization of changes for which the poem agitated; or, in Chartists' eyes, to militate for more.

The Mask's combination of punctual address and posthumous publication is just one of many performative contradictions that have made it a lightning rod for critical debates about aesthetics and politics and the pretenses of poetry at historical consequence. The poem issues an impassioned call for working-class Englishmen to risk their lives in further acts of mass assembly and nonviolent protest, but does so from the isolated safety of Italy and of aristocratic privilege; it offers an astonishing analysis of systematic working class exploitation, yet shies away from Revolutionary violence; it promulgates, in a near-Burkean vein, "The old laws of England" as "Liberty's" best defense, but does so in double-edged words that make their complicity with "Oppression" unmistakable.[11] *The Mask of Anarchy*'s formal contradictions make plain "the ideological bind of proffering poetry as the thing to be 'done' in political crisis," Susan Wolfson argues.[12] And as William Keach has pointed out, those who have sought to rescue the poem's political efficacy have had to downplay such formal contradictions.[13] Or, I would add, they have turned them to good through a negative aesthetic formalism keen to see the poem's political message self-destruct in the interest of a purer invitation to futurity.

On 16 August 1819, some sixty thousand people gathered on St. Peter's Field in Manchester to demonstrate for expanded suffrage and parliamentary reform: "for the purpose," as their handbills put it, "of taking into consideration the most effectual legal means of obtaining a Reform in the Representation of the House of Commons." As the principal orator, Henry Hunt, began to address the audience, the local Manchester militia, backed by a regiment of British hussars, forced its way through the packed assembly to arrest Hunt and disperse the crowd. "Hacking" and "hewing" as they went (according to eyewitness accounts quickly disseminated in the newspapers and repeated in Shelley's poem), the local police and national military forces together killed a dozen people and left hundreds wounded.[14] Writing to the Earl of Derby on behalf of the Prince Regent in a widely publicized letter, Home Secretary Henry Addington, Lord Sidmouth—whom Shelley's poem would depict as a crocodile-riding personification of *Hypocrisy*—commended the area militia for their "prompt, decisive, and efficient measures for the preservation of the public tranquility."[15]

Public outrage against "our should-have-been protectors" was equally prompt and decisive.[16] Seizing on the fact that the perpetrating militiamen included veterans of the British victory at Waterloo, the left-wing press stressed the link between foreign military policy and domestic police repression by dubbing the event the "Peter-loo Massacre," and the resulting media frenzy amounted to what James Chandler has shown to be a watershed in English historical and media consciousness. An "intensifying cycle" of governmental repression, journalistic condemnation, and further crackdowns on freedoms of press and assembly produced a heightened national "self-consciousness about the historical *state of the representation*," in a sense that yoked parliamentary conditions to the print media public sphere.[17] The quick and widespread perception of "Peterloo" as an event of national historic magnitude was reliant upon a press-mediated "experience of national simultaneity" that made palpable the "hyperactive public sphere" in which Shelley, for instance, could participate at a distance of a thousand miles.[18]

Since Shelley's poetic response, as Steven E. Jones observes, "stages a series of shifting representations that work to question the ground of representation itself—in the political as well as the semiotic sense," readings of *The Mask of Anarchy* need to account for the poem's rhetorically disjunctive parts. I survey them briefly before taking up each at some length and advancing a new argument for their coherence as increasingly materialist exercises in allegory and pedagogy. *The Mask of Anarchy* begins with a hundred lines of vicious political satire: in Shelley's wicked spin on Renaissance courtly *masque*, a crew of rampant allegorical "Destructions" reenacts the violence of Peterloo. This "ghastly masquerade," the poem's first phase and form of political allegory, bodies forth the vices "Murder," "Fraud," "Hypocrisy," and "Anarchy" as particular ruling, Regency reactionaries: "I met Murder on the way—/He had a mask like Castlereagh" (lines 27, 5, 14, 24, 30, 5–6). Foreign Secretary Robert Stuart, Viscount Castlereagh, is accompanied by "Fraud," decked out as Lord Chancellor Eldon, and "Hipocrisy" in the guise of Home Secretary Henry Addington, Viscount Sidmouth. "Anarchy," vamping on a blood-splashed horse, resembles no one so much as the Prince Regent himself ("as if his education/Had cost ten millions to the Nation" [lines 76–77]). They trample the assembled "adoring multitude" to "a mire of blood" (lines 40–41) and generally ransack the national purse, population, and conscience, until a "maniac maid" calling herself "Hope" ("she looked more like Despair") lays her body down in the street before the onrushing crew (lines 86–88).

This act of passive resistance triggers the flagrantly aesthetic segue mentioned at this chapter's outset: "A mist, a light, an image" (line 103) arises and passes over the course of seven stanzas, leaving the marauding allegorical oppressors dead and dispersed in its wake. Shelley's *figura-ex-machina* "Shape" (line 110) clears the scene of its maskers: when the misty curtain lifts, an utterly different poetry seems to take their place. Set off in quotation marks but assignable to no particular speaker in the poem, an earnest and stirring oration to the "Men of England" arises, as if the Earth itself had "turned every drop of blood/By which her face had been bedewed/To an accent unwithstood" (147, 143–45). Discoursing on the question "What is Freedom?," this terrestrial voice first denounces in detail the "Slavery" of wage-labor exploitation and systematic poverty, then advocates a program of positive "Freedom" in the form of "clothes, and fire, and food," "Science, Poetry and Thought" for "the trampled multitude," and finally urges them on to further acts of nonviolent, mass assembly, despite the casualties such assemblies will continue to incur (lines 156, 205, 221–22, 254). A famous refrain both opens and closes the oration, thus bringing the poem to an end, too: "Rise like Lions after slumber/In unvanquishable number/Shake your chains to earth like dew/Which in sleep had fallen on you—/Ye are many—they are few" (lines 151–55, 368–72).

b. Ghastly Masquerade

As nearly every reader acknowledges, the fact that abstract vices *impersonate* Regency powerbrokers in Shelley's opening "ghastly masquerade" betrays something unusual in the poem's allegorical operations. "I met Murder on the way—/He had a mask like Castlereagh" is a personification worth pausing over. "The irony," as Steven E. Jones puts it, "is that the reality is abstract evil;

its appearance merely takes the form of persons."[19] But the case is stranger still, since in fact affording readily graspable, personal bodies to abstractions is the routine work of personification allegory, and was accomplished the instant murder was bodied forth with a capital "M." What startles here is rather the way Shelley *presupposes* a first-order, allegorical hold on abstract Vice—that is, the way he takes "Murder" as his starting point—and hijacks that already-allegorical sign as a vehicle for a concrete, historical personage: "Castlereagh." As Theresa Kelley puts it in a far-reaching history of allegory that credits Romanticism in general, and Shelley in particular, with holding allegorical figures in subversive proximity "to real people and events," *The Mask* "humanizes abstraction with something like figural vengeance."[20] It is as if what readers risk missing, what they might fail to perceive without figurative help, are real-historical persons, whom Shelley's variety of political allegory summons by their proper names to ad hominem accountability. The poem's opening presumes a public so habituated to abstraction that biographical specificity has to be stagily donned, like a "mask."

To worry, then, as Marc Redfield does in a fine-grained reading of the scene, that "[w]hen we imagine Murder to be incarnate or grounded in Castlereagh, we make a mask into a person and fall into the naiveté of taking a mask for a face, literalizing a prosopopoeia," betrays a significant political difference between this mode of deconstructive rhetorical reading and the poem's express concern for the casualties that will continue to compound—"They are dying whilst I speak" (line 171)—as long as contemporaries *fail* to "literalize" in just this way, fail to ground abstract social and moral ills in the faces of present political elites.[21] The accusation of "literalism" itself had a political valence in 1819: the inaugural issue of the radical newspaper *The Medusa; or, Penny Politician*, for instance, opened with a satirical address "TO THE PUBLIC, *Alias* The '*ignorantly-impatient Multitude*'" that pretended to chastise its working-class readership, in the voice of "the Statesmen at Whitehall, the Judges on the Bench—all the Parish officers in the Nation," for the crass literalism of their demands for happiness:

> O YE factious, seditious and discontented crew! . . . are you determined . . . to '*feel, feel, feel* and *touch, touch, touch*,' before you will allow your happiness to be real? . . . how unacquainted you are with the wonder-working powers of *imagination*!—Can you not believe that your hunger and thirst are gratified, unless you *eat* and *drink?* Can you not believe that you are clothed and warm, unless you are *covered* from the inclemency of the season?—O, what political unbelief is this![22]

Clearly, from such editorial perspectives, literalism could constitute a critically potent form of "political unbelief." As Anne Janowitz observes in the seminal study that puts Romantic lyricism back in touch with decades of interventionist communitarian poetry in the radical press, *The Mask* borrows from this tradition to turn "the usurpation of the figurative by the literal . . . into an asset, not a liability."[23]

The stakes of political allegoresis, of getting the right reading on contemporary political pageantry, are indeed set high at the outset of *The Mask of Anarchy*, which also sets up pointed disparities of access to the necessary interpretive tools and resources. In a poem that pays a good deal of attention to pedagogical politics, part of what makes the opening masquerade amusing *for readers* is the high-speed education we receive in the second meanings traditionally "hidden" in allegorical double-speak. In this anything but "dark" conceit, that is, the reading audience learns what's behind the mask ("Murder") *before* we've even glimpsed the mask ("Castlereagh"). And while the rampant political elites within the poem's diegetic frame are depicted as consciously equipped to reap benefits from their duplicitous performance, the "adoring multitude" (line 41) that forms their audience in the poem is beset by failures of interpretation. While readers, for instance, are instructed in advance that Chancellor Eldon's public sentimentalism screens an industrialist's exploitative agenda—"His big tears, for he wept well / Turned to millstones as they fell" (lines 16–17)—the ordinary persons who surround their Lord Chancellor within the poem are blindsided by second meanings that do Peterloo-style damage:

> And the little children, who
> Round his feet played to and fro,
> Thinking every tear a gem,
> Had their brains knocked out by them.
> (lines 18–21)

The masquerade thus casts an information deficit it helps perpetuate as a defining condition of the "multitude's" vulnerability. Were it possible, moreover, for readers to maintain the line between the literal and figural levels that traditionally powers allegorical dialectics, we would rest assured that children playing on the literal level would be literally safe from the figural millstones that the allegorist has insinuated behind Eldon's tears. But the blunt force trauma invoked in the last line renders such a reading untenable by virtue of its all-too-literal reference to Peterloo, where troops reportedly "dash[ed] the heads of infants against stones"; instead, it nearly suggests that better political literacy might have saved the children.[24] Here, the allegorical relay between the politician's sentimental performances of charity and the deadly millstones intimates less a gap between literal and figural levels than a causal chain that links the "brains knocked out" in Manchester to the differently debraining, slow violent "round" of mindless labor in the mills and the tradeoff between literacy and wage work.[25]

In his magisterial study of allegory and violence in Renaissance literature, Gordon Teskey defines "allegory's primary work" as the attempt to "force meaning on beings who are reduced for that purpose to substance."[26] Allegory thus performs and conceals the violence that erecting "a structure of meaning" does to what Teskey alternately calls "life," "bodies," the "organic substratum," and "the material world."[27] The formulation so neatly describes the overt plot of the first part of *The Mask of Anarchy*—which follows allegorical personifications as they trample multitudes of people into an undifferentiated "mire of blood" while chanting their allegiance to "God, and Law, and King"—that we are clearly dealing with an allegorist acutely aware of the dynamic Teskey describes. Why this might be the case for a poet writing "on the Occasion of the Massacre at Manchester" comes into focus as Teskey specifies the function of allegories for the nation-state, which work to incorporate multiple lives into a unified body politic: here, the particular violence veiled concerns "the experience of the body, in the clearing or *agora* of political life, living under the threat of destruction."[28] In other words, national allegory typically conceals exactly the political risk brazenly exposed at Peterloo.

The Mask of Anarchy, then, seems to seek a mode of allegorical figuration that would do something other than disclose its own complicity in the signifying violence that reduces persons to mere matter and mere life; a mode that would cease, as Teskey had it, to take "bodies as the material basis of an order of signs."[29] Pushing past its opening masquerade, *The Mask of Anarchy* attempts several further tellings of Peterloo, each time allowing the poem's "multitude" to have perceived—and indeed, to have helped produce and interpret, rather than merely fallen victim to—the political meaning of the historical events at Manchester. By the poem's end, the opening masquerade's allegory of "Anarchy" will have taken on the campy, anachronistic air of a pageant that staged Regency politics in Renaissance drag not only at the level of formal convention (*masque*) but also at the level of the political intelligence borne. For as Kir Kuiken has argued, the poem is bent on producing figures adequate to the changing ground of modern political sovereignty.[30] But whereas for Kuiken this means finding a figure suited to the "radical groundlessness of modern sovereignty itself," I see Shelley as drawing ever more attention to the grounds that sustain figures for political belonging.[31] This increasingly materialist movement transforms the initial idiom of political theology and moral condemnation toward an analytics of power and social theory that is indeed better suited to sovereignty's nineteenth-century incarnations.

c. Events Take Shape

Between the grim "Pageant" that opens the poem and the rousing political oration with which it closes—and, spatially, between the "prostrate multitude" and the allegorical "Destructions" they have so far failed to recognize—there intervenes an extravagantly poetic figure (lines 151, 126). Shelley ultimately calls this figure, as he so often does, simply "a Shape." Built up gradually over nearly fifty lines, this vague and misty trope disrupts the allegorical clarity and celerity, that swift and tidy pairing of abstract Vices and proper names, that has supported readers until now:

> When between her [Hope] and her foes
> A mist, a light, an image rose,
> Small at first, and weak, and frail
> Like the vapour of a vale:
>
> Till as clouds grow on the blast,
> Like tower-crowned giants striding fast
> And glare with lightnings as they fly,
> And speak in thunder to the sky,
>
> It grew—a Shape arrayed in mail
> Brighter than the Viper's scale,
> And upborne on wings whose grain
> Was as the light of sunny rain.
>
> On its helm, seen far away,
> A planet, like the Morning's, lay;
> And those plumes its light rained through
> Like a shower of crimson dew.
>
> With step as soft as wind it passed
> O'er the heads of men—so fast

> That they knew the presence there,
> And looked,—but all was empty air.
> (lines 102–21)

When the Shape has passed, the crowd looks up to find *Anarchy* "dead," his cohort "[ground] / To dust" (lines 131, 133–34), and the passive resister, Hope, "walking with a quiet mien" (line 129).

The unexplained revolutionary efficacy of this shamelessly aesthetic trope is a scandal that has elicited rich critical debate around the political pretenses of poetic utterance, or, as Marc Redfield puts it, "the politics of literariness itself."[32] As Steven Goldsmith observes, "the passage seems to imply that tyranny is a linguistic fabrication dissipated the moment language returns to its inherent troping process." And there is much in Shelley—and especially in his aesthetic advocates—that affirms Goldsmith's diagnosis of their shared, unexamined faith in "linguistic empowerment [as] a universal solvent."[33] Reading the Shape as an emblem of aesthetic experience uncontaminated by the violence of "the empirical world," for instance, Matthew Borushko writes of this passage that "mist and light—substantives representative of nothing other than themselves—give way to an image" that "possesses only indeterminable or indescribable content."[34] But this is a peculiar assessment of a manifestly bloodstained Shape that owes its substance and its cause to the preceding violence: a *"crimson* dew" precipitating from Peterloo's "mire of blood." Nor can it be said, as Borushko does, that its "content" remains "indescribable," unless one is prepared to ignore the remaining two-thirds of the poem, which amount to its extended exegesis.[35]

Especially in comparison with the decisive allegorical figurations that precede it in the poem, the Shape is a poor emblem for the nonreferential feats and fiats of poetic (canonically, lyric) autonomy. For it is in fact assembled gradually, hesitantly, from carefully sourced elements of the scene, and kept rather studiously explicable as a chance weather effect. Assessing changes to this passage across Shelley's drafts, Steven E. Jones discovers a poet at pains to tone down the distinction between figural and meteorological process:

Apparently, as he revised, Shelley gradually rejected similes, choosing instead to blur the visible outlines of the figure so that it is visualized only vaguely, as superimposed on the sky—or even as consisting of the sky's chaotic and unpredictable weather, a mythopoeisis of meteorological processes.[36]

Indeed, Jones's passage mimics Shelley's own, revising a first characterization of the Shape as "superimposed" *on* the sky in order to tentatively acknowledge that this Shape might be made *of* and *by* it. Indeed, even as the apparition takes on the stronger contours of a conventional and poetically fashioned figure in the poem—even as the Shape gets poetic "wings" and intertextual allusions—its plausible explicability as an atmospheric and optical reworking of the material evidence, of the lingering "steam of gore," persists.[37] The "grain" of the Shape's wings is also given as "the light of sunny rain," the apparent "glare" of its armor is like flashes in a gathering storm cloud, the beaming tiara that shines through "those plumes" does so "Like a shower of crimson dew" (lines 113, 117). The "crimson" color ties the matter of this chancy, particulate image—its "grain," as line 112 has it—to blood lingering as "red mist" in air, or settling into the ground ("blood is on the grass like dew" [line 192]).[38]

The first figure in the poem whose occurrence is equivocally *described* rather than decisively imposed by poetic speech, this Shape is also the first allegorical persona "heard and felt" by everyone, "prostrate multitude" included. All "knew the presence there" (line 120). In a further, pointed reversal of the previous allegorical mode, abstract thoughts ("Murder," "Fraud," "Hypocrisy") do not tendentiously *precede* particular embodiments ("Castlereagh," "Eldon," "Sidmouth"), but are rather, in the case of this as yet anonymous and amorphous "Shape," triggered by the sensation of physical movement: "Thoughts sprung where'er that step did fall" (line 125). Hence the Shape's meaning, its becoming-sign, entails a tender-empirical process: writing that "A sense awakening and yet tender / Was heard and felt—," Shelley rigorously activates both the sensory and the semiotic senses of *sense*. Here fresh meaning breaks or dawns with a tingling sensation, and the "abstraction" of allegory retains its etymological sense of something drawn off or away (*abs* + *trahere*) from another, by distillation or chemical sublimation rather than absolute distinction in kind. Compounded up from the scene of the damage, then, this Shape presents an alternative etiology for political allegory as a "frail" but real production of beings and substances at the scene of a revealing political catastrophe.

For it is not that there is no longer allegory here, but that allegory has ceased to operate to "force meaning on beings who are reduced for that purpose to substance" (Teskey). This Shape "rose" and "grew" from below, out of the very bodies and environments that the political-allegorical regime it replaces had reduced to "organic substrate." Operating in the habitually abstract milieu they nicknamed "the German Ideology," Marx and Engels describe a strategic reversal in the genesis of meaning that deftly captures what Shelley is up to in this phase of the poem:

> [H]ere we ascend from earth to heaven. That is to say, we do not set out from what men say, imagine, conceive, nor from men as narrated, thought of, imagined, conceived, in order to arrive at men in the flesh. We set out from real, active men, and on the basis of their real life-process we demonstrate the development of the ideological reflexes and echoes of this life-process. The phantoms formed in the human brain are also, necessarily, sublimates of their material life-process, which is empirically verifiable and bound to material premises.[39]

It is such a process of material sublimation, rather than aesthetic sublimity, to which Shelley's Shape belongs as a changed image of political collectivity, even of "the people"—massive, fragile, aggrieved, in armor, in pieces, love-crowned—that might be one "ideological reflex" of Peterloo at the level of affect and consciousness. Imaged in the "Shape" is a change in the national atmosphere, a political feeling that Shelley's contemporaneous prose elaboration, *A Philosophical View of Reform*, called a "sentiment of the necessity of change."[40]

One advantage of seeing the Shape in the terms Marx and Engels provide is that it is openly, inescapably "ideological," and that the very gesture which seems to "reduce" ideation to "material life-process" simultaneously affords to these "phantoms formed in the human brain" a tenuous material reality. Slighter, certainly, than other things, mental phantoms are nonetheless of the same stuff: multiplied among many heads, Shelley thinks, "opinions" compound into nothing less than the "cement" of our "social condition"; conversely, when "the equilibrium between institutions and opinions" is out of kilter, this mental-material medium is capable of "discharging its collected lightening."[41] Such an approach, in which Marx and Engels converge improbably with Wordsworth in an obsessive commitment to "men in the flesh," is indeed how Shelley seems to answer Wordsworth's critique of allegorical personification as an obtrusively abstract, "mechanical device of style" obnoxious to the modern Poet's resolution to "keep the Reader in the presence of flesh and blood."[42] Shelley's resolutely flesh-and-blood allegory owes little to a difference in kind between corporeal matter and meaning and does little to enforce it; this figure's flickering, unfinished significance has other origins and consequences.[43]

It will come as no surprise at this point that this "Shape," which dissolves *The Mask*'s hard-edged allegory into a tenuous mental-material apparition of dubious, collective provenance, wears a Lucretian emblem on its forehead. In contrast to Anarchy's starkly emblazoned brow ("I AM GOD, AND KING, AND LAW!" [line 37]), the Lucretian muse Venus, morning star, subtly illumines and permeates the militant Shape: "On its helm," yet "far away," she opens cosmic distances internal to the figure and atomizes its armor into a dazzling, "sunny rain" (lines 114, 109). (Is this not what all the "Shape's" attributes— "mail," "scale[s]," "plumes"—have had in common all along: their evident texture of interlinking elements?)[44] Shelley in fact signaled his shift to a Lucretian semiotics in the previous stanza, when he compared the increasing size and definition of the "image" to clouds coalescing into massive likenesses of giants: "as clouds grow on the blast,/Like tower-crowned giants striding fast" (lines 106–7). As *De rerum natura* has it: "We sometimes see clouds quickly massing together on high [*concrescere in alto*] and marring the serene face of the firmament. . . . often giant's countenances appear to fly over and to draw their shadow afar" (4.134–37).

I have had cause, in each previous chapter, to refer to this area of *De rerum natura*: Lucretius's materialist explanation of thought and sensation as occasioned by the atom-thin *simulacra* that pour from the surfaces of things, bearing fractions of bodies into the distance and into perceptibility. But in the passage adapted in *The Mask of Anarchy*, Lucretius is explaining a particular case: simulacra that derive from no single thing, but instead arise spontaneously (*sponte sua giguntur*), when floating figures shed from *many* sources accidentally cohere (*concrescere*) in air. As John Mason Good put it in the 1805 translation of *De rerum natura* that Shelley owned: "Borne through th'aërial realms in modes diverse,/Their forms for ever shifting," simulacra, "combining strange," can form a "spontaneous" apparition of any kind of thing (4.137–41).[45] What matters for *The Mask of Anarchy*, that is, for the possibility of communicating the history and significance of the events at Peterloo without miming the state's authoritarian regime of signification, is the possibility of a spontaneous collective image of a multitude.

Tapping into the Lucretian possibility that significance can coalesce without the forceful imposition and supervision of rhetorically savvy poets, readers, or politicians—that a rival "image of the people" can derive from, index, and resemble them in their new visibility, vulnerability, and power—Shelley shifts allegorical authority from the speaker and maskers to the interaction between the many participants that made the massacre as natural and historical event. Coalescing "O'er the heads of men," in a simulacrum that mingles their images with the airborne evidence of the damage, the "Shape" renders back an estranged, fleeting, collective likeness of "the Occasion of the Massacre of Manchester" (line 119).

Approached from this angle, Shelley's political poem seems less aesthetically extravagant: an attempt not so much to produce as simply to register and "reduplicate," as he would say, an amorphous but palpable alteration in the political atmosphere—"a sense awakening and yet tender"—in the wake of a catastrophic event of police brutality. Shelley offers a Shape that would be the involuntary and collective product of this enraging and instructive loss, documenting the feeling of feeling differently among those recently forced to know, directly or at a distance, the kind of violence to which certain lives can be exposed in the interest of "secur[ing] the public tranquility." In this context, the "tenderness" in tender empiricism, the tenderness that characterizes the "sense awakening" in the aftermath of the massacre, acknowledges a shocking lesson in political vulnerability.

Noticing the newly descriptive, documentary, constative approach to the figural genesis of the Shape in the poem enables the question of its aesthetics and politics to be formulated differently. If the poem is attempting to grasp a change it does not purport to call into being, "communicating and receiving" a difference rather than making one, it alters the way the question of aesthetics and politics might be framed.[46] Instead of asking what political difference an artwork can make, the poem prompts us to ask what kind of artwork a political difference can make. *The Mask of Anarchy*, that is, depicts the massacre at St. Peter's Field as changing the conditions of signification in which it is possible to communicate as an artwork, enabling a shift from one kind of national allegory to another.

Nor is it disqualifying, any longer, that the poet Shelley, or his dreaming speaker, communicates this "spontaneous" collective image through the mediation of letters, newspapers, and rumors that reach him "from over the Sea" and in allusion to other poems. "Spontaneous" in this Lucretian sense means unauthorized and multiply derivative, not *im*mediate. It seems, moreover, that even participants in the occasion have a belated, environmentally and intersubjectively mediated experience of perceiving what has happened: *The Mask* mimes not only the slow-motion concrescence of the shape but also its gradual psychosomatic apprehension, telling these through an attenuated series of sensory and mental verbs—they knew, they looked, they thought, they looked, they heard and felt—that follow on the Shape's heels, but miss a frontal view or total grasp. Steven E. Jones has resituated Shelley's Shape in a popular symbolic idiom that *The Mask of Anarchy* shared with graphic satires and optical and theatrical amusements; he also makes the crucial point that reform assemblies like the one at Manchester were "already symbolic events, awash with red caps of liberty," costumed allegorical goddesses, significant props and clothing, colorful banners and broadsides.[47] Shelley's tentative simulacrum of the people, then, participates less in the unanchored whirl of images theorized in postmodernity than in a "semiotic milieu" that passes from dense to rare and back again, filling out the gamut from bodily gesture to loaded word and linking plastic, graphic, and verbal art to embodied political performance.

Now that we have some grasp of how the Shape's genesis differs from the poem's previous allegories, it remains to be seen how its allegoresis differs, and what different kind of political intelligence such a figure bears. The stanzas that draw out the kind of "Thoughts" to which the Shape gives rise indeed offer an alternative account of the passage between visibility and invisibility, sensuous figure and abstract thought, that allegory traditionally premises and promises:

> As flowers beneath May's footsteps waken
> As stars from Night's loose hair are shaken
> As waves arise when loud winds call
> Thoughts sprung where'er that step did fall.
> (lines 122–25)

Waves attest to wind, stars to invisible night, blossoming flowers to the warmth of the season. The epistemology of these analogies is classically Lucretian in the sense explored in chapter 3: they derive the reality and force of imperceptibly subtle bodies, up to and including "thoughts," from perceptible ones that dynamically index their subliminal activity.[48] Thus the allegory in each line personifies not the archetypal natural forms with which it begins (flowers, stars, waves), but the "negative" spaces that surround them, shifting attention to the action of transparent or invisible backgrounds (the wind, the season, the night) that bear or draw those figures into perceptible relief.

Thoughts, the last line contends by analogy, are like this: bright indices of broad contexts with which they are physically in touch. And so also are we asked to reconceive the "abstract" thoughts allegory bears in shapely bodies: by the logic of the stanza's analogies, thoughts are allied *with* such shapes/figures (flower, star, wave), and not their seizure by a radically alien power. Instead, the alterity and duplicity of allegory consist in the difference between that which is drawn into relief and its darkened or recessive ground; these stand in a relation of metonymic contiguity, of physical and causal contact and context, that Shelley's practice constantly proves to be reversible.[49] Thus the interplay between visible and invisible and the "other-speaking" long recognized as basic to the allegorical mode need not derive from and attest to, let alone "exacerbate," an irreducible difference between signification and matter.[50] Shelley handles allegorical shapes as figural ostentions of a wider field that eludes simultaneous apprehension.

d. Material Poetic Justice

Given this understanding of the provenance and valence of *The Mask*'s second allegorical mode, the departure into oratory that constitutes the poem's third and last phase seems less disjunctive. Rather than capitalizing on a tendentiously lyrical moment of aesthetic free play to launch a political program, the oratory intensifies what is emerging as the poem's series of turnings toward the material (back)ground, each more scrupulous than the last. After the Shape has passed, and after the concomitant "sense awakening and yet tender / Was heard and felt—," the poem actually sets about vocalizing the particles of bloodstained turf that must have supported each prior allegorical enactment:

> A sense awakening and yet tender
> Was heard and felt—and at its close
> These words of joy and fear arose
>
> As if their own indignant Earth
> Which gave the sons of England birth
> Had felt their blood upon her brow,
> And shuddering with a mother's throe
>
> Had turned every drop of blood
> By which her face had been bedewed
> To an accent unwithstood.—
> As if her heart had cried aloud:
> (lines 136–46)

What comes to narration by way of this extraordinary prosopopoeia, which routes violently dispersed blood through the ground to distill coherent syllables and enchain them in a syntax, is a "background" that is causal and contextual, in addition to being phenomenal. Thus the poem next offers the system of circumstances that supported the violence of Peterloo, making possible and indeed probable the apparently scandalous fact of a massacre. The earthly words an-atomize the indistinct, composite, collective "feeling" conveyed by the Shape as an atmospheric afterimage of Peterloo, breaking down and spelling out the "sense" it was "heard and felt" to index as a nonverbal sign. At stake in this transition is a limit case not, as in many prior readings, for the political (in)efficiency of lyric fiat, but rather for the ability of materialist allegory to articulate affect and structure: to demand and to facilitate a passage between sensory-affective and discursive-analytical knowledge.[51]

And the fact is, once Shelley's poem has summoned Lucretius to relax the agon between matter and meaning at the level of figuration, the resulting allegoresis sounds a lot like historical materialism. The oratory postpones the high question "What is Freedom?" until the "day to day" "slavery" of subsistence wage-work has been adequately grasped:

> "'Tis to work and have such pay
> As just keeps life from day to day
> In your limbs, as in a cell
> For the tyrants' use to dwell
>
> "So that ye for them are made
> Loom, and plough, and sword, and spade,
> With or without your own will bent
> To their defence and nourishment."
>   (lines 160–67)

Capturing the instrumentalization and reification of wage laborers with much-admired "proto-Marxist" precision, the speaking ground goes on to deliver a thoroughgoing critique of the historically new system of working-class exploitation that "take[s] from Toil a thousand fold / More than" the "tyrannies of old" ever "could" (lines 177–79).[52] She indicts the extraction of surplus value through wage-work, the alchemy of the money form ("Paper coin" and the "Ghost of Gold" [lines 180, 176]), the division of mental and manual labor ("Science, Poetry, and Thought" ought to have a place in the worker's "cot" [lines 254–56]), the rule of law as a cover for class interest (lines 230–33), producers' alienation from the place, products, and profits of their labor (lines 201–4), fresh extremes of expropriation and income inequality, and religion as a psychological surrogate for social justice (lines 234–37).

If the bloodied ground of Peterloo could speak, it would define Freedom first in terms of the satisfaction of basic material needs ("Thou art clothes, and fire, and food / For the trampled multitude" [lines 221–22]), after which "Justice," "Wisdom," "Peace," and "Love" might be spoken of.[53] Indeed, aside from a keen grasp of what others would soon name "the proletariat," the most Marxian feature of the speech might be its ardent promotion of the unglamorous, material wants of human animals as worthy of a high discourse on "Freedom": its insistence, like Marx and Engels's in the *German Ideology*, "that men must be in a position to live in order to be able to 'make history.'"[54] In this sense "eating and drinking, a habitation, clothing and many other things" constitute "historical act[s]" and merit serious historiography.[55]

To present-day readers, the cited stanza's picture of somatic "life" scrupulously managed for "tyrants' use" seems just as unmistakably to prefigure Michel Foucault's analytic of biopower. By virtue of long-term alterations in the logic and technologies of sovereignty, he argued, in early nineteenth-century Europe, "the biological came under State control." Against the classical political theology of sovereignty typified in the monarch's right to kill a citizen-subject (to "make die" or "let live"), Foucault outlines the (ongoing) "biopolitical" power of governments to "make live" and "let die": to furnish and withhold the requisites of modern life (medicine, hygiene, employment, housing, nutriment, safety, insurance) from groups of people considered in their species character as biological populations.[56] Structurally, *The Mask of Anarchy* rather exactly expresses this shift: while in the Elizabethan-style opening masquerade, a sovereign monarch rules by threat of immediate "Death" and plunder, the earthly oration replaces his ritualized display of the power to kill with a chronicle of the "slow violence" of systematic deprivation to which certain lives are exposed—starvation, sickness, anxiety, cold, homelessness, vagrancy, political marginalization—and the institutions (workhouse, prison, military) that variously "let die" this population, whose access to food, rest, and shelter the poem compares unfavorably to that of other classes of humans and animals.

"Slow violence" is Rob Nixon's more recent name for the gradual and unseen "violence of delayed destruction" that the ravages of poverty share with ecological devastation: "a violence that is neither spectacular nor instantaneous, but rather incremental and accretive, its calamitous repercussions playing out across a range of timescales."[57] The "attritional violence" afflicting the planet and the poor—indeed, affecting them together as long as socioeconomic poverty entails disproportionate exposure to environmental damage—takes place in marginalized and insensible tempos and spaces, Nixon argues, and issues in "long dyings" that scarcely register as violent at all.[58] In *The Mask of Anarchy*, the earth's exposition of "slow violence"—of the sites of "daily life / Where is waged the daily strife / With common wants and common cares" (lines 279–81)—helps account, in retrospect, for the peculiar saliency of the act of passive resistance that put an end to the opening masquerade. Refusing the death-by-attrition afforded to her set, Hope / Despair "lay down in the street / Right before the horses' feet," provoking the authorities to take her life in a single act of spectacular violence rather than by incremental and invisible deprivation (lines 98–99).[59] The wager of "Hope" here is that in a context where state sovereignty no longer derives from the right to "make die," forcing the government to kill in public will jeopardize its sovereign legitimacy.

The point of enumerating such "proto"-Marxian, -Foucauldian, and -ecocritical insights is not to vindicate Shelley's political poetics through the prestige of future theories, but rather to notice the perhaps surprising congruity *The Mask of Anarchy* demonstrates between the sensuous naturalism of Lucretian semiotics and structural, historical, and theoretical critique. This poem does not permit readers to rest content with the affective sensation of history that critics are learning to appreciate as a virtue of Romantic poetic mediation, but mandates that feeling's regrounding in an analytics of power. Three hints from Walter Benjamin's study of Baroque allegory will enable us better to grasp why this might be so. First, he observes, allegory is born "by virtue of a strange combination of nature and history" attuned to history's physical dissolution into the setting. Second, the allegorical mode of expression is inherently and not incidentally didactic, epitomized not only by the semantic density of Baroque visual emblems, but also by heavy-handed discursive exposition that "drags the essence of what is depicted out before the image." Third, the mode ultimately subordinates beauty to knowledge: allegorical artworks court their own critical "mortification" or dissection, sharing with scientific understanding the pleasure of connecting "detail" to "structure."[60] How these features fit together in an inherited set of allegorical, didactic poetic tactics put to punctual use in *The Mask of Anarchy* becomes clearer in light of some eighteenth-century precedents for the poem's historicizing form of neo-Lucretian atomism.

e. Atomic Prehistories for *The Mask of Anarchy*

To read Shelley's turns toward the earth and atmosphere in *The Mask of Anarchy* as gestures that communicate, rather than suppress, historical eventfulness is to draw Shelley's poem into unexpected kinship with the georgic line that Kevis Goodman has brilliantly rehabilitated within eighteenth-century and Romantic writing, a genre that Mary Favret and Tobias Menely have since taken to the skies.[61] Building on Raymond Williams's notion of a "structure of feeling," Goodman theorizes georgic poetics as a technology of mediation set to modulate the potentially overpowering sensations of historical experience to the limits of life, sympathy, and sense.[62] And yet, she argues, to accommodate the "touch of the real" to readerly limits is also and inevitably to provide an "aperture" through which its "unfixed elements" rush in, registering as sensory and affective disturbances to the poems' pleasurable norms of mediation and meaning.[63]

Virgil's *Georgics* were, among many other things, a brilliant first act in the adaptation history of *De rerum natura*, and to read poems in this joint tradition with specifically Lucretian attention is to notice treatments of "history in solution" that are ham-fistedly forthright in comparison: moments when the poets, like Williams and Goodman themselves, summon an atomist vocabulary and ontology to speak directly of "history in solution" as "that immanent, collective perception of any moment as a seething mix of unsettled elements."[64] From the neo-Lucretian perspective I have attempted to reinhabit, historical influence enters not as nonsense, pain, noise, or occlusion, against the grain of poems at work to mitigate its impact, but as an acknowledged coauthor at the level of figuration and object of didactic exposition at the level of theme.[65] This is the genealogy of a materialist conception of history that aims not only to take material concerns as its object, but also to acknowledge their force in its making, shaping, and writing.

James Thomson's *The Seasons*, for instance, compounds the "angry Aspect" of airborne "*Plague*" from a "copious Steam" of elements emanating from colonial environments and the carnage of the slave trade. By the time this "dire *Power*" is said to "Walk," "stain'd / With many a Mixture," its appearance and motion express the proximate edge of a complex of imperial, biological, and meteorological culpability spun out over a thousand lines. She comes complete with a footnote directing readers to Richard Mead's medical theory of contagion, "*Matter* capable of conveying Mischief to a great distance."[66] That is, in what we might break down into the two-step movement of neo-Lucretian didactic poetry (though in fact their order is reversible and overlapping), historical nature enters as the poem's welcome (also: inevitable) collaborator, through figural strategies that heighten the shaping pressure of extradiegetic circumstance; second, such figurations are objects of expository analysis, de- and recomposed in narratives that tease out the provenance of parts and embed them, more or less shamelessly, into a systematic picture and lesson plan—what Thomson calls "this complex stupendous scheme of things."[67]

In the Lucretian tradition, the material recombinatory that is nature can be relied upon to put forward again—piecemeal, elsewhere, in alien configuration—the material evidence of social violence (and, to be fair, what Erasmus Darwin calls "past Delight").[68] The phenomena of nature, in this sense, are not historically innocent, but keep posing material incitements to justice conveying incidentally "Mischief to a great distance" from the historical scene of the crime. And Thomson is not alone. William Cowper, in "Yardley Oak," approaches the hulk of the eponymous ancient tree as a figural ready-made whose body submits centuries of "unrecorded facts" to the poet's present perusal. Since the oak has been intricately sculpted by time's "sly scythe"—"an atom and an atom more / Disjoining from the rest"—the poet's art is an "inquisitive" reconstruction of the story from the "scoop'd rind" and roots "crook'd into a thousand whimsies":

> By thee I might correct, erroneous oft,
> The Clock of History, facts and events
> Timing more punctual, unrecorded facts
> Recov'ring, and mis-stated setting right.

The poem stands to refine our argument about material allegory in so far as it stresses the imperfect coincidence between the oak's figural and indexical insistence and the didactic "discourse" extrapolated therefrom. For although the oak is "well-qualified by age / To teach," its lack of "voice" permits the speaker to "perform / Myself the oracle," and "discourse / . . . such matter as I may."[69] Not a pure impasse, however, Cowper depicts the resulting "discourse" as physically occasioned and supported by its object—"seated here / On thy distorted root" (from which Shelley seems to borrow that "old root which grew / To strange distortion" that was once Rousseau in *The Triumph*)—even as he marks a narrowing into glottocentric expression that avows the partiality of the performance. Shelley likewise marks the conversion of "every drop" of blood lost on St. Peter's Field into "an accent unwithstood" with an "as if" that emphasizes its conjectural elaboration. Yet the action seems different from pure fantasy: a bottom-up retelling of the history of Peterloo in collaboration with materials that solicit, support, and co-produce, rather than resist, their formulation into words. The new accent is *un*withstood, Shelley's unusual term "accent" itself stressing words' shaping by the body that says them.

Taking the risk of speaking *for* unspeaking elements that do not disclose their provenance in human language, these poems' discursive, expository turns help us to apprehend the tradition from which Shelley borrowed in order to launch his particulate, partisan form of didactic allegoresis in *The Mask*. Marking this risk, the didactic poets proceed with a kind of forensic conjecture bent on voicing the continued presence and aftereffects of events that otherwise go unsaid.[70] The "words of joy and fear" that arise through this process in *The Mask of Anarchy* are frequently said to arise from nowhere, such that critics write, for convenience's sake, of the voice of "Liberty," "Britannia," or "the Orator."[71] But the "accent's" historical and literary-historical derivation is clear in the tradition of didactic poetic atomism to which Shelley's poem, at this juncture, belongs: the exhortation is an accentual-syllabically reconstructed story of droplets of violently spilled blood as they threaten to disappear, and reappear, in cycles of weather, water, and soil.[72] Set off in quotation marks, this speaking, like that of Cowper and Thomson before it, extrapolates a back-story—both perceptive and prejudicial—of tributary causes and materials for the palpable change that it has already depicted as a "Shape."

The aptitude of Shelley's "Shape" for structural analysis is an obvious affordance of its Lucretian composite form, whose heterogenous parts and sources produce an ineluctable shadow of temporal self-difference.[73] Such shapes are pulled apart at the seams by divergent back-stories that require a kind of double-speak, an allegory, to be told at once.[74] In the Romantic era, Erasmus Darwin's popular didactic epics epitomized the allegorical potential, or demand, of neo-Lucretian recombinatory naturalism. In poems whose great theme is "the perpetual circulation of matter" between the "changeful" forms of nature and art (botanical, mineral, animal, technological, political), diverse atomic histories tend to grin uncannily through present faces:

> With ceaseless change how restless atoms pass
> From life to life, a transmigrating mass;
> How the same organs, which to day compose
> The poisonous henbane, or the fragrant rose,
> May with to morrow's sun new forms compile,
> Frown in the Hero, in the Beauty smile.[75]

Darwin's stanza is a compact demonstration of a poetic materialism that frankly admits the processes under investigation to be significant forces in their own figural representation. It also models the kind of perceptual strabismus apposite to the atomist examination of forms: a double-perception alert at once to a present thing ("henbane," "rose") and to the "restless" atomic trajectories that chance to actuate and destroy it.[76] Equally material and real, these two perspectives are hard to hold at once, though the one everywhere evidences the other. The prosopopoeia that choreographs them together into a "smile" or a "frown"—a personal expression of atomic "mass action"— amounts to a scientifically and poetically correct rendition of the two-way causal link between these magnitudes in the event of person-ification which makes a human "Beauty" or "Hero." Though Darwin's personifications have taken a lot of hits, the poems share credit and blame for them with the other kinds of compositions they describe.[77]

In each previous chapter we saw poet-empiricists attempt to cultivate and propagate an attitude of "tender" receptivity to the figural expressions of things. Yet in each of this section's examples, Shelley, Thomson, Cowper, and Darwin make a further move, spinning discursive poetry from the figural heterogeneity of present bodies, whether dense as a tree stump or rare as a cloud. To shift focus toward this additional phase or mode, which not only registers and transmits complex shapes in their sensory and affective turbidity, but also teases them apart, is to begin to attend to the affordances of *De rerum natura*'s poetics as specifically *didactic* in genre. And it is also to notice that the didactic poet's task is thus basically, inherently political, as pedagogy is political: full of conscious and unconscious choices about which particular histories to see and to tell, and which explanatory structures and contexts to furnish for their assimilation.

In rehabilitating poems in this tradition, it is fair to say that Romanticists have been wary of their lesson plans, tending instead to stress their claim to some of the sophisticated work on affect and disruptions of instrumentality traditionally monopolized by lyric expression. What follows can be taken as a preliminary brief for confronting the stodgy pedantry and verbosity of these poems, instead: for thinking about them as "literature of knowledge," in Thomas De Quincey's pejorative phrase, rather than of affective "power"—or rather, as literature of impassioned pedagogy that articulated structure and feeling in ways that agitated mainstream Romantic and Victorian criticism.[78]

f. Getting Didactic

The didactic oratory that arises in *The Mask of Anarchy* after the passage of the "Shape"—the last and longest of the poem's distinctive phases—has not fared particularly well in Shelley criticism. Susan Wolfson's influential new formalist reading found the poem's formal features consistently subversive of the oratory's political pretenses, revealing a self-addressed "fantasy of political performance" whose apparently circumstantial failure to reach an audience in 1819 was in fact well prepared at the level of form.[79] Wolfson's argument hinges on the observation that "the poetic forms that make Shelley's political point do not translate into oration." Relying on a dream-frame that never closes and textual puns that must be *read* and not heard, Shelley ultimately conveys "a masque in the mind of a poet dreaming about being a political orator and projecting this figure as fantastic epipsyche."[80]

Yet calling into question Wolfson's presumption that "political poetry" is "poetry that does what it intends, realizing within an immediate social field the presence-to-self of an intention," Marc Redfield diagnoses such political efficacy as "arguably *the* primal fantasy" of Romantic aesthetic nationalist ideology: "a vision of the nation as an assembled body, gathered into the presence and presence-to-self of an orator."[81] From his perspective, then, the poem's political value resides precisely in the formal subversions that for Wolfson marked political shortfalls: through them, a deeply ideological positive program self-destructs, thankfully leaving "nothing more than the blank fact of its own material occurrence" and making way for "the real political work" that "lies ahead . . . in the extratextual futurity of political struggle."[82] Here Redfield corroborates Forest Pyle's point that the value of Shelley's poetry inheres "not in its reference to the present or the empirical, but in its blank opening onto futurity."[83] And all build on Robert Kaufman's trenchant interilluminations between Shelleyan lyric and Adornian aesthetic autonomy, defending the political value of artworks that work by "determinate negation of empirical reality" over nominally "committed" ones that act as "vehicles for what the author wants to say."[84]

Several premises of this consensus ought not to be accepted without further scrutiny: that the poem's political value pertains to its *future*, as a prerequisite to an emancipatory politics that "lies ahead"; that an "assembled body, gathered into the presence . . . of an orator" denotes protofascistic aesthetic nationalism, rather than mass demonstrations in general, or Peterloo in particular; that materiality is tantamount to blank unmeaning.[85] But first, what is striking in both the new-formalist challenge to *The Mask*'s politics and its negative-aesthetic defense is the shared premise that the "form" of Shelley's poem is the part withdrawn from its nominal contents, audience, and "social field." Whether exposed as self-absorption, or vindicated as the withdrawal from social determination requisite to "the new," form here tacitly means lyric form, and lyric construed in a historically specific way. Audible behind these critics' agreement is a high Romantic argument crystallized in

John Stuart Mill's brilliant 1833 antithesis between true poetry and "oratory," or "eloquence": between "feeling confessing itself to itself," as if in "utter unconsciousness of a listener," and rhetoric out to win friends and influence people. Poetry risks becoming mere eloquence wherever it acknowledges an audience, a world, and a "desire of making an impression on another mind."[86]

But Mill was openly levying this definition against nonlyrical, narrative, descriptive, and didactic works still very much professing to be poems. His concerted invalidation of the patent incident-, object-, and audience-dependencies of these forms is what made "What Is Poetry?" an important salvo in the long nineteenth-century process of "lyricisation."[87] To make a formal failure, fortunate or otherwise, of *The Mask*'s pretended oration, then, is to accept the polemical marginalization of the poetic genres for which the written illusion of oratory was perfectly good form, indeed a generic expectation. "This is not poetry to be overheard," writes Alexander Dalzell in his theory of classical didactic, with a wink toward Mill's ongoing power to rule his subject unpoetical: "The didactic poet shares with the teacher and the preacher a particular kind of communication, and what is significant about that relationship is that it is always directed towards an audience."[88]

Dalzell's sketch of what might be called the didactic "radical of presentation," which he elaborates with *De rerum natura* in mind, rather instantly depathologizes *The Mask of Anarchy*'s oratory: the didactic voice is pedagogical (about the message) rather than confessional (about the messenger); the reader is expected to identify with the addressee, not the speaker; the explicit addressee and the implicated readership do not exactly coincide; the poem offers a systematic account of a subject.[89] To tell the story of Romantic poetic science as a genre study would require another book (and writer!), but consider for a moment how many paradigmatically Romantic poems might be implicated in the basic, triadic structure of didactic address: a fictive speaking situation that projects "the subtle relationship of teacher, student, and subject,"[90] rather than the self-address of soliloquy or a strictly unrealizable apostrophic dyad. The cases multiply if we allow the didactic dynamics of authority and identification to be under contention in a period as signally obsessed with *Bildung* as the Romantic one: a cursory glance at the English Romantic lyric canon through the lens of didactic form evinces pedagogical variations ranging from identification with a soliloquist who turns out, at the last moment, to have been a teacher ("Greater Romantic Lyric") to scenes of dialogical instruction concerned to make of the poetic classroom a space "large enough to contain incommensurate minds" (lyrical ballads).[91] Already in the classical archetypes, the teacher presents in the role of another's student; in their most closely neoclassical didactic epics, Erasmus Darwin and Shelley heighten this tendency until the first-person resides in a learner who frames instruction received from another authority.[92]

Romanticists come by an antipathy to the didactic honestly. After all, it was Romantic-era writers and critics who turned "didactic poetry" into "a byword for mediocrity or a simple contradiction in terms," as David Duff has nicely glossed the situation, with none other than Shelley supplying the slogan "Didactic poetry is my abhorrence" in the Preface to *Prometheus Unbound*.[93] The complications begin in the latter half of Shelley's sentence, however, and continue into his paraphrase of Horace on the poet's duty to "amuse and instruct."[94] (And the complications culminate, more widely, in the multimodal "coordinated assault on a variety of readerships" through which Shelley was attempting to move the national sentiment in 1819, making *Prometheus*'s promise of "beautiful idealisms" to gratify "the highly refined imagination of the more select classes of poetical readers" sound more like flattery of those classes' self-image than consistent aestheticist principle.)[95] But the readiest way to reframe the complex issue of Romantic antididacticism, which implicates the heated reception and canonization controversies nicknamed "Pope" and "Wordsworth," is with recourse to the *OED*.[96] There it becomes clear that when a new, lightly pejorative usage of *didactic* came into use in the late eighteenth century, it worked not to dismiss instruction and instructive literature wholesale, but, on the contrary, to dispute in new detail about pedagogical *method*. *Didactic* now contrasts "formal" or "rote" pedagogy unfavorably with methods that foster "greater involvement or creativity on the part of those being taught"; it appears in tandem with "dry" and "cold."[97] Romantic antididacticism in poetics should be similarly reframed: an objection not to teaching but to "bad" teaching, and a question less of whether poetry should teach than what and how.

Duff's illuminating work on the subject makes recognizable a poetry of purposive instruction that persists by other names and with new psychological sophistication in the Romantic period: for instance, in the widespread vocabulary of awakening, defamiliarizing and retraining readerly faculties of mind and feeling, not least in prefaces that, from *Lyrical Ballads* to *Prometheus Unbound*, forthrightly defend the "purpose" of a given poem. Some of what scholars elsewhere take for obvious should be brought to bear on the purported demise of didactic, he argues: in post-Revolutionary contexts keenly convinced of the ideological influence of literature, the better-known, anti-instrumentalist disengagement strategies of aesthetic autonomy competed against equally powerful instrumentalist attempts to redouble or reform the ideological powers and responsibilities of literature, to argue its merits in terms of social and political utility, and to "defend the place of poetry in the modern circle of knowledge."[98] This last phrase makes Shelley's contribution particularly obvious.

But there is also a pattern worth noticing in the way lyricizing polemics begin to construe didactic as poetically untenable in the period, and this relates less to audience relations and reader/pupil psychology than to that third dimension of the didactic radical of presentation: the subject matter. Taking in this pattern means leaving Shelley's poem a little longer in order to listen to some brilliant hostile witnesses against didactic poetry that help restore the more-than-academic threat posed by the genre. Above all, Thomas De Quincey's antididactic polemic turns out to be pitched against a peculiarly Lucretian combination of knowledge and passion: a materialist form of corporeal inspiration, threatening because it is neither personal nor divine.[99] Pathological in De Quincey's judgment, this self-dispossessing mode of mortal solidarity and terrestrial attachment exemplifies for Shelley, I suggest, the possibility of a poetry whose passion is history's, not his.

g. As Nature Teaches

"The poetry is not in the object itself, nor in the scientific truth itself," Mill argues, "but in the state of mind in which the one and the other may be contemplated." An "incident" worthy of story, "an object which admits of being described," or "a truth which may fill a place in a scientific treatise," he argues, may very well "*also* furnish an occasion for the generation of poetry" (narrative, descriptive, and didactic, respectively).[100] Yet in all these cases, such events, objects, and truths are strictly incidental. This rather exactly prohibits the form of poetic realism extended in *De rerum natura*'s figural physics, in Darwin's, and in Shelley's, where minds are not the only source of poiesis, which tends rather to evidence contact between different natures through figural distortions proper to neither.

De Quincey takes the case directly to Lucretius in the course of his famous antididactic distinction between the "the literature of *knowledge*" and "the literature of *power*." "The function of the first," he explains "is—to *teach*; the function of the second is—to *move*."[101] Unlike Mill, then, De Quincey is perfectly comfortable construing poetry as a species of the rhetorical power of persuasion (what Mill disparaged as "eloquence" and "oratory"). Yet beginning with a delicious comparison of *Paradise Lost* to "a cookery-book," De Quincey radicalizes Mill's distinction between "external objects" ("knowledge," "items," "information," in De Quincey's vocabulary) and the "human passions, desires, and genial emotions" in which poetry essentially consists.[102] "Didactic poetry" is for De Quincey an oxymoron whose "predicate destroys the subject."[103] Pointing out that Virgil's dubious farming recommendations are obvious pretexts for more properly poetic delights—"It is an excellent plea for getting a peep at the bonny milk-maids to propose an inspection of a model dairy"—he saucily denies that "the element of instruction enters *at all* into didactic poetry." Teaching is a "foil" for poetry's excursive passions and pleasures, the "prosy thread" that "holds together the laughing flowers which go off from it to the right and to the left."[104] (No room here for the Lucretian possibility that laughing flowers might be figurally instructive.)

Hence, De Quincey argues, the illustrative failures of Pope's *Essay on Man*, and—what amounts to the same thing—*De rerum natura*. Both didactic poems are equivalently undermined by the systematic demands of a subject too serious to serve as a light tie between the extravagant flowers of rhetoric:

> To evade the demands in the way that Pope has done, is to offer us a ruin for a palace. The very same dilemma existed for Lucretius, and with the very same result. . . . both [poems are], and from the same cause, fragments that could not have been completed. Both are accumulations of diamond-dust without principles of coherency.[105]

As a Lucretian might point out, de Quincey's logic here is seriously figural, a repudiation that is philosophical and metaphorical in equal measure of the decadent incompletion of atomistic form. These poems are "accumulations of diamond-dust" whose inorganic shimmer betrays the worldliness of (neo-) classical materialism, their fragmentary status bearing out its nontranscendent commitment to "ruin." In other words, the passage begins to reveal that, far from being irreconcilable, the poetic and didactic exigencies of such poems might be all too consistent—but consistent in what looks, to De Quincey, like spiritual destitution.

That the problem with didacticism might be a question of morality and metaphysics becomes clearer as De Quincey softens his initial antithesis between serious teaching and passionate persuasion to account for the obvious, indisputably admirable counterexamples of *Paradise Lost* and *The Prelude*. The passage is worth citing at length for the way it reveals one powerful strain of Romantic and Victorian antididacticism as a campaign *for* a rival poetic curriculum:

Poetry, or any of the fine arts (all of which alike speak through the genial nature of man and his excited sensibilities), can teach only as nature teaches, as forests teach, as the sea teaches, as infancy teaches, viz., by deep impulse, by hieroglyphic suggestion. Their teaching is not direct or explicit, but lurking, implicit, masked in deep incarnations. To teach formally and professedly, is to abandon the very differential character and principle of poetry. If poetry could condescend to teach anything, it would be truths moral or religious. But even these it can utter only through symbols and actions. The great moral, for instance, the last result of the Paradise Lost, is once formally announced; but it teaches itself only by diffusing its lesson through the entire poem in the total succession of events and purposes: and even this succession teaches it only when the whole is gathered into unity by a reflex act of meditation; just as the pulsation of the physical heart can exist only when all the parts in an animal system are locked into one organization.[106]

A literary synthesis between knowledge and power, it seems, is possible and desirable after all. But it must proceed by way of organic form, which insinuates the right "truths moral or religious" throughout the whole of an artwork, generating a living totality that "lock[s] into one organization" the "genial" minds and sensible hearts of author and reader. This specific, organicist understanding of what "nature" is and means explains how De Quincey, borrowing language from the *Prelude*, can incidentally provide a very fine description of Lucretian pedagogy—"Poetry . . . can teach only as nature teaches, as forests teach, as the sea teaches"—in the process of declaring it impossible. When he writes of moral and religious truth "masked in deep incarnations," it is no exaggeration to hear organic form connecting creature to Creator through the symbolic vehicle of the creation. As "incarnations," the passage's forests, seas, and infants are creaturely vehicles for divine truth; through organic form "the genial nature of man" participates in the Creative logic of symbolic expression.

The part of Wordsworth's *Prelude* to which De Quincey alludes clarifies what it means to "teach as Nature teaches" in this sense:

> In progress through this verse my mind hath looked
> Upon the speaking face of earth and heaven
> As her prime teacher, intercourse with man
> Established by the Sovereign Intellect,

> Who through that bodily image hath diffused
> A soul divine which we participate,
> A deathless spirit.
>
> ([1805] 5.231, 5.11–17)

By a prosopopoetic logic that could not be more different from that of *The Mask*'s terrestrial oratory, here earth gets a "speaking face" by sanction of "the Sovereign Intellect," entering into "intercourse with man" as an authorized vehicle for divine communication: a "bodily image" through which the incorporeal intelligence addresses our own immortal soul. Such logic is particularly liable to figural disruption: in an eloquent reading that inaugurates his study of *lyric* pedagogy, Brian McGrath leans into the "as" in Wordsworth's "teach as Nature teaches" to show how the difference between grammar and rhetoric surprises this aim, such that "Poetry teaches the difficulty—if not impossibility—of learning to read it, which is not to suggest that it teaches how to overcome this difficulty."[107]

Wordsworth's "philosophic Song," which Coleridge envisioned as a counter-epic against Epicurean tendencies, indeed harbors plenty of uncanny, unauthorized figurations that do not fit the lesson plan.[108] For us, however, the passage incidentally but instructively (!) marks out the difference between conversing with the faces of things under the aegis of what is often called Wordsworth's "animism" or "pantheism" (a keyword in the eighteenth-century Spinoza controversy), and doing so under Lucretian materialist auspices.[109] The first inspirits the natural world, "diffus[ing]" a "soul divine" throughout all the visible members of "earth and heaven"; the subtle heresy risked here is that of rendering the "Sovereign Intellect" superfluous beside an animate, immanently divine nature whose activity might not require transcendent sanction. The second flatly denies the existence of such a sovereign, "Untenanting creation of its God," as Coleridge puts it.[110] The Lucretian tradition maintains that bodies are creatures not of God but of earth and sky; they go on figuring without divine underwriting, appearing and transpiring without incarnating incorporeal meanings or plans.[111]

In De Quincey's context, to "teach only as nature teaches" reinforces poiesis as a force of *human* nature—"genius is a voice or breathing that represents the *total* nature of man"—and casts the relation between truth and phenomena as one of symbolic incarnation.[112] To say that poetry "can teach only as nature teaches" in a neo-Lucretian sense would mean, on the contrary, that artistry cannot exit nature; that in poetry human language is bent by other natures' teaching; and that such lessons do in fact merit "direct and explicit" pedagogical breakdown for human learners. (Recall the pedagogical redundancy in the Lucretian phrase we saw Goethe and Knebel working on in chapter 3: *uti docui, res ipsaque per se/vociferatur*—as I have taught, and the thing itself exclaims—or, in Good's translation, as proves "many a clamorous fact.")[113]

De Quincey is in fact quite sensitive to these differently "natural" features of Lucretian didactic, which ultimately make *De rerum natura*, in his judgment, not so much bad as insane. Though the essay on Pope is no place to admit it—there Lucretius's poem functions as a synonym for *An Essay on Man* and the extirpation of poetic "power" by the demands of "knowledge"—De Quincey elsewhere stresses precisely "the extraordinary *power* of Lucretius."[114] Far from representing a desiccated and plodding didacticism, *De rerum natura*, he comments in an essay section on Keats, bears witness to a "powerful nature" that "precipitated its own utterance, with the hurrying and bounding of a cataract"; "the style is not only stormy, but self-kindling and continually accelerated."[115] Lucretius, paragon of the "literature of *knowledge*," in fact evinces no shortage of the *power* that was supposed to be its antithesis, that impassioned "function . . . to *move*" that characterized true poetry in De Quincey's estimation.[116]

Yet De Quincey has a problem with *De rerum natura* that makes its passion a pathology, and not a power. Lucretius's "frenzy of an earth-born or hell-born inspiration" admits no human or divine source for its movement:

> divinity of stormy music sweeping round us in eddies, in order to prove that for us there could be nothing divine; the grandeur of a prophet's voice rising in angry gusts, by way of convincing us that prophets were swindlers; oracular scorn of oracles; frantic efforts, such as might seem reasonable in one who was scaling the heavens, for the purpose of degrading all things, making man to be the most abject of necessities as regarded his causes, to be the blindest of accidents as regarded his expectations.[117]

Lucretius's materialist rapture is an "earth-born" passion of and for the nature of things that overwhelms the poet's skill and the human exception and yet refuses transcendence, denies having escaped the order of necessities and chances. For De Quincey, this amounts to an intolerable combination of knowledge-power.[118]

De Quincey calls Lucretian poetry "an insanity affecting a sublime intellect." The materialist degradation of all things to body must be a symptom of bodily sickness; it proves Lucretius to have been writing "under a real physical disturbance," intoxicated, "delirious."[119] Such writing is something of which De Quincey, like Coleridge, is more than entitled to speak—and also to resist, especially to resist glamorizing. But tracking the diagnosis of Lucretian pathology in their poetic theory reveals that one purpose and effect of valorizing lyric and demonizing "didactic" in the nineteenth century was to deny the possibility and the legitimacy of a mode of ecstatic knowledge whose resolutely worldly transports were not assignable to personal feeling or divine inspiration. "It was an *aestrum*, a rapture, the bounding of a maenad, by which the muse of Lucretius lived and moved," De Quincey writes with penetrating disapproval: a poetic bearing compelled and propelled by the objects of knowledge and driven to communicate this movement, rather than its own, properly personal and human feelings; a power of corporeal *affection* that draws its explicit polemics from such bodily testimony and submits its knowledge, ultimately, back to corporeal judgment.[120] The bleak scene with which *De rerum natura* finally breaks off—a depiction of the plague of Athens that seems out to countenance the risk of contagion that physically and philosophically attends a materialism of contingent, corporeal vulnerability—chronicles the dissolution of every social feeling but one: the refusal to "abandon the bodies," the very last words of the fragmentary epic (6.1286).[121] Audible here is perhaps a prototype of the Marxian historical-materialist commitment to beginning and ending in the joy and suffering of worldly beings: a vehement denial of the afterlife that, in De Quincey's eyes and others', risks the irony of a substitute faith.[122]

h. Pedagogy of Knowledge-Power

For Shelley, who clearly tended to seek, rather than suppress, the irreligious and self-dispossessing liabilities of writing, the Lucretian example of a poetry of "knowledge-power," a genius of world-affection rather than self-expression, would have offered up something extremely important: a possibility of poetic form as a passion of worldly *history*, which Shelley's poetry of and around 1819 is everywhere attempting to attain.[123] *The Mask*'s unrelenting turnings toward the ground, I think, represent a suitably contemporary form of such "earth-born," impassioned knowledge: a mode of expression keen not to originate in the poet, not to abandon the bodies, and not to assign otherworldly causes and purposes to their suffering. What this yields, in early nineteenth-century England, is a stirring depiction and exposition of socioeconomic violence and biopolitical precarity.

In fact, in the place where Shelley's *A Defence of Poetry* takes the measure of the Latin didactic tradition—"Lucretius is the highest, and Virgil in a very high sense, a creator"—it is to link the creativity of both to a culture whose "true Poetry" consisted in its historical and political forms. "The true Poetry of Rome lived in its institutions," Shelley argues, in the "rhythm and order in the shews of life" through which republican politics were enacted. In this case, "The Past, like an inspired rhapsodist" assumes the role of poet: the "dramas" of the Roman republic and empire "are the episodes of the cyclic poem written by Time upon the memories of men."[124] The living conduct of Roman political history approached poetry in its "rhythm and order," Shelley thought; Latin didactic, it seems, offered an example of how to let living history co-author a poem.[125]

The Mask is not simply or exclusively a didactic poem, and genre reassignment is no end in itself. But lyricization, no less lately than with De Quincey, has operated as something of a containment strategy for poetic materialism, resulting, as we have seen, in the truly astonishing conclusion that Shelley's *Mask*, for good or ill, lacks "an analysis of how material change might be realized in the historical moment of 1819."[126] Switching out the lens of negative aesthetics for the (equally formal) one of positive didactics makes a different kind of assessment possible. To ask the question about poetry and politics from the point of view of didactic means, most basically, to consider the poem in terms of the passionate reception and transmission of *knowledge*: that is, to ask about the politics of its teaching in multiple senses, including the politics it teaches and the politics of its pedagogy.

Because the knowledge at issue is knowledge of Peterloo, this line of inquiry opens immediately onto a more precise question about the teaching of radical history as a potentially political act. This is an act that *The Mask of Anarchy*—despite poetry's much-respected inability to make things happen, and only with the help of a host of circumstantial and institutional supports— keeps making people commit. I mention this not to use didactic form to turn the problem of poetry and politics into a reflection internal to "the profession," but rather to suggest that the stronger than usual contextualization-effects of *The Mask of Anarchy*, which induce literature teachers to give their preparation and class time to the study of a subject matter that exceeds the poem, are a feature of didactic form in its fundamental heteronomy. While it has proved both possible and generative to consider lyric address in terms of the triumph of the (fictive, written) event of vocalization over the referents and interlocutors summoned to confirm it, "didactic," by contrast, would seem to name the (fictive, written) address that wishes to see them prevail.

Far from suppressing *The Mask of Anarchy*'s verbal ambivalences, grasping the didactic aspect of the poem's address means, first and emphatically, *affirming* the consensus among deconstructive and negative aestheticist readings regarding the self-evacuating character of its "lyric" utterance. The oratory that forms the bulk of the poem culminates in a hundred-line sequence of *fiats*—"Let a great Assembly be," and so on—that exhort the "Men of England" to further acts of nonviolent protest in quintessentially performative poetic grammar that regularly undercuts its own emancipatory power. This is how, for instance, the poem's thundering rhetoric promises to deliver "oppression's" death sentence in the penultimate stanza:

> "And these words shall then become
> Like oppression's thundered doom
> Ringing through each heart and brain,
> Heard again—again—again—"
>   (lines 364–67)

The ambiguous possessive "oppression's doom" works equally to call for oppression's demise and to perpetuate ("again—again—again") the "doom" it presently metes out on "each heart and brain."[127] And indeed, this portion of the poem leaves systematically unclear whether its performative power augurs oppression's end or is doomed to repeat it. Such double-binds are classic Shelley, arguably a poetic specialty. It is of this stanza's double-edged repetitions that Redfield observes that "the text collapses reflexivity into the mechanical iteration of an inscription," leaving "nothing more than the blank fact of its own material occurrence."[128] But in *The Mask of Anarchy*, performatives are falling short in a manner that is anything but "blank." It is not so much that these fiats undermine their capacity to enact a desired future apart from their own utterance, a "failure" that students of lyric have long since learned to recuperate as the negative essence of lyric voicing. More specific and strange than collapsing toward contentlessness, these fiats collapse toward the constative, toward a simple statement of the case of what happened "on the Occasion of the Massacre at Manchester."[129]

In what amounts to *The Mask*'s third telling of the events, here a "great Assembly" (line 262) of working-class and pauperized "Women, children, young and old" (line 277) and their allies (lines 283–90) gather on a "spot of English ground / Where the plains stretch wide around" (lines 264–65) to demand the rights purportedly guaranteed them by "The old laws of England" (line 331) by nonviolent and orderly means ("measured words" [line 297], "folded arms and steady eyes" [line 344]). Nor, more chillingly, does Shelley's chorus of jussive "lets" quit rehearsing what has already happened as the assembly became a massacre and a galvanizing national scandal: "Let the charged artillery drive / Till the dead air seems alive" (lines 306–7); "Let them ride among you there, / Slash, and stab, and maim, and hew" (lines 341–42); "And that slaughter to the Nation / Shall steam up like inspiration" (lines 360–61). To my knowledge, only Eric Lindstrom, in a book-length study of Romantic *fiats*, has stressed this oddity in *The Mask*: a "logic of retrospective command over world-historical events that have already transpired," as he puts it, reading Shelley's gesture as a Nietzschean "*amor fati*."[130]

But we might also recognize the closing oration as a feat of historical pedagogy, an act of conjuring or recalling the events of Peterloo in the minds of student-readers. Here the incantatory force of "let" does not so much (fail to) legislate a future of freedom as induce a form of historical memory of an episode in the fight for it, operating as a "let's say" that enlists an audience in summoning to mind an event hard to believe from the outset and perhaps increasingly easy to forget or never know. Repeating the violence of Peterloo, then, is an act not only of lyric self-incrimination but also of didactic instruction, even indoctrination, with all the political difficulties this formulation implies. On this reading, *The Mask of Anarchy* becomes a script for the vivid and polemical recollection of Peterloo, interpolating readers as learners and teachers in the production and transmission of (what may or may not succeed in being) radical, popular counterhistory. One recalls, in this context, that the paradox of a pedagogy of change rather than continuity—"it is essential to educate the educator"—was one of Marx's key motives for mandating the transformation of theory into "revolutionising practice."[131]

What might revolutionary pedagogy look like, and to what extent do the transformations of didacticism charted in Shelley's poem participate in this possibility? Paulo Freire took up Marx's charge in the late 1960s, developing a "pedagogy of the oppressed" that sought to make education a "co-intentional" practice of freedom. He stressed the basic point that "the oppressed must be their own example in their struggle for redemption," which "is not a gift bestowed by the revolutionary leadership, but the result of their own *conscientização*."[132] Yet he also insisted that despite appearances, coming to critical consciousness is not a self-reflexive process, and cautioned against the "risk of shifting the focus of investigation . . . to the people themselves." Instead, as in the didactic poetic tradition we have been following, a third object takes a strange preeminence: teacher-students and student-teachers must be together engrossed in the investigation of a specially gathered and prepared object, a "codification" in which many elements of their "thematic universe" intersect.[133] The method of the dialogical educator, Freire argues, is to absorb from learners a "generative theme" and to "re-present" it back to them "not as a lecture, but as a problem." The educators' major task is to produce an image, sketch, or photograph—let us call it a "Shape"—the contemplation of which allows "previously inconspicuous," "background" conditions of the historical situation to "stand out" as phenomena.[134]

In *The Mask of Anarchy*, the "generative theme" is "Peterloo" as gathered from many sources and rendered back toward the people who made it, in a "codification" that offers "the situation within which they are submerged" as a perceptible problem.[135] We saw above how, in interaction with the Shape, "thoughts arose" that were quite rigorously and multiply thoughts of the backgrounds to the massacre. And this happened, in Shelley's poem, as a pointed alternative to a top-down mode of allegorical instruction that gave comically little room for thought and was hard to distinguish from the de-braining violence of the government's authoritarian reaction.

i. The Power of Assembly

It is well known that *The Mask of Anarchy*, particularly its climactic stanzas, was taken up and reprinted in the Chartist and Owenite socialist movements in the late 1830s and 1840s. Yet as Anne Janowitz has shown, this uptake is better understood as a dialogical *return* of *The Mask* to its sources in the communitarian poetry of the radical and ultraradical press in the teens and twenties.[136] The very same formal features that triggered Wolfson's worry that *The Mask* evinced a "self-addressed" and "rarefied political poetics"—its fictions of public ventriloquism, of "fighting tyranny with allegorical signs," of demystifying corrupt power in the mode of a dream—are borrowed from interventionist poems so unrarified that they have only recently garnered critical attention.[137] Janowitz and Raphael Hörmann have since separately identified the radical Spencean poet (and hairdresser and musician) Edward James Blandford's "A REAL DREAM; *or, Another Hint for Mr. Bull!*" as a key intertext for Shelley's *Mask of Anarchy*.[138] Published in April of 1819 in *The Medusa; or Penny-Politician*, which seems to have been among the papers Shelley's correspondents sent him in Italy, Blandford's allegorical dream tour of "titled pomp's all-splendid raree-show" offers an undeniable prototype for *The Mask* in form and didactic message, from "blood-stained vice assuming saintly grace" to the final figural turn that vocalizes mute witnesses to the depradations of "peculation, fraud, and falsehood."[139] Blandford's May poem, "Glorious Poverty," even features "a glorious Sidmouth-saint,/Whose glori-

ous murders fame shall paint,/With all his bloodhound crew!"—a hypocriti-
cal human-canine posse that makes more than a cameo in Shelley's poem.[140]

Yet relocating *The Mask of Anarchy* within the idiom of radical popular
song—an "interventionist poetic, a poetic fused not to aesthetic autonomy but
to poetic and political collective agency," as Janowitz puts it—raises important
new questions about the politics of its teaching.[141] The new contexts propel us
from a first didactic reading of the poem as a matter of historical instruction,
toward a second that evaluates the political theory it promulgates. For reading Shelley's popular songs alongside *The Medusa*'s makes clear the political
radicalism that it was then possible to communicate in verse, and Shelley's
distance and difference from it. Blandford's dreaming speaker frankly calls
the starving "million" to revolutionary violence—"'Rouse men,' cried I, 'and
for your freedom fight!'"—while Shelley scholars have been at pains to show,
at best, *The Mask*'s ambivalent "recognition that liberatory resistance must include the physical destruction of a genocidal oppressor."[142] It is certainly true,
as Keach points out, that Shelley's call to "Rise like Lions after slumber/In
unvanquishable number" transfigures "the most familiar of royal heraldic de-
vices into an image to make all royals and royalists tremble"; but also that, as
others observe, the advice to "Shake your chains to Earth like dew/Which
in sleep had fallen on you" seriously underestimates the weight of the chains
and the force that might be necessary to remove them (lines 153–54).[143]

The most important, and still unanswered, critique of this program comes
from Goldsmith, who argues that it sells short Shelley's trenchant diagnosis of
the material basis of working-class disempowerment by lapsing into a politics
of purely linguistic emancipation:

> "Let a vast assembly be,
> And with great solemnity
> Declare with measured words that ye
> Are, as God has made ye, free—
>
> "Be your strong and simple words
> Keen to wound as sharpened swords,
> And wide as targes let them be
> With their shade to cover ye."
> (lines 295–302)

Here, Goldsmith argues, Romantic political aesthetics and its deconstructive
advocates alike perpetuate an unacknowledged legacy of apocalyptic formal-

ism: a tendency to locate freedom in the power of words (formerly, *the* Word) alone, attributing to them an independence from historical suffering and an emancipatory power to end it, and performing, again and again, "the transformation of matters of power into matters of language."[144] In context, moreover, Shelley's ultimate substitution of words for weapons and confidence that the multitude's linguistic agency will suffice to spell "oppression's doom" coincided with the merely moderate English reformist agenda, which was similarly limited to demanding the *verbal* freedom of adequate representation in Parliament. Here "the literary and the political converge in a faith in the work of representation"—to the possible exclusion of the work of material freedom.[145] The point is well taken, and was made in 1819, too. As one *Medusa* editorialist put it, "To have reformed the Commons' House of Parliament upon liberal principles, to give every honest man a vote, every man of fair character (as honest of themselves) a free voice as an elector" might once have sufficed—but now "the fall of our tyrants" must mean "retributive JUSTICE": "we must have a far more extensive rummaging into social rights, a more general search after national property," than parliamentary reformists are "willing to talk about."[146]

Of Shelley's call for a "great Assembly," Goldsmith rightly asks, "what is an assembly if not a place of speech and debate, a place where those who are represented have their say?"[147] But recent mass movements and occupations are inspiring different answers to this rhetorical question, including answers that value the power of assembly precisely for its noncoincidence with freedoms of speech and state-sanctioned forms of parliamentary representation. "If we consider why freedom of assembly is separate from freedom of expression," Judith Butler argues, "it is precisely because the power that people have to gather together is itself an important political prerogative, quite distinct from the right to say whatever they have to say once the people have gathered."[148] Never exactly containable by its government sanction (or suppression) as a legal "right," the power of people to assemble in fact tends to stage the distinction between popular sovereignty and that of the state.[149]

Visibly laying claim to the public through embodied "actions in concert" that do not reduce to the verbal demands they may or may not articulate, people who assemble, Butler argues, issue a "bodily demand" that tends to foreground the conditions of "livable life" over those of speech and representation:

> [W]hen bodies assemble on the street, in the square, or in other forms of public space (including virtual ones) they are exercising a plural and performative right to appear, one that asserts and instates the body in the midst

of the political field, and which, in its expressive and signifying function, delivers a bodily demand for a more livable set of economic, social, and political conditions no longer afflicted by induced forms of precarity.[150]

The Mask of Anarchy, I have tried to show, pursues at every level, from its minute allegorical tactics to its generic affiliation, a mode of corporeal signification capable of bringing into relief the political causes and contexts of evitable suffering, quick and slow. What we have been tracing in ontological and poetic dimensions through Shelley's mobilization of classical figural materialism, in other words, Butler identifies at the heart of a performative theory of democracy. The agreement bespeaks another: the congruity between Shelley's Lucretian semiotic and the risk and power of popular assembly that Peterloo vividly signified in action and event.

Political assemblies, Butler argues, enact "not a unitary will, but one that is characterized as an alliance of distinct and adjacent bodies" whose repertoire of embodied gestures, actions, and inactions together exert a claim to public space and to being "the people" that "can never be fully codified into law."[151] Irreducibly plural, relational, embodied, context-dependent, and transient—signifying freedom from necessity by way of the same "power" that opens them to harm—the expressive political power of assemblies belongs, in other words and in another dimension, to the genre of materialist assemblage. Shelley's four tellings of Peterloo in *The Mask* evince new coherence in this light: a vicious send-up of the affinity between authoritarian politics and a mode of signification that "exacerbate[s] the antipathy of the living to the significant"; a pointed switch to an alternative semiotic mode that openly owes its material and its meaning to the bodies assembled and generates a composite image of their force and fragility; a didactic breakdown of the material backgrounds (socioeconomic and biopolitical) that conditioned the manifestation and its affections; a political affirmation of assembly in all its scary, practical liability and power.[152] This last *A Philosophical View of Reform* simply avows as "the necessity of exciting the people frequently to exercise their power of assembling."[153]

It was already as an assertion of bodily significance in the midst of a deadly political field that Shelley offered his first figure for nonviolent resistance in *The Mask of Anarchy*. Hope/Despair "lay down in the street/Right before the horses' feet" in order to provoke the authorities to take her life in a single, visible act of violence rather than through decades of biopolitical neglect. What the allegorical regime of the first part of the poem pictures as a singular act of heroic martyrdom, the closing oratory retells as collective action.

Reviewed in this light, a whole repertory of embodied gestures, bearings, and motions distinguishes Shelley's sense of assembly from purely linguistic agency and purely parliamentary aims, and still more from pure passivity or pure negation. Above all, the oratory stresses the significance of "assembling" in itself: of showing up, in "vast" and intimidating numbers ("From the corners uttermost," "From every hut, village, and town"), and of bringing normalized, quotidian suffering into view (out from hiding in "the haunts of daily life / Where is waged the daily strife / With common wants and common cares"). Repeating the word "common" and equivocating between "others' misery" and one's "own," the passage makes clear that to assemble is to have transformed elite "compassion" and political exasperation into cross-class physical solidarity. A "phalanx" choreography is recommended: "Stand ye calm and resolute / Like a forest close and mute," with "folded arms and looks which are / Weapons of unvanquished war" (lines 319–22).

Whatever "strong and simple words" are uttered, and the poem does not say which, they form but one part of a message that is also delivered "mute": a message of superior force—but more specifically, of superior force contained, for the moment, in an attitude of ethical superiority: "With folded arms and steady eyes, / And little fear, and less surprise / Look upon them as they slay / Till their rage has died away" (lines 344–47). What follows in Shelley's poem fills out a platform of nonviolent resistance that has less to do with the purportedly inherent nonviolence of "the aesthetic" as Borushko has argued it, than with a tactical repertoire familiar, today, from the desegregation movement in the U.S.-American South, the anticolonial struggle in India, and more recent occupations of public space.[154] It plays out the high-cost wager that "the blood thus shed" by people whose visible, physical insistence in certain spaces provokes outright violence "will speak" in multiple ways as media amplify the scandal: catching authorities red-handed in uses of excessive force that erode their legitimacy in the eyes of a broader public, exciting mainstream sympathy and shame, performing what a later theorist describes as a "moral jiu-jitsu" that converts enforcers and enablers to the cause of the demonstrators.[155]

At this place in *The Mask of Anarchy*, "The old laws of England," are summoned in a peculiar and telling way. The demonstrators are to "Let . . . stand" these laws, whether "Good or ill" laws, "between" their bodies and those of law enforcement as they advance toward a "Hand to hand, and foot to foot" face-off.[156] To many readers, the recommendation has seemed to backpedal toward an anxious advocacy of civil obedience that evinces Shelley's class interests, as well as a near-Burkean reverence for traditional juridical forms (the

Laws' "reverend heads with age are grey") and the belief, bad faith or naïve, that the law will suffice to shield assembled people from harm (since when?). But in a poem that has already declared England's "righteous laws," frankly, "sold" (line 231), the stanzas in fact seek to turn the laws' pliancy to power to popular advantage, emphasizing the way juridical codes mean differently in the mouths and backed by the collected power of a multitude they have not previously been forced to represent. In these changed acoustic circumstances, the Laws' "solemn voice" begins to "echo," rather than to prescribe, the assembly's meaning: the "solemn voice" of juridical precedent "must be / Thine own echo—Liberty!"

Shelley, of course, is more famous for a different juridical-poetical agenda, known equally for its grandiosity and its double-edge. "Poets are the unacknowledged legislators of the World," he would write in the *Defence*: they lack acknowledgment not only in the sense of being stunningly underappreciated but also in that they are themselves unknowing, legislating to the precise extent that they neither know nor feel what their words "express," "inspire," "sing to battle," or "mov[e]." It is less frequently remembered that in *A Philosophical View of Reform*, where these famous final sentences of *A Defence* were first worked out, their nearly perfect expression of aesthetic purposiveness without purpose was followed up by a quite different juridical-poetic plan of action. In the section on "Probable Means," Shelley suggests that such *un*acknowledged legislators ("poets, philosophers, and artists") write petitions to their more *acknowledged* legislators, taking part in a scheme of mass civil over-obedience aimed at paralyzing government through the logistically overwhelming exercise of every allowable form of minor complaint and appeal. Memoranda would "load the tables of the House of Commons," the courts would be flooded with "an overwhelming multitude of defendants," revenue collection and legal administration would grind to a halt.[157] The plan gives yet another, surprisingly corporeal meaning to *The Mask*'s seemingly tame recommendation to "Let the Laws of your own land, / Good or ill, between ye stand" (lines 327–28). Shelley might be envisioning a wadding of paper so thick it impedes the normal course of governmental abuse. In *A Philosophical View of Reform*, that is, on-paper and in-person strategies alike collude in "generally conveying the sense of large bodies and various denominations of the people in a manner most explicit to the existing depositories of power."[158]

In this chapter, I have tried to offer a didactic poetic reading of *The Mask of Anarchy* by taking seriously the entwined problems of the politics of its pedagogy and the politics it teaches. A closing reflection on the political meaningfulness of continuing to teach the poem and the historical events to which, as good didactic, it makes teachers and even its own text give place, may be in order. On a didactic reading, the continued relevance of *The Mask of Anarchy* cannot be attributed to an aesthetic autonomy from its context that frees it to be equally and differently meaningful in another. Its pertinence and accessibility will attest instead to affinities between the context of composition and that of reception, such that loving a poem might mean willing its obsolescence rather than its eternal life: pulling, in the case of *The Mask of Anarchy*, for the change in circumstances that would render its pictures of quick and slow state violence quaint or moot. In some ways, we have wandered far from *De rerum natura*. But we have also stayed broadly true: love knowledge, it says, hate violence, teach. And whatever happens, refuse, as the last sad words of the materialist epic have it, to "abandon the bodies."[159]

In its "strange combination of nature and history," didactic poetry as well as the form of materialist allegory with which it is entwined, I have tried to suggest through this exploration of *The Mask of Anarchy*, knows that the past obtrudes "punctually" into the present. It converts this pressure into an address that, punctual in a different sense, bears forcefully on some present purpose.

Coda: Old Materialism, or Romantic Marx

History is the true natural history of man (on which more later).

KARL MARX, *Economic and Philosophic Manuscripts of 1844*

"Human sensuousness is therefore embodied time."[1] This, concluded Marx, dissertating at twenty-two on "The Difference between the Democritean and Epicurean Philosophy of Nature" (1841), is the brilliant upshot of Epicurus's ancient atomist materialism in its Lucretian poetic transmission: it shows, "at last," how the living body forth their times . . . and not, as he would soon come into the habit of adding, just as they'd like.[2] At the outset of a materialism whose credentials as "historical" and whose aspirations as "practical" are generally undisputed, that is, one finds the unfamiliar style of material corporeality whose occasional reappearances in Romantic science and poetry have been the subject of this study: fractious, composite bodies, made up of myriad smaller bodies in motion, perpetually falling into pieces and perhaps into an observer's eyes.

To draw his early conclusion about "embodied time," Marx zeroed in on the very Epicurean and Lucretian technicality that had fascinated Goethe and Shelley before him: the theory of the *eidola* through which bodies produce, at once, the inseparable possibilities of sensory and of temporal experience. "At last," Marx concludes of the outward-turning "sensuous activity" through which beings decay into perceptibility, "the temporal character of things and their appearance to the senses are posited as intrinsically one."[3] Marx stresses, in particular, the Epicurean argument that "Time does not exist by itself" (*tempus item per se non est*), apart from the perceptible, perception-generating traffic in *eidola*: "from things themselves follows the sense of what has been done in the past," of "what thing is present," and of "what is to follow," as Lucretius explains, "nor may we admit that anyone has a sense of time by itself separated from the movement of things and their quiet calm."[4] There is a very real sense in which the kind of subjects and circumstances who, by the time of "The Eighteenth Brumaire," are collaborating to make history—"Humans make their own history, but they do not make it out of free pieces, not out of circumstances they choose themselves, but under those directly encountered, given, and passed down"—can do so because they are presumed to corporeally partake in a Lucretian materialization of "embodied time."[5]

In the 1960s, amid the controversy surrounding the rediscovery of early Marxian writings like the dissertation on Epicurus, Louis Althusser influentially dismissed them as "the Marx *furthest from Marx*."[6] Ideological, humanist, subjectivist, Hegelian—one might as well say "Romantic"—they were divided from Marx's coming materialist science of history by "an unequivocal 'epistemological break'" that Marxist critique ought not to cross and was not obliged to reconcile.[7] Romantic these writings are, albeit (also) in some minoritarian senses this book has tried to recover: they subscribe to an avowedly "sensuous" and "poetic" empiricism as scientific method, they take an interest in "objective activity" and the life of "man's *in*organic body," they confront the complex entwinement of "an historical nature and a natural history," and they find, in *De rerum natura*, exquisitely fit means of grasping the biological reproduction of life as suffused with the social circumstances of production.[8] In the Romantic poetic perspective opened by this study, we can see that Marx was eager to situate historical subjects within a materialist philosophy of nature not, as Althusser worried, to uncover there a permanent and universal "*essence of Man*," alienated by conditions of modern labor, but rather, much like Blake, to depict humans as active and passive embodiments of historical time, even at the level of the bodies they take for natural: as shapers and bearers of social circumstance, dependent for their very life to the weight and the call, as Marx would say famously in "The Eighteenth Brumaire," of "dead generations."[9]

Against present critical habits, which understand "naturalization" as the mystification that strips social constructs of their contingency, Marx devoted much of the 1844 Manuscripts to demonstrating that "History itself is a *real* part of *natural history*," a view that commentators from Friedrich Engels to Alfred Schmidt and John Bellamy Foster have shown he never relinquished.[10] Such is the way Marx framed *Capital* itself, the Preface to the first edition avowing "My stand point, from which the evolution of the economic formation of society is viewed as a process of natural history."[11] One principal task of "the materialist conception of history" first adumbrated by name in *The German Ideology* was to target and historicize the very "antithesis between nature and history" that, Marx and Engels argue, had so far excluded the literal "production of life"—through the "labor" of *both* work and procreation— from historical understanding.[12] Thus a self-consciously "historical" materialism joined the old materialisms of nature (and the new pre-Darwinian and Darwinian sciences of life) in telling a single natural and social history of production: a history whose natural science and political economy were understood to occupy the same, mundane side of an opposition structured against "extra-superterrestrial" accounts of the form and purpose of biological and social life.[13]

What this history might look and sound like, we began to witness in the bloodstained "accent" of the Earth in Shelley's *The Mask of Anarchy*. That accent signified at an extreme of natural-historical materialism that late twentieth-century criticism has been reluctant to match, an extreme at which the (initially) posthuman, ecologically motivated "new materialisms" and the classically humanist "old" Marxist struggle for socioeconomic justice might circle back and rejoin one another. Attributing an "accent unwithstood"—not resisted and not limited to speech—to the paradigmatic *mater* and *materia* of the Earth, Shelley's natural-historical materialism limns this convergence through a materialist semiotics that defies "the axiomatic resistance of matter to mind" in which many flavors of idealism, materialism, and vitalism have unwittingly concurred.[14] Heeding poetic modes of affording semiotic action and passion to matter, in violation of the taboo against anthropomorphism (which covertly presumes and perpetuates the human exception), those who identify as human can cease to be surprised by the rigorously unpredictable responses of other natures in their vulnerability, artistry, and power—terrestrial species and forces no longer mistakable for the predictable backdrop to human adventures in freedom and yet, by the same token, perhaps amenable to new forms of strategic confabulation.[15] Likewise, material semiotics might name the way "new" materialisms rise to the challenge of broaching the classically political problems (freedom, representation, institutions, power) initially left behind in their own recoil from discursively turned cultural theory.[16] It is interesting, in this climate, to recall that "old" materialism in the Marxian vein saw itself as providing the first-ever "*earthly* basis for history" (*eine* irdische *Basis für die Geschichte*) and that among the early writings that have seemed so very un-Marxian are four hundred pages of poems.[17]

I close this book amid a full-fledged "new materialist turn," an unsteady affiliation not least because the ancient, renewable tradition traced in my book's pages found materialist solidarity in transience and in senescence. This makes it impossible not to ask, of each emphatically "new" thing, *Who are you call-*

ing old? If the answer, this time, has been the Marxist materialism of history, the deconstructive materialism of the letter, or New Scientific and Enlightenment materialisms typed as overly passive, mechanical or inert, I hope there are hints in this book about the value of all of these old agendas for the task of contemporary materialist criticism. Such criticism would seem to need to reconnect the human and nonhuman depredations of late capitalism without letting the omnibus categories of "neoliberalism" or "the anthropocene" eclipse the political differences lived in morphologically different bodies and without mistaking the theoretical extension of agency down a *scala natura* (that there are few Neoplatonists left to defend) for politics achieved. It might consider that among the assets of texts historically fictionalized as literary, figurative, or poetic has been to let other natures twist human languages to their purposes through styles of passivity and receptivity different from absence, deprivation, or negation.

Some two decades after designating a hard break between "ideological" and "scientific" portions of Marx's oeuvre, Althusser, recovering from a break of his own, began a luminous return to the old atomist natural philosophy about which the young Marx had written his "splendid piece of nonsense."[18] In the posthumously reconstructed writings on "the underground current of the materialism of the encounter," Althusser discovers in the Lucretian *clinamen* the purest expression of a suppressed tradition of "aleatory materialism" that he wishes to illuminate within, before, and after Marxian materialism: a materialism "not of a subject (be it God or the proletariat), but of a process, a process that has no subject yet imposes on the subjects (individuals or others) which it dominates the order of its development, with no assignable end."[19] Far more threatening than "the various materialisms on record," he argues, whose compatibility with necessity and teleology leaves canonical philosophy's idealist "priority of Meaning over all reality" intact, the Epicurean plot that dared to put chance before law and before logos authorized a conception of historical process "haunted by a *radical instability*," by rigorously unpredictable events that cohere from co-present elements and take their own agents by surprise.[20]

Although he never revised his judgment about flagrantly Romantic, early Marx, Althusser's own late turn to Epicurean natural philosophy may constitute an implicit invitation to do so; an invitation, moreover, from a thinker fiercely protective of Marx's structural and economic science and not apt to recontain it in anthropology or philosophy. For Marx, who would complain in the "Theses on Feuerbach" that hitherto existing materialisms had ceded the active modality of "the thing, reality, sensuousness" to idealism and ide-

ation,[21] the Epicurean theory of the *eidola* gave "active form" to nature's material compositions, the "human" among them. Marx explains this way, zeroing
in, as I mentioned, on the same feature that captivated Goethe and Shelley
before him:

> The *eidola* are the forms of natural bodies, which, as surfaces, detach them-
> selves like skins and transfer these bodies into appearance. These forms
> of things stream constantly forth from them and penetrate into the senses
> and in precisely this way allow the objects to appear. . . . For it is precisely
> because bodies appear to the senses that they pass away.[22]

This is the basic way that "*sensuousness*" amounts to "*embodied time*" for
Marx and the materialist philosophy of nature he summons here: having sensations means experiencing, as a palpable alteration in one's own body, the
matter of others' serial loss. By way of the *eidola*, the atomic attrition that
is "withdrawing [things] from our eyes" (*ex oculisque vetustatem subducere
nostris*) is the very same process that showers them with visions: "at last," as
Marx puts it, "the temporal character of things and their appearance to the
senses are posited as intrinsically one."[23] For what bodies are making as they
fall away from themselves and into appearance, Marx stresses, is time, in its
character as a collective, irreversible, corporeal experience.

By "not returning into themselves," by transpiring, via the *eidola*, into others' senses, Marx writes, beings "have their sensuousness outside themselves,
as another nature."[24] For Marx, then, being sensuous—the active form of appearing that things can be said to "have" only in so far as they lose it, to have
in so far as they are taken up by *another* nature—is a fundamentally relational
kind of corporeality, distributed between bodies that give portions of their
being to the materialization of their times. It is through "sensuousness" so
understood that the dissertating Marx affirms sensation as the criterion of
knowledge, discovering, in Epicurean materialism, the variety of empiricism
fit to make "sense-perception" into "the basis" of "*true* science," as the 1844
Manuscripts and the *German Ideology* seek to do.[25] Empiricism, then: not because sense-based knowledge is invulnerable to temporal contingencies, but
precisely because it varies with them. And not because the empirical method
expresses the special privilege of the human knower or progresses toward her
technological erasure, but precisely because situated, species-specific figurations are, to paraphrase Goethe and Donna Haraway, the only available objectivity: the inevitably partial means by which nature, as Marx puts it, "hears,"
"smells," and "sees itself."[26]

In this transcorporeal mode of experience, Marx notices, impressions are shaped by the narrows of species-peculiar perception without losing their dynamic connection to the natures they represent.[27] And when Marx and Engels polemically differentiate their materialism from eighteenth-century "mathematical" and "mechanical" materialisms of Enlightenment and bourgeois revolution, they do so by proclaiming their allegiance to the history of *empiricist* experimentalism: to moments, as they write of Bacon in *The Holy Family*, when "matter in its sensuous poetic glamour smiles upon the whole man" (*Die Materie lacht in poetisch-sinnlichem Glanze den ganzen Menschen an*).[28] This smile, like so many of the prosopopoeiae in this book, attributes "poetic glamour" to the matter and not the man; or better, the figural smile bespeaks a complicity at the level of substance. Marx and Engels here celebrate a conduct of knowledge whose very figural felicity ("poetic glamour") amounts, as we heard Herder put it in the introduction, to a "seal of truth": a stylish style of knowing whose pleasure and allure bespeak the reality of contact inseparable from the bodies involved, testifying to knowers' nonviolent affection as and by (other) material things.[29] For Marx and Engels, too, it seems, a basic acceptance of matter as poetic (which is not to say alive) conditions a thoroughgoing historical materialism, which would be hypocrisy were it to take material concerns as an object without dignifying them as co-producers of its knowledge.

If in the 1841 dissertation, Marx's form of poetic empiricism seems to concern a kind of philosophical "time" devoid of history, Marx soon cast sensuous time as the embodiment of historically specific relations of production and exchange. In conformity with their training under a system of private property, for instance, "organs" of "seeing, hearing, smelling, tasting, feeling, thinking, being aware, wanting, acting, loving" operate as senses "of *possessing*, of *having*" and "*direct*, one-sided *gratification*."[30] Chastising Feuerbach for an ahistorical view of the sensuous nature of and outside persons in *The German Ideology*, Marx insists that "the sensuous world . . . is not a thing given direct from all eternity . . . but a product of industry" and the "unceasing sensuous labor" of "generations"—so much so that had this history been interrupted for a moment, Feuerbach would be surprised to find "his own perceptive faculty" suddenly "missing."[31] Marx leaves no doubt that the time embodied in human sensuousness has *historical* shape, stressing that the morphology of the sensuous body responds to the pressure of present labor relations and archives those of the past. "The *forming* of the five senses is a labour of the entire history of the world down to the present,"[32] he writes in the 1844 Manuscripts. Blake arguably pioneered the insight in depicting millions of male and female laborers hammering, weaving, forging, binding, and ornamenting human sense organs ("The sons of Ozoth within the Optick Nerve stand fiery glowing / And the number of his Sons is eight millions & eight"); or Lamarck, in the morphology that saw a being's body shape as the effect, not the cause, of "its way of life, and the circumstances in which those individuals, from whom it derives, found themselves."[33]

Marx's old, Romantic materialism, in its uptake of positively ancient materialism, was also alert to the heterogeneity of embodied temporality: to the lapsed and distant figures that haunt any perceived present, exerting a material pressure from the past within any current "atmosphere of sensation." As Lucretius put it: "in one moment of time perceived by us . . . many times are lurking" (*tempore in uno, / cum sentimus . . . / tempora multa latent*) (4.794–96). It was Derrida who was most sensitive to Marx's "spectropoetics," to the anachronistic proliferation of specters involved in his attempt to bind historicism to the actual circumstances of the living. The resistance motivating Derrida's deconstruction relies on Marx's purported "hostility towards ghosts," treated as ideological mystifications to be banished from and by practical materialism. Yet for Derrida, the specters that haunt every ontology of presence exert an inescapable and indispensible ethical demand: making felt "*the non-contemporaneity with itself of the living present*"—a recognizably Romantic feeling by now—these figures "recall us to anachrony" and to an ethos of "respect for those who are no longer or who are not yet *there*, present and living."[34] Recognizing the Lucretian derivation of the spectral sensations Derrida rightly experiences reading Marx, however, allows us to recognize historical materialism as an open "spectropoetics," an ethos heedful of the anachronistic insistences of the past because, and not in spite of, its materialist commitment to the actual material circumstances of the living.

This is because, as some readers may have intuited, the Lucretian corporeal poetics of sensation and time that Marx and some of his Romantic historicist predecessors found so useful was also a physics of phantoms. The wandering *eidola* are first introduced in *De rerum natura*, in fact, as an alternative to ghosts and spirits: "lest by chance we should think that spirits [*animas*] . . . or ghosts [*umbras*] flit about amongst the living" (4.37–38). But the theory that follows is an exceedingly strange strategy for diminishing the fearful belief in ghosts: "the wonderful shapes and images of the dead" that terrify us waking and sleeping, Lucretius assures us, are not ghosts or spirits, but wonderful shapes and images of the dead. Wayward simulacra more durable than their referents, these visitations *are* fractions of the deceased body, and come bearing it into our eyes. *Don't be afraid*, Lucretius seems to say, *the ghosts you see are real.* At issue, therefore, is a materialist tradition so credulous of and hospitable to the still-circulating vestiges of past generations that "ghosts" are rejected as inadequate to the reality with which "dead generations," as Marx put it, "weigh like a nightmare on the brain of the living."[35] In this sense, the Derridean ethical commitment to specters, to "learning how to let them speak or how to give them back speech," echoes the Lucretian injunction to attend to the slight and nonliving, but significant and real, notices that are "striving ardently to fall upon your ears" and the figural forms that harbor them in language. Understanding the sensations upon which the experience of life depends as anything but raw or immediate, as making and taking their time from the vestiges of past lives and distant realities, is a very ancient materialist lesson. Romantic materialisms, in particular, put this lesson to use to grasp and articulate the temporal complexity of historical experience and to train readers to sense the lighter address of past and passing things: those "vanishing apparitions" which, Shelley says, "haunt the interlunations of life."

Paying attention, as nineteenth-century readers did, to Lucretius's costly and nonsynchronous sense of embodied time—to the kind of happening both lost and given, in Goethe's and Shelley's poetry, through the images that belatedly reach our eyes; to the ways that present bodies, in Erasmus Darwin, Cowper, and Thomson, ostend divergent material histories that demands redress—helps to correct the presentism, "affirmative vitalism," and even redemptive theology of which some of the latest neo-Lucretianisms have been justly accused. "Reconstituted, material life persists, swerves, becomes afterlife," Steven Goldsmith points out of some recent "new materialist" Lucretian revivals: "There are resurrections everywhere. Atomism keeps an afterlife alive even for atheists."[36] Yet wherever philosophy attains this "triumphal quality," it has avoided the demanding peculiarity and power of Lucretian consolation.[37] While elements that have made a life together may go on to make others, such reconstitutions have no subject: no person, sentience, or perspective to bridge what John Dryden faithfully and freely translated as "a pause of Life, a gaping space," "Which yet our dark remembrance cannot trace."[38] Atoms in their "solid simplicity" partake in many lives and never die, but neither do they ever live.[39] For those who live, loss of life is absolute: any "afterlife" is a rigorously *other* life, which is what keeps each life grievable within the poem's ethical purview.

De rerum natura in fact breaks off in a scene of uncompensated grief: the loudly lamenting survivors of a plague-ravaged Athens brawl rather than leave the bodies of their dead, victims of a contagion against which no honest theory of material contingency can ever immunize itself. It is important that Deleuze, keen to see Lucretius "denounc[e] everything that is sadness" and "mak[e] of thought and sensibility an affirmation," needs to disavow the poem's inconsolable end. It must have been forged by Lucretius's enemies, he suggests, because it is so terribly *sad*. "I fantasize about writing a memorandum to the Academy of the Moral Sciences to show that Lucretius' book cannot end with the description of the plague, and that it is an invention, a falsification of the Christians who wanted to show that a maleficent thinker *must* end in terror and anguish."[40] Much of the affirmative vitalism of the new materialist turn, I think, is guided by Deleuze's determined philosophical happiness, which deserves to be assessed in its own right.

Lucretian materialist consolation works by other means: not through the promise of resurrection, but by casting mortality, literally and immediately, as the substance and condition of any experience at all. For, as Marx, and Shelley, and Goethe have variously observed by Lucretian lights, it is only by way of their passing that bodies regale each other in sensations: "for assuredly we see many things cast of particles with lavish bounty [*largiri multa*]" (4.72). Emphasizing the lush, inadvertent largesse in mortal loss, Lucretius reminds readers who fear death that their very experience of life—of sensing and being sensed, feeling and being felt, understanding and being understood—is co-extensive with their own and others' dissolution. Pledged to the "living storm" through which they "Mix with each other in tempestuous measure" in Shelley's anti-*Triumph*, mortals "live mutually upon another" (*inter se mortales mutua vivunt*) (line 141; 2.76). Without our frangible, mortal softness, there would be no sight, touch, pleasure, pain, or knowledge: nothing but an insensibility that would differ little, after all, from death. Every chance for material happiness and material justice in this tradition hinges on a mutual commitment to mortality set against the ultimately violent fantasies of invulnerability and of life without loss.[41]

One final sense in which Lucretian poetic physics, in its Romantic returns, prepared the vocation assumed by Marxist materialism and a challenge to the ones that follow: the role of an antidisciplinary science attuned to the realities that expertise renders unrepresentable. Early on, Marx whimsically cast communism as the society that would let him "hunt in the morning, fish in the afternoon, rear cattle in the evening, criticize after dinner . . . without ever becoming hunter, fisherman, shepherd, or critic."[42] In *The Political Unconscious*, Fredric Jameson draws on an earlier, structuralist Althusser to reinstate, more soberly, Marxism's claim as "an interdisciplinary and a universal science"—a science unmistakably Romantic in its determination to "break out of the specialized compartments of the (bourgeois) disciplines." At stake in Marxist interdisciplinarity, as Jameson expressed it, is "the strategic choice of a particular code or language, such that the same terminology can be used to analyze and articulate two quite distinct types of objects . . . or two very different structural levels of reality."[43] Jameson's observations capture with telling precision the eighteenth- and nineteenth-century effects of implementing Lucretian poetics: an unconstraining idiom of relation wherein "linking language events and social upheavals or economic contradictions" requires no special pleading because they all partake of, and take place in, the same "reality."[44] (Such is the reality, for instance, of Erasmus Darwin's epics, in which calcareous and siliceous mineral formation, technologies of porcelain manufacture, and abolitionist exhortation bump up against one another in a poetic medium that both shows and tells the reality of their interconnection.)[45]

But the irreducible character of the older materialist medium as poetry makes all the difference, lashing its science to impropriety, partiality, and transience against any pretense at totality and teleological unfolding. In place of what Jameson called "the unity of a single great collective story," then, we are returned to what Shelley called "the chaos of a cyclic poem," whose unfinishable openness is the only guarantor of the truth or freedom it spells.[46] Though the plot, one might hope, still tells "the collective struggle to wrest a realm of Freedom from a realm of Necessity," the luxury and burden of thinking that Freedom's protagonists are humans, extricating themselves from a ground of natural determination, are denied.[47] The word *Anthropocene* indicates the growing realization that the "collective story" will have to be written as natural history and with other natures; in fact, this story has been so written all along, in witting and unwitting collaboration with nonhuman materials and forces by which it will be subsumed.[48]

To the extent that several technical, hard-to-reconcile senses of "matter" still compete in contemporary criticism—those of the text and letter, of history and production, of the body and biology—materialism's ancient and Romantic indisciplinarity still has something to offer. Lucretian materialist poetics have proven their irrepressible aptitude to communicate things as they have been all along—"simultaneously real, like nature, narrated, like discourse, and collective, like society," as Bruno Latour puts it in his anthropology of Western modernity—against partitions of knowledge that render these dimensions mutually unrepresentable.[49] For what Lucretian matter enabled, in its Romantic deployments, was the un- and then counter-disciplinary attempt to secure between signs, natures, and exchanges, as between every entity and process, a minimal substance of relation and susceptibility.[50] From Montaigne to Shelley and Louis Althusser, this materialism offered its poetic science of tiny simple elements and complex composite shapes as fit to articulate the relation between apparently incompatible modes and scales of analysis, and validated figuration as the real relay between them.

For us, Romantic materialist figuration models the refusal to concede the real to the genres that presently monopolize its representation. In a present joined, we have lately learned, to the Romantic era from the perspective of the earth, its counterdisciplinary techniques of worldly poiesis might help historicist and materialist criticism learn to write, again, as natural history—which will not be to write apolitically, or apoetically, at all.

ACKNOWLEDGMENTS

It takes an improbably splendid and supportive milieu to foment a book about the poetry of milieux and the collaborative forms of knowledge they can gently engender. This one is a composite image of the teachers, friends, families, and places that sustained it over many years; trying to do justice to my good influences burst my footnotes and made me wish for a first-person-plural way to write. Kevis Goodman and Steve Goldsmith fostered this project from its first scholarly inklings, not only through their brilliant teaching and guidance, but also through their almost debilitatingly fine examples of what a book of literary criticism can be. Speaking of, my gratitude to Anne-Lise François for taking me on as a (strictly useless) research assistant is growing still: her ways of thinking and writing exert an indispensible counter-action against my most stubborn critical tendencies. Joanna Picciotto somehow took my weird science seriously from the first and interrogated it with the brilliant intensity of her own literal imagination; I owe many epiphanies to the electric conversations that make me feel like a peer while giving me an education. I have realized too slowly just how much this project learned its ethical bearing from Judith Butler and its intellectual-historical bent from Martin Jay, both of whom took time they did not have to give precision to my wild thoughts and prescribe reading cures with care. I am indebted to Tim Hampton's Rabelaisian teaching and Niklaus Largier's mystic arts, meanwhile, for my first, indelible instruction in nonmodernity. The questions and comments of all these teachers, echoing in my mind for years, drove this project right up to its searching last chapter.

The book's emergent ideas were elaborated in delirious interchange with

Lily Gurton-Wachter, my co-conspirator in the Romantic arts of receptivity; David C. Simon, our early-modernist correspondent; Selby Schwartz, who taught me teaching as a life of gracious experimental imagining; and Tom McEnaney, who was becoming my partner in all things. Those ideas would have been unthinkable had we not chanced to share Berkeley with many ingenious spirits and friends—Andrea Gadberry, Katrina Dodson, Javier Jimenez, Mike Allan, Tristram Wolff, Kathryn Crim, Annie MacClanahan, Ted Martin, Ben Tran, Allison Schachter, Daniel Immerwahr, Joshua Weiner—and people like Penelope Ismay, Dan Clinton, Jasper Bernes, and Dean Krouk, who dropped phrases in seminar that were milestones in my education. I chanced into early contact with some giants in Romantic literature and science— Theresa Kelley, Noah Heringman, and Rob Mitchell—whose encouragement and subsequent collaborations (whether in Bloomington, Houston, Western Ontario, Durham, or Berlin) sparked and supported this book in more ways than I can count or know.

Sweet Science was transformed by a postdoctoral year on biopolitics at the University of Wisconsin–Madison, thanks to the gracious hospitality of Sara Guyer, Rick Keller, and Megan Massino at the Center for the Humanities, and to Theresa Kelley (again), who loaned me a down comforter and her large and sparkling mind. I treasure conversations begun there with Monique Allewaert, J. K. Barret, and Alex Dressler (who is not responsible for the errors in the book's classicism but is responsible for pricking its political conscience). Thanks to fellow fellows Ellery Foutch, Melissa Haynes, Elizabeth Johnson, Jessica Keating, Mathangi Krishnamurthy, Michelle Niemann, Trevor Pearce, and Lynette Regouby for sharing their splendid projects and their skills at provisional community.

Cornell University's generous institutional support made finishing this book feasible; its people made the finishing joyful and new. I am especially grateful to Jenny Mann, Liz Anker, Rayna Kalas, Jeremy Braddock, Masha Raskolnikov, Margo Crawford, Mary-Pat Brady, and Diane Berrett Brown for their on-demand mentorship and the kind of substantive curiosity and interchange one only dreams about across wide differences in fields. And to the ladies of literature and science, Jenny Mann (again), Elisha Cohn, and Courtney Roby, who leant their keen minds to some of the manuscript's most reckless and unformed parts. Thanks to Harry Shaw for reading and responding with care to an earlier version of this text; to Rick Bogel, Mary Jacobus, Andy Galloway, Karen Pinkus, Philip Lorenz, and Tracy McNulty for their sincere interest and bright hints; to Cynthia Chase for her candid and constructive criticism; and to the working group on *Besetzung*: Paul Fleming, Antoine

Traisnel, Johannes Wankhammer, Nathan Taylor, and Bruno Bosteels. And thanks, especially, to Roger Gilbert and Jonathan Culler for supporting this project from writing sample to publication and making a home for it in Cornell English, where Romanticists dream to land. Here students in my materialism, sensation, Romanticism, and "life of things" seminars showed me much about what I purported to know, and did so with a kind of enthusiasm that allayed this humanist's nagging fear of irrelevance.

A year at Cornell's Society for the Humanities under the generous stewardship of Tim Murray was decisive for the book's final development. Thanks to him and my fellow fellows in the Society's "sensation" year, Ann Cvetkovich, Munia Bhaumik, Ida Dominijanni, Anna Watkins Fisher, Saida Hodžić, Benjamin C. Parris, Dana Luciano, Verity Platt, Annette Richards, Munia Bhaumik, Christine Balance, Maria Flood, and Rebecca Kosick, and Andrew McGonigal, who showed me through their brilliant work that, among many other things, mine is more feminist, queer, American, and affective than I have known. This project also benefited from ACLS-Mellon and DAAD fellowships, the hospitality of the Max-Planck-Institut für Wissenschaftsgeschichte, Berlin, and the opportunity to workshop its chapters at the University of Missouri–Columbia Science Studies Colloquium (special thanks to Noah Heringman and Stefani Engelstein), the Indiana University–Bloomington Eighteenth-Century Workshop and Poetologies of Knowledge Colloquium, and the Clark Library at the University of California, Los Angeles.

Marjorie Levinson and James Chandler generously workshopped my manuscript at a late stage: their perceptive, imaginative, and encyclopedically knowledgeable feedback is a gift this book alone cannot answer and one that holds open the way to further things.

In their readiness for bouts of speculative naturalism, fringe artistry, and experimental cookery, Anna Fisher and Antoine Traisnel lightened the heavy task of composition. I owe to them, with Merike Andre-Barrett, Alex Livingston, Patty Keller, Naminata Diabate, Elisha Cohn, Theo Black, Catherine Taylor, Stephen Cope, Leslie Brack, Mike Askin, Athena Kirk, Andrew Moisey, Eli Freedman, Julia Chang, Louis Hyman, and Kate Howe, the ongoing sensation that Ithaca, in defiance of technical measure, is a warm and expansive place.

For difference-making interchanges you may or may not remember, thanks to Orrin Wang, Miranda Burgess, Rebecca Comay, Khaliah Mangrum, Harris Feinsod, Branka Arsić, Eric Santner, Jonathan Sachs, Heather Keenleyside, Timothy Cambell, Peter Manning, Tilottama Rajan, Andrew Piper, Lauren Berlant, Ethel Matala di Mazza, Eva Geulen, Jill Frank, Ian Duncan, Celeste

Langan, Lezlie Namaste, Sam Baker, Brian McGrath, Kate McCullough, Helmut Müller-Sievers, Finnicus McEnstein, Michel Chaouli, Daniela Helbig, Dalia Nassar, Charles Wolfe, Hinrich C. Seeba, George Hutchinson, Ella Diaz, Dan Hoffman-Schwartz, Barbara Natalie Nagel, John H. Zammito, Peter Hanns Reill, Johannes Türk, Rebecca Sprang, Nancy Armstrong, Len Tennenhouse, William Keach, and Tom Mitchell.

My gratitude to Alan Thomas and Randy Petilos of the University of Chicago Press for accepting and realizing this book, and to the extremely generous press readers, Jon Klancher (again!) and Terry Kelley (yet again!), whose insights were indispensable to its final form. I am especially indebted to my copyeditor, Susan Tarcov, for her myriad artful adjustments and questions, and to my assiduous and ingenious research assistant, Noah Lloyd. I owe the deft index to Derek Gottlieb and thank Michael Koplow for ferrying all the parts into type.

Last and most, I would like to thank the intercontinental Lee-Haubrich-Goldstein-McEnaney clan, who made me know that de- and re-composition could be a beautiful thing. My father Michael Goldstein's love of ideas, my mother Ingrid Haubrich's energetic devotion to work and art, my stepfather Larry Lee's wisdom and equilibrium, my "little" sister Anya Goldstein's amazing breadth of heart, all gave me, differently, the models and the room to pursue an interminable and impractical thing. What still astonishes is the love they—with my friends Kate, Mike, Pete, Diana, Selby, Michael, Jen, Sarah E.—gave when the work made me quite unlovable.

This book is for Tom McEnaney, whose touch is in every sentence, my senses, and whatever sense I make.

Earlier versions of two chapters appeared in the following publications: chapter 2 as "Obsolescent Life: Goethe's Journals on Morphology," in *European Romantic Review* 22, no. 3 (2011); and chapter 4 as "Growing Old Together: Lucretian Materialism in Shelley's 'Poetry of Life,'" in *Representations* 128, no. 1 (2014).

NOTES

D R N : Lucretius, *De rerum natura*

L A : Goethe, *Die Schriften zur Naturwissenschaft*, Leopoldina-Ausgabe

M A : Goethe, *Sämtliche Werke nach Epochen seines Schaffens*, Münchner-Ausgabe

M E C W : *Marx-Engels Collected Works*

M E G A : *Marx-Engels-Gesamtausgabe*

M E R : *The Marx-Engels Reader*

INTRODUCTION

1. Blake, *Four Zoas*, "Night the Ninth, Being The Last Judgment," 120.40, *Complete Poetry and Prose*, 390. Unless otherwise noted (as here), Blake citations refer to the Blake Trust / Princeton edition of *Blake's Illuminated Books*, 6 vols., general editor David Bindman.

2. Blake, *Milton A Poem* 44 (43).5, 44 (43).15.

3. Blake, *Four Zoas*, "Night the Ninth," MS 139.1-10, *Complete Poetry and Prose*, 407.

4. Stefani Engelstein focuses generatively on the term, too: "Blake called for a new science," she argues, "a con-science, that would link the search for knowledge with ethics by recognizing and encouraging the *flexible* interaction of bodies no longer envisioned as discrete individuals." *Anxious Anatomy*, 17.

5. When it comes to "REASON, and SCIENCE," wrote Hobbes (in a terrific simile), "*Metaphors*, and senseless and ambiguous words, are like *ignes fatui*" whose end is "contention, and sedition, or contempt"; for Locke, "the artificial and figurative application of words" is among the "perfect cheats" to be avoided "in all discourses that pretend to inform or instruct"; Bacon's *Parasceve* banishes "all that concerns ornaments of speech, similitudes . . . and such like emptinesses"; and Sprat famously railed against "the mists and uncertainties these specious Tropes and Figures have brought upon our knowledge" (*Leviathan* 1:5.116-117; *Essay* 3.10.34, Fraser ed., 2:146; Bacon and Sprat, quoted in Walmsley, *Locke's Essay*, 107). Yet plain

style programs never fit their caricature as unambiguous or naïve regarding the role of rhetoric in thinking and sensation: see Kroll, *Material Word*, for English Restoration culture's plastic and pragmatic—neo-Epicurean—understanding of language; Jenny Mann on the generative tension between the Latin rhetorical instruction and an English vernacular literature in the Renaissance; Walmsley, *Locke's Essay and the Rhetoric of Science*, especially illuminating on historicity and decay in the Lockean theory of language and mind; Law, *Rhetoric of Empiricism*; and Caruth, *Empirical Truths and Critical Fictions*, for empiricism's underrecognized sophistication (intended and unintended) regarding the rhetoric of perception.

6. Blake, *Four Zoas*, "Night the Ninth," 138.14–15, *Complete Poetry and Prose*, 406.

7. Horace, *Ars poetica* (c. 16 BCE) 343–44; Lucretius, *De rerum natura*, trans. W. H. D. Rouse, rev. Martin Ferguson Smith, 1.933–50 and 4.8–25 ("amarum" [1.940, 4.15], "tristior" [1.944, 4.19]). I refer most frequently to this standard Loeb edition and prose translation, hereafter cited parenthetically in the text. Chapters 2–5 foreground the important new verse translations to English and German with which Goethe and Shelley were working, that of John Mason Good (1805) and Karl Ludwig von Knebel (1821), respectively. Elsewhere, I draw upon verse translations in circulation in the early nineteenth century, especially Good and Thomas Creech (1682), but also John Dryden's 1685 variations and John Evelyn's book 1 (1656). I regret not adopting Lucy Hutchinson's near-complete translation from the 1650s as my anachronistically modern standard: David Norbrook brought her *Order and Disorder* into print for the first time in 2001; and see the splendid facing-text scholarly edition he, Reid Barbour, and Maria Cristina Zerbino have made since (2012). Russel M. Geer's physics-minded translation, *On Nature* (1965), is a great help; I use it where noted, and recommend A. E. Stalling's amazing fourteeners to any English speaker willing to accept the poem as her contemporary (*The Nature of Things*, 2007). For an overview of the long translation history, see Hopkins, "English Voices of Lucretius"; on the Romantic period see Priestman, "Lucretius in Romantic and Victorian Britain," and Otto, "Lukrez bleibt immer in seiner Art der Einzige." And see Priestman's *Romantic Atheism* and Vicario's *Shelley's Intellectual System*, for the stakes and contexts of Good's facing translation and Gilbert Wakefield's new scholarly edition of the Latin text in the early nineteenth century.

During the last phase of working on *The Four Zoas*, Blake was living in Felpham and enthusiastically learning Latin, Greek, and Hebrew under the (sometimes unwanted) guidance of his employer and patron, William Hayley; Hayley himself had just completed *An Essay on Epic Poetry*, and Blake explicitly worked up *The Four Zoas* in classical (and Miltonic) epic style; see Ackroyd, *Blake*, 214. Blake provided many engravings for Hayley's *Life and Posthumous Writings of William Cowper, Esq.*, which, quoting Cowper's appropriation of the Lucretian honeyed-cup topos, may have been one of many prompts for Blake: "if we give them physic, we must sweeten the rim of the cup with honey," Cowper wrote regarding "Table-Talk" in a letter Hayley discusses in the *Life*; see Cowper to Bull, 24 March 1782, in Hayley, *Life*, 4: 241–42 and 216–17.

8. Wordsworth, 1802 Preface, Wordsworth and Coleridge, *Lyrical Ballads*, 258–60.

9. *DRN* 1.72 and 1.29. See Loeb note *a* on the propriety of translating *religione* as "false religion" or "Superstition" (1.64); "savage works of war" is *fera moenera militiai* (1.29).

10. *DRN* 1.7–8, 29–40. I allude to Blake's "Earth" and "Enitharmon" (Los's female counter-

part) in the passage and to *De rerum natura*'s famous opening tableau: Mars draped "vanquished" and vulnerable in Venus's lap, "shapely neck thrown back," as "gaping" he drinks "sweet coaxings" from her lips (*suavis ex ore loquellas*, 1.39). (This introduction has much more to say about taking in sweet knowledge by mouth.) Blake's poem also continually punctures the power of Urthona's "reign," especially in line 4's unmarked caesura and enjambment, which lets stand a contradiction to this figure's "rise" and "strength" (lines 7–8): "Urthona is arisen in his strength *no longer* now" (line 4, emphasis added). On what it might mean for subjects of knowledge to assume unrisen postures of leaning and lying, see Adriana Cavarero, whose "postural ontology" tracks the centrality of uprightness to philosophical rectitude from Plato forward. "Steadily balanced on its vertical posture, the philosophers' subject does not lean, does not bend, does not incline," she concludes: the "bending self" is a rarity in philosophical writing, "an 'I' that, by assuming an oblique position, leans out of the vertical axis that allows one to stand erect on one's own basis as a perfect autonomous subject." "Recritude," 221, 226.

11. Something similar happens in *Milton*, where the End Times (the apocalyptic destruction of fallen, historical, standardized Time) are announced two ways: once in Los's heroic, prophetic register ("Six Thousand Years / Are finshd. I return! Both Time & Space obey my will" [21 (20).16–17]), and once by a bird and an herb. Apocalypse waits, in that book, on the "Wild Thyme" and the mounting of a "Lark," as though one could also break from fallen history on a *lark*, by letting *time* go *wild* (45 [44].29–30). For an early intimation of this book's interest in what "strangely anachronistic concepts of nature still covertly prevail" in critical theoretical unmaskings of "naturalization," see Beer, "Has Nature a Future?"

12. I quote from Thomas Creech's popular 1682 translation, *Of the Nature of Things, in Six Books, Translated into English Verse. Explained and Illustrated, with Notes and Animadversions, Being a Compleat System of the Epicurean Philosophy*. See above, note 7.

13. *DRN* 2.216–24. Theoretical engagements with *De rerum natura* have tended to focus on this feature. (Jacques Derrida's, Louis Althusser's, Michel Serres's and Gilles Deleuze's engagements with Lucretius all center on the *clinamen*, that "Swerve" Stephen Greenblatt has made famous again in his eponymous study of Renaissance humanism.) Throughout this study, I hold the atomic *clinamen* to the more subsidiary role I think it plays in *De rerum natura* as an abductive inference from a visibly tropic universe (see chapter 3). Keeping this proportion helps to counter a philosophical bias toward "firsts," even among bodies (here, atomic "first bodies" [*primordia rerum*]), that affects even the poem's most sympathetic and antimetaphysical readers. For those seeking it, the clinamen appears in its biological manifestation in chapters 2 and 4, through important earlier modern controversies surrounding "equivocal generation" with which *De rerum natura* was tightly associated into the nineteenth century.

Still, a few brilliant philosophical notes on the swerve are helpful from the outset. Althusser observes that with the aleatory *clinamen*, Lucretius audaciously permitted chance to precede reason, law, form, and meaning: chance chanced to generate what occupies the site of origin in rival classical philosophies. (This in a cosmos whose chances, as biosemiotician Jesper Hoffmeyer might put it, can be "habit-forming.") Michel Serres observes that the Lucretian account of the rain of atoms and its interruption by the *clinamen* rather reverses the usual expectations regarding the genesis of meaning or sense from chaos. For Lucretius, "univocity"—the laminar flow of atoms in parallel—amounts to a non-sense in which "nothing stands in relief, nothing appears." Here signification requires deviation from, not institution of, order, and Serres suggestively connects the point to the issue of figuration as a general feature of the real that concerns me in this study: "since existence only appears in and by deviation from equilibrium, physically speaking, I'm willing for this deviation to be the primary space in which every metaphor finds its place and time. The *clinamen* is transport in general." Among Romanticists, Miranda Burgess has undertaken a far-reaching and splendid study of "transport in general"— that is, in vehicular, affective, psychoanalytic, and figural dimensions. See Serres, *Birth of Physics*, 144–51; Hoffmeyer, *Biosemiotics*, 90–95; Althusser, "Underground Current," *Philosophy of the Encounter*, 168–69 (an essay to which I return in chapter 2 and the Coda); and Burgess, "On Being Moved," "Transporting *Frankenstein*," and "Transport: Mobility, Anxiety, and the Romantic Poetics of Feeling." And for an elegant overview of "Lucretius in Theory" from the perspective of gender and sexuality, see Goldberg, *Seeds of Things*, chapter 2.

14. *DRN* 2.102. In *The Book of Los*, Blake makes the notoriously causeless and world-producing Lucretian *clinamen* part of Los's early history (as Milton does of Satan's): "Falling, falling! Los fell & fell / Sunk precipitant heavy down down", "thro' the void" (4.27–28, 4.32). Then, though, "contemplative thoughts first arose"—apparently Lucretian thoughts—for "His downward-borne fall. chang'd oblique" and "in process of falling he bore / Sidelong on the purple air" (4.40, 42, 46–47).

15. Shelley, *Triumph of Life*, line 157.

16. "Contact" and "contingent" both derive from the Latin *con-tingere*, to touch together.

17. *De rerum natura* was itself first produced and received in the later Roman republican and early imperial period that Philip Hardie calls "singularly obsessed with time and with the construction of models of temporal and historical process." *Lucretian Receptions*, 63. See Toohey, *Epic Lessons*, 90–92, for a glimpse of the extremity of civic, political, and military strife in Lucretius's age, and the essays in Gale, *Lucretius*, for an introduction to *De rerum natura*'s literary and social contexts. And see Sachs (*Romantic Antiquity*, 3–36) for the particular prominence of Rome in the neoclassical imaginary of a British culture increasingly aware of its own transition to empire.

18. I am grateful to Kevis Goodman for her help with this line.

19. See Daniel Tiffany's *Toy Medium*, which likewise takes seriously the consubstantiality between physical and poetic matter. But for Tiffany, this commonality resides in the aporia that atomism and poetry purportedly posit between empirical phenomena and their real substructure: in lyric and atomist imagery, he argues, "what matters about the world . . . often defies intuition," bearing witness to an impasse between epistemology and ontology (2, 171). Such interpretations have a venerable history in Romantic and post-Romantic aesthetics, especially those of sublimity: atomism, James I. Porter argues, "seems practically designed to elicit feelings of sublimity, of fear, and awe," threatening to "annihilate phenomenological meaning" in a "radical negation of all that is and has sense" ("Lucretius and the Sublime," 168). But there is an important irony to the recurrent contention that atomism alienates and frightens the subject from the world of sense, since this materialist philosophy presents itself as a technology for diminishing fear (*terrorem animi*, 1.146), not least through the radical validation of sensory experience (see, esp., chapter 3). Indeed, in the Epicurean tradition, sensation is the basic and irrefutable criterion of truth, from which any insight into "hidden matters" (*occultis de rebus*)

must be derived and to which it must be referred (see *DRN* 1.423–25 and 1.699–700). This is not a case of breach between being and appearing. At stake, Deleuze rightly observes, is "a gradation which causes us to pass from the thinkable to the sensible, and vice versa" ("Simulacrum and Ancient Philosophy," 275). Is this contest of interpretation one of genre? Tiffany's study of the radical disjunction between the two senses of "sense" in physics and poetry is a study of the lyric. Yet the fullest exposition of Epicurean atomism that came down to the moderns had a poetic form, and that form was didactic and epic. Understanding the way atomist materialism purports to relate sensible and insensible realities may require attention to the premises and possibilities of this poetic form, something I attempt in chapter 5.

20. In fact, for Jacques Derrida, whose writings can be particularly susceptible of this kind of caricature, an encounter with Epicurean atomism provides an opportunity to observe that his notion of "the mark or trace" is broader than human language, intending linguistic signification as one system among others: "If I say mark or the trace rather than signifier, letter or word, and if I refer these to the Democritian or Epicurian [*sic*] *stoikheion* in its greatest generality, it is for two reasons. First of all, this generality extends the mark beyond the verbal sign and even beyond human language." Language is "but one among those systems of *marks* that all have as a proper feature this curious tendency: to increase *simultaneously* the reserves of random indetermination *and* the powers of coding or over-coding. . . . Whatever its singularity in this respect, the linguistic system of these traces or marks are, it seems to me, just one example of this law of destabilization" ("*My Chances*/Mes chances," 360, 345–46). See Christopher Johnson, *System and Writing in the Philosophy of Jacques Derrida*.

21. Priestman, "Lucretius in Romantic and Victorian Britain," 289. The first Lucretian moment to which Priestman alludes would be that of the pan-cultural atomist revival in seventeenth-century culture generally, and especially in the New Scientific, corpuscular imaginary. (See Kroll, *The Material Word*; Wilson, *Epicureanism at the Origins of Modernity*; Rogers, *The Matter of Revolution*; Picciotto, *Labors of Innocence*). Priestman's "second moment" points to the marked early nineteenth-century bump in translations, editions, and imitations by nonconformist intellectuals and antiquarians who continued to promulgate an agenda of rationalist Enlightenment into the post-Revolutionary period when their ideology came "under ever more furious attack as godless, pro-French and unpatriotic." But counting "moments" may be inadequate to the task of tracking the diffuse, tendentious, and piecemeal afterlives of what has been an illicit text for most of postclassical culture—an "underground current of the materialism of the encounter" in Althusser's formulation—which is now resurfacing, for reasons that have as much to do with our moment as any past one, in scholarship on periods far outside either of these moments. The reconstruction and dissemination of *De rerum natura*, Gerard Passannante argues, may have taught Renaissance cultures the very idea of subtle intellectual influence. *Lucretian Renaissance*, especially chapter 4.

22. On this "axiomatic resistance of matter to mind," see especially Marjorie Levinson, teasing out a suggestion by T. J. Clark, in "Pre- and Post-Dialectical Materialisms," and chapter 5, below.

23. Wordsworth, Note to "The Thorn," Wordsworth and Coleridge, *Lyrical Ballads*, 288; Goethe, *Sämtliche Werke*, MA 17: 532.

24. Tilottama Rajan also summons the term "re-visionary empiricism"; see her description

of Blake's early works in *The Supplement of Reading*, 234. And see Dalia Nassar's philosophical sketch of the (to some) oxymoronic notion of "Romantic Empiricism" in its German, post-Kantian context and ecocritical potential ("after the 'End of Nature'"). As philosophical hallmarks of the mode, Nassar points to an ontological conviction that humans participate in natural processes, an epistemological conviction that "knowledge is only possible if the known object (nature) is not ultimately beyond or distinct from the knowing subject," and a practical and methodological commitment to careful observation. "Romantic Empiricism after the 'End of Nature,'" 300.

25. Though my strategy in this study is to dislodge the preeminence of this kind of organicism by looking outside it, Joan Steigerwald's work particularly well exemplifies the complementary strategy of nuancing it from within. See her forthcoming essay "Degeneration: Inversions of Teleology"; the special issue she edited on "Kantian Teleology and the Biological Sciences," in *Studies in History and Philosophy of the Biological and Biomedical Sciences*; her "Rethinking Organic Vitality in Germany at the Turn of the Nineteenth Century"; and "Treviranus' Biology: Generation, Degeneration and the Boundaries of Life."

26. Goethe, "Bedeutende Fördernis durch ein einziges geistreiches Wort" (Meaningful Progress through a Single Witty Word), *Zur Morphologie* 2.1, MA 12: 306–9. See chapter 3 below.

27. Herder, "On Image, Poetry, and Fable," 358–59, 365; "On the Cognition and Sensation of the Human Soul," 188–90; see *Werke in zehn Bänden*, 4: 329–93 and 633–77, respectively. For a fuller elaboration of Herder's work on sensation and figuration in these two essays, see my "Irritable Figures," 273–95.

28. Nietzsche, "On Truth and Lying in a Non-Moral Sense," 148.

29. Herder, "On the Cognition and Sensation of the Human Soul," 180–181, 242. Compare Simon Jarvis, "Thinking in Verse," 98–116, and *Wordsworth's Philosophic Song*, 1–23.

30. "All Religions Are One" (c. 1788), the "Argument," 3: 46. For a recent discussion of Romantic experimentalism in light of New Scientific legacies (and including Blake's aphorism), see the recent conversation in *Wordsworth Circle* 46, no. 3: Richard Sha, "Romantic Skepticism about Scientific Experiment"; Robert Mitchell, "Romanticism and the Experience of Experiment"; and Amanda Goldstein, "Reluctant Ecology in Blake and Arendt."

31. Herder, "On Image, Poetry, and Fable," 362; "On the Cognition and Sensation of the Human Soul," 188–189.

32. Picciotto, *Labors of Innocence*, 323–24.

33. Allewaert, "Toward a Materialist Figuration," 68; Allewaert draws on eighteenth-century empiricism, contemporary new materialism, and Emily Dickinson's poetry to advance a theory of figuration "that is in no sense exclusively driven by human beings and that is in every sense material and relational," 63. And see Natania Meeker's *Voluptuous Philosophy*, the first and only study, to my knowledge, to recognize the centrality of figuration to Lucretian materialism, a dimension Meeker pursues in all its estranging implications for the relation between literature, knowledge, and subjectivity across the French eighteenth century.

34. Robert Mitchell, *Experimental Life*, 64. While Mitchell construes this "ontological" approach to sensation as opposed to an epistemological one—one concerned to evaluate sen-

sations' contribution to knowledge formation—I explore the way sensation grasped in terms of corporeal interactivity authorizes empirical knowledges of a kind invalid or irrelevant in previous epistemology.

35. Heringman, *Romantic Rocks*, 25, citing Paul de Man's skepticism about the reliability of reference (to which I give significant attention in chapter 3). *Romantic Rocks* is this study's clear precedent in discovering, in the materiality and textuality of the Romantic "rock record," a historically specific form of Romantic materiality stationed "precisely between the two materialities recently competing for the objects of Romanticism, that of the letter and that of history." Heringman, *Romantic Rocks*, 20.

36. Wordsworth, 1802 Preface, Wordsworth and Coleridge, *Lyrical Ballads*, 257.

37. For the latter, see esp. Appel, *Cuvier-Geoffroy Debate*.

38. Latour, *We Have Never Been Modern*, 12.

39. On vitalism: Robert Mitchell *Experimental Life*; Packham, *Eighteenth-Century Vitalism*; Gigante, *Life*. On Spinoza: Levinson, "A Motion and a Spirit" and "Of Being Numerous," and Förster, *Twenty-Five Years of Philosophy*.

40. As James Secord points out in "The Crisis of Nature," to posit a "break" at 1800 gives too much weight to the deliberately modernizing polemics of certain researchers eager to distinguish themselves from prior and ongoing practices of natural history: when scholars "locate an epistemic break at 1800," Secord argues, "they simply underwrite the definitional strategies of Cuvier and his colleagues" (449). In chapter 5, I bring this point to the particular "break" that affords an authentic history to nature, in contrast to the purportedly ahistorical, spatial taxonomies of a previous "natural history."

41. I am thinking of Emily Rohrbach's argument in *Modernity's Mist: British Romanticism and the Poetics of Anticipation* (which teaches us to hear "missed" and "mystery" in "mist"), and of Tobias Menely's work on the human-generated hazes that obscure the baseline of seasonal temporal reckoning and the atmospheric prognostication in Cowper's *The Task*. See Menely, "Present Obfuscation" (and chapter 4). Rohrbach draws an opening contrast between the immersive Keatsian position—"We are in a Mist"—and the Wordsworthian perspective *"above* the mist" achieved in the *Prelude*'s "Snowdon Pass" episode, but tracking a specifically Lucretian kind of atmosphere, turbid with *figures*, makes a motif of modern immersion recognizable in Wordsworth and Coleridge, too—significantly, in their nightmarish depictions of crowds. Consider, for instance, Bartholomew Fair in *The Prelude*, where "the comers and the goers face to face— / Face after face after face" becomes a dizzying whirl in which it is impossible to tell whether they are "allegoric shapes, female or male, / Or physiognomies of real men" (7:172–73, 179–80); or the "fiendish crowd / Of shapes" that torture the sleeper in Coleridge's "The Pains of Sleep" with the specifically Lucretian threat of "deeds" that confuse the distinction between action and passion: "all confused I could not know / Whether I suffered, or I did" (16–17, 28–29).

42. Schaffer, "Indiscipline and Interdisciplines"; Klancher, Introduction to *Transfiguring the Arts and Sciences*; Craciun and Calè, Introduction, 1–13. And see Heringman's reconstructions of capacious and permissive umbrella disciplines like natural history and the "science of antiquity," which fostered the emergence of more specific empirical sciences and persisted in and alongside them: *Romantic Rocks*, 19–27, and *Sciences of Antiquity*, 3–4, for example.

43. Valenza, *Literature, Language, and the Rise of the Intellectual Disciplines in Britain*, 10,

26–27, and 139–72, esp. 144. Valenza's study is particularly valuable for its angle on nineteenth-century specialization as a shift from rendering diverse objects "conversible" within a broadly literate public sphere, toward discovering discourses specifically fit to specific objects of knowledge, to the exclusion of inexpert readers. She also suggestively explores Romantic poetic complicity in narrowing "literature" from written things in general that require only general learning to read them, toward certain kinds of prestigious texts that require a literary reading—defined, in particular, against writing and reading for information.

44. 1802 Preface, Wordsworth and Coleridge, *Lyrical Ballads*, 258–60.

45. On now unfamiliar senses of "passion" as coming upon one from without, passing between persons, and conditioning personhood itself in processes of personification that precede and license it, see Pinch, *Strange Fits of Passion*. For a beautiful example of how this insight inflects the intersection of Romantic medicine and aesthetics, see Kevis Goodman's reading of *Lyrical Ballads* as offering a "historical epidemiology, in the root sense of epidemiology—the study of what is 'upon people' (*epi* + *demos*)." Goodman, "'Uncertain Disease,'" 217.

46. Amid the growing literature on sensibility, irritability, and vitalism in literature and medicine, see especially Elizabeth A. Williams, *Cultural History of Medical Vitalism* and *The Physical and the Moral*; Riskin, *Science in the Age of Sensibility*; Engelstein, *Anxious Anatomy*; Vila, *Enlightenment and Pathology*; Reill, *Vitalizing Nature*; Zammito, *Kant, Herder and the Birth of Anthropology*; and Packham, *Eighteenth-Century Vitalism*.

47. The pluripotent vitalities in Denise Gigante's *Life: Organic Form and Romanticism* and Saree Makdisi's *William Blake and the Impossible History of the 1790s* (see chapter 1 below) are recent representatives of this tendency in Romantic criticism. Goethe, Blake, and Shelley's sense of their political and ethical insufficiency receives long-distance corroboration in queer theory. See, for instance, Ann Cvetkovich on femme writings that make visible and challenge the stigma on passivity by "stress[ing] the power and labor of receptivity." In a phrase that could have been written about Goethean "tender empiricism" (chapter 3 below), Cvetkovich chronicles how "vulnerability in these writings takes on positive meaning as a desirable and often difficult achievement." *Archive of the Feelings*, 58.

48. A sensitivity most certainly learned from Anne-Lise François's *Open Secrets: The Literature of Uncounted Experience*.

49. Note to "The Thorn," Wordsworth and Coleridge, *Lyrical Ballads*, 288. It is important to acknowledge Wordsworth's suggestion, in a note to both the 1800 and 1802 Prefaces, that criticism ought to replace the misleading "contradistinction of Poetry and Prose" with "the more philosophical one of Poetry and Science" (1800)—modified to "Poetry and Matter of fact, or Science" in 1802—and a "strict antithesis" between Prose and Metre (254). Yet, as I explore at some length in chapter 4, the 1802 expansion that works out the question "What is a Poet?" in complex interplay with the "Man of Science" does not uphold this proposed distinction. It argues instead for forms of ineradicable knowledge in poetic experience and ineradicable passion in scientific experimentation that preserve each in its purported opposite (not to mention throwing down a gauntlet to the "Man of Science" to collaborate in making visible the relations subtending poetic and scientific routines of observation and description). On an increasing professional rivalry motivated by profound philosophical and social *similarities* in the public role, reading audience, and humanitarian aspirations of poets and men of science, see Catherine E. Ross, "Professional Rivalry of William Wordsworth and Humphry Davy."

50. The bibliography on Romantic literature and science has happily grown beyond the possibility of survey. Some of the groundbreaking studies that have helped restore canonical English and German Romantics to their scientific contexts and expand the canon past these figures, but are not individually engaged in this introduction, include Bewell, *Romanticism and Colonial Disease*; Beer, *Darwin's Plots*; Richardson, *British Romanticism and the Science of the Mind*; Fulford, Lee, and Kitson, *Literature, Science and Exploration in the Romantic Era*; Holland, *German Romanticism and Science*; and Chaouli, *Laboratory of Poetry*.

51. Noel Jackson, *Science and Sensation in Romantic Poetry*, 7–8. Yet John Tresch's *The Romantic Machine* and Helmut Müller-Siever's *The Cylinder* certainly succeed in imaginatively inhabiting this lapsed reality.

52. Noel Jackson, *Science and Sensation in Romantic Poetry*, 14.

53. McLane, *Romanticism and the Human Sciences*, 2–3.

54. Wordsworth's slight caveat about the word *science* in the Preface—"If the time should ever come when what is *now called* Science . . . shall be ready to put on, as it were, a form of flesh and blood" ([1802], 260, my emphasis)—registers this shift in usage, which gained irreversible momentum in his time. As S. Ross put it in a now classic essay, the meaning of "science" narrowed from any knowledge (especially if systematic and accurate), to natural and physical science, whose growing prestige "explains why it could thus arrogate to itself the word previously used for all knowledge." See "Scientist: The Story of a Word," 3, 7. Wordsworth's alternation between the now quaint-sounding circumlocution "Man of Science" and lists of specialties—"the Chemist, the Botanist, or Mineralogist" (260)—speaks to another felt paradox in the new social organization of knowledge, whereby what was common or general to modern "men of science" was their particularity and specialization. The word *scientist* was first proposed in jesting analogy to *artist* in the *Quarterly Review* in 1834, for lack of "a general term by which we can designate the students of the knowledge of the material world collectively"; its legitimacy would be debated for a century, with *philosopher* still very much in use in the period 1800–1850 (Ross, 9ff.). I return to this part of Wordsworth's Preface in chapter 4, and briefly in chapter 5. See also Jarvis, *Wordsworth's Philosophic Song*, 1–33.

55. Evidence from the history and sociology of science makes clear that "science" as a kind of knowledge and social role was undergoing every bit as much redefinition—or, perhaps better, invention—in the Romantic period as "literature" was: this was an era of contested and emergent, rather than institutionally secured, disciplinarity and professionalization. Andrew Cunningham and Perry Williams go so far as to promote the early nineteenth-century over the seventeenth-century "Scientific Revolution" as a "candidate for the *origins* of science"—by which they mean the origin of its now presumptively secular, salaried senses. See their "De-Centring the 'Big Picture,'" 422, my emphasis. And see Berman, *Social Change and Scientific Organization*, for the English Regency-era production of professional, expert research as a tool for managing (and manufacturing) social crisis—a sense of modern science that Shelley contests in chapters 4 and 5, below. On the place of poetry and the "human sciences" in this changing field, see again Klancher, *Transfiguring the Arts and Sciences*; McLane, *Romanticism and the Human Sciences*; and Shaffer, "Indiscipline and Interdisciplines." Valenza is particularly forceful on the disciplinary insecurity of poetry attempting to "become a specialized field whose paradoxical task was to defend common life against the advance of specialization," *Literature, Language, and the Rise of the Intellectual Disciplines in Britain*, 35.

56. Daston and Galison, *Objectivity*, 17–53.

57. S. Ross, "Scientist: The Story of a Word," 4, 7, dating this long-term transformation in the sense of "science" to the period 1620–1830.

58. Raymond Williams, *Culture and Society*, 47.

59. Ibid., 34–35.

60. Klancher, *Making of English Reading Audiences*, 136.

61. Meeker, *Voluptuous Philosophy*, 7.

62. For the modification after the dash, see François, *Open Secrets*; Khalip, *Anonymous Life*; and Terada, *Looking Away*.

63. See Kroll, *Material Word*, and, in the Renaissance of Lucretius within Renaissance studies, see especially Lezra, *Unspeakable Subjects*; Jonathan Goldberg, *Seeds of Things*; and Passannante, *Lucretian Renaissance*, which sees the philological reassembly of *De rerum natura* from atomized textual remains as both typifying and theorizing the very project of a "renaissance," 1–14, 78–119. For the French eighteenth century, see Meeker's extraordinarily important *Voluptuous Philosophy*.

64. Mann's study closes with Samuel Shaw's play *Words Made Visible* (1679), which declares, "The whole life of man is a Tropical Figurative Converse, and a continual Rhetorication"—a pronouncement from which one could well imagine Herder taking off. *Outlaw Rhetoric*, 218.

65. Böhme, *Alternativen der Wissenschaft*, 15. One especially relevant alternative Böhme explores is the Aristotelian perspective, which presumes that humans experience bodily substances in their capacity *as* bodily substance, prompting Böhme to propose an "affected science" (*Betroffenenwissenschaft*), predicated on affection by, rather than detachment from, the object or problem of knowledge (112–18, 20–22).

66. Kelley, *Clandestine Marriage*, 13. I am indebted to Kelley's splendid exploration of plants' "material invitation to figure." Kelley is careful to caveat the likeness between plants' activities and those of the botanists they inspire with "the force of a Kantian *as if* that keeps the difference between matter and figure in play"—plant matter's resistance to every form of epistemic capture being a major argument of the book (13). Worrying instead about the critical habit of relying on matter to offer resistance and the kinds of commonality the Kantian "as if" guards against (by fictionalizing), in this study I experiment in letting go the ethics of radical and absolute alterity, exploring instead the impropriety of figuration as itself an index and guardian of difference, albeit among different styles and kinds of corporeality, rather than between matter and its ontological others.

67. As I suggest intermittently (particularly in chapter 1, and in notes throughout), versions of this conception are enjoying a revival in the contemporary life-scientific disciplines, which have called into question molecular genetics' model of ontogenetic development and phylogenetic evolution as controlled by a contextually invulnerable inheritance of genetic information: developmental systems theory, epigenetics, and biosemiotics.

68. Kant, *Critique of the Power of Judgment*, §65, 246. For Kant the claim is regulative when it comes to living beings, though this did not stop contemporaries from interpreting it as ontological (see Zammito, "Lenoir Thesis Revisited," and "Teleology Then and Now.") Nor, as Jennifer Mensch has recently argued, did it stop Kant from recognizing in autotelic organicism a causal logic suited to authorize rational freedom in his critical system. See Mensch, *Kant's Organicism*, and chapter 3, below. I take the expression "vitalizing nature" from Peter Hanns Reill's sensitive and important book on the subject, *Vitalizing Nature in the Enlightenment*.

69. Kant, *Critique of the Power of Judgment*, §§61, 65 [233-34, 244-47, ed. Guyer].

70. See Charles I. Armstrong's detailed analysis of this selective appropriation in *Romantic Organicism: From Idealist Origins to Ambivalent Afterlife*. Armstrong points out that Coleridge's importation of Jena organicism expunges its most volatile and self-critical component—the theory of the fragment—as if by "immunising repulsion" from his explicit theoretical work (51).

71. Notes for a lecture given in the 1812-13 series at the Surrey Institution, *Coleridge's Criticism of Shakespeare*, 51-53.

72. Ibid., 52.

73. Gigante, *Life*, 3.

74. Foucault, *Order of Things*, xxii-iii, 50, 217-18, 238-39, 272-76.

75. See essays collected by Allistair Hunt and Matthias Rudolf in the special *Romantic Circles* PRAXIS issue, "Romanticism and Biopolitics."

76. Foucault, *Order of Things*, e.g., xxi.

77. See Guyer's luminous "Biopoetics, or Romanticism." On the theological and political entailments of divergent theories of life, see Jacyna, "Immanence or Transcendence."

78. Coleridge, *Statesman's Manual*, 28. Yet there persists, in the most valuable studies of Romantic biology we have, a near-unanimous tendency to synonymize Romantic biopoetics and organicism: beyond Gigante's *Life*, see Müller-Sievers, *Self-Generation*; Armstrong, *Romantic Organicism*; Gasking, *Investigations into Generation*; Roe, *Matter, Life, and Generation*; Reill, *Vitalizing Nature*; Richards, *Romantic Conception of Life*; Packham, *Eighteenth-Century Vitalism*; Mensch, *Kant's Organicism*; Cunningham and Jardine, Introduction; Knight, "Romanticism and the Sciences"; Hamilton, "Romantic Life of the Self"; Zammito, "Teleology Then and Now" and "Lenoir Thesis"; and Pfau, "'All Is Leaf.'"

79. Coleridge, *Statesman's Manual*, 28, 22, Appendix E, 108-9.

80. Coleridge, *Collected Letters*, 4: 761.

81. Coleridge, *Statesman's Manual*, Appendix E, 108.

82. Passannante, *Lucretian Renaissance*, 14. A technically similar formulation about "pervasive influence" turns up in proximity to Lucretius in the Preface to Shelley's *Laon and Cythna* (one of Shelley's early formulations of what Chandler, above all, singled out as his lasting preoccupation with the relation between the person and the age): contemporaries "cannot escape from subjection to a *common influence* which arises out of an infinite combination of circumstances belonging to the times in which they live; though each is in a degree the author of the very influence by which he is pervaded." See Chandler, *England in 1819*, 79-85, 483-554; and Shelley, preface to *Laon and Cythna*, *Works*, 92, my emphasis.

83. Coleridge, *Statesman's Manual*, Appendix C, 88.

84. Coleridge, *Philosophical Lectures*, 195.

85. Thelwall, *Essay Towards a Definition of Animal Vitality*, 7-8. Yasmin Solomonescu offers a ground breaking study of the neglected polymath: *John Thelwall and the Materialist Imagination*.

86. Thelwall, *Essay*, 7-8.

87. Coleridge, *Statesman's Manual*, Appendix C, 89. Against this, Coleridge's "living and spiritual philosophy" permits "two component counter-powers" to "interpenetrate each other, and generate a higher third."

88. Good, Appendix to *DRN*, lxxxvi–lxxxvii. Matter, Good continues, "[i]f arranged, therefore, in one mode . . . discloses the power of magnetism; in another, that of electricity, or galvanism; in a third, that of chemical affinities," right up through botanical and animal irritability and sensitivity, and human intellection (lxxxvii).

89. *DRN* 1.818–19, 909–10; 2.761–2, 1008–9.

90. Indeed, these four difficulties tend to come thick on each other's heels in *De rerum natura.* If composite emergence "seems absurd," chides the Creech translation, consider how, though laughing now, you are not of "*Laughing* things compose[d]": so, "If *Wise* and if *Discursive* Things can rise / From *Seeds,* that neither *Reason,* nor are *Wise*; / What hinders then, but that a *Sensible* / May spring from *Seeds,* all void of *Sense,* as well?" 2.951–56.

91. Coleridge, "Hints towards the Formation of a More Comprehensive Theory of Life," *Coleridge's Poetry and Prose,* 597; Thelwall, *Essay,* 8.

92. For this and the previous points about Epicurean causation, compare Jonathan Kramnick's lucid account in *Actions and Objects from Locke to Richardson,* 61–98.

93. Thelwall, *Essay,* 19, 13, 38.

94. Ibid., 39, 12.

95. Lamarck, *Recherches sur l'organisation des corps vivans* (1802), 58.

96. Of the material theory of mind expressed in *DRN,* Good affirms that Lucretius "boldly predicted and accurately ascertained . . . the sources of all nervous communication," amounting to "the very philosophy of the present day" (1: lxxxix), an opinion Shelley confirms in the preface to *Laon and Cythna,* which calls *DRN* "that poem whose doctrines are yet the basis of our metaphysical knowledge" (*Works,* 95).

97. Khalip, *Anonymous Life,* 11, 14.

98. Shelley, *Defence of Poetry,* §37, *Shelley's Poetry and Prose,* 530.

99. Goodman, *Georgic Modernity and British Romanticism,* 3.

100. Koselleck, *Futures Past,* 239.

101. Barad, "Posthumanist Performativity," 801. See, on this point and in general, Diana Coole and Samantha Frost's very fine introduction to their collection *New Materialisms: Ontology, Agency Politics,* esp. 3.

102. See Bennett's suitably lively fusion of Spinoza (as read by Deleuze), Bergson, Lucretius, Darwin, and others in *Vibrant Matter: A Political Ecology of Things.*

103. Levinson, "Of Being Numerous," 649, "Pre- and Post-Dialectical Materialisms," 113.

104. Goodman, *Georgic Modernity and British Romanticism.*

105. Butler, *Precarious Life.*

106. See Allewaert, "Toward a Materialist Figuration," 63–64.

107. On Goethe, see chapter 3; Keats to Woodhouse, 27 October 1818, *Complete Poems and Selected Letters,* 501.

108. See, for instance, de Man, "Resistance to Theory," in the volume by that name, 11, and chapter 3 below.

109. I borrow from the title of Duncan Kennedy's study, *Rethinking Reality: Lucretius and the Textualization of Nature,* to which I return in chapter 3.

110. Meeker, *Voluptuous Philosophy,* 11.

111. Auerbach, *Scenes from the Drama of European Literature,* 16; Deleuze, "Simulacrum

and Ancient Philosophy," 266. Brooke Holmes has written a lucid exposition of Deleuze's essay, arguing that from Deleuze and Lucretius's joint vantage "thought . . . is not a way of looking at the real, but is, rather, part of the real itself, and its nature" ("Deleuze, Lucretius, and the Simulacrum of Naturalism," 326). That Deleuze's reading of Epicureanism centers on the identity of nature and "power" in Holmes's account sheds some light on the way a Deleuze-mediated new materialism has given shorter shrift to atomism's intensive theorization of vulnerability and dependence.

112. Asmis, *Epicurus' Scientific Method*, 30.

113. Judith Butler, *Bodies That Matter*, ix; Latour, *Pandora's Hope*, 1.

114. Judith Butler, *Bodies That Matter*, 8–9.

115. Ibid., 33; *DRN* 1.59–60.

116. Latour, *Pandora's Hope*, 127, 153. Latour uses the very Lucretian expression *concrescence* "to designate an event without using the Kantian idiom of the phenomenon," explaining that "concrescence is not an act of knowledge applying human categories to indifferent stuff out there but a modification of all the components or circumstances of the event," 305.

117. Herder, "On Image, Poetry, and Fable," 365 [*Werke in zehn Bänden*, 4: 645].

118. In a closely related passage, Herder likens sensory apertures to customs offices: "pass[ing] through the gates of a different sense," an object is "given a different stamp, in keeping with the different customs prevalent there and the different use to which it is put" ("On Image, Poetry, and Fable," 359). Such revisionary empiricism thus reveals its sensitivity to the force of culture, convention, and even state power within the most minute and apparently unmediated of bodily processes. And yet, even as he attributes active, figurative power to the perceiver, Herder does not render perception a matter of unilateral imaginative projection ("On the Cognition and Sensation of the Human Soul," 212).

119. Herder, "On the Cognition and Sensation of the Human Soul," 189. Haraway's answer to the question of "how to have *simultaneously* an account of radical historical contingency for all knowledge claims and knowing subjects, a critical practice for recognizing our own 'semiotic technologies' for making meanings, *and* a no-nonsense commitment to faithful accounts of a 'real' world, one that can be partially shared and that is friendly to earthwide projects of finite freedom, adequate material abundance, modest meaning in suffering, and limited happiness"; see "Situated Knowledges," 579.

120. See Goldstein, "Irritable Figures."

121. Gurton-Wachter, *Watchwords*, 31.

122. Such as versions of "first principle" in their alliterative negations: *nullam rem e nilo gigni divinitus umquam* (no thing is ever by divine power produced from nothing), 1.150; *natura neque ad nilum interemat res* (nature does not reduce things to nothing), 1.216; *nil posse creari/de nilo neque quod genitum est ad nil revocari* (nothing can be produced from nothing and what has been made cannot be brought back to nothing), 1.543–44.

123. *DRN*, 2.875–76.

124. Martha Nussbaum concludes that in Lucretius's transmission of Epicurean philosophy, finitism, "yielding to human life," trumps Epicurus's emphasis on the attainment of philosophical invulnerability. *De rerum natura* tends rather to foster than to eradicate "ties of interdependence and mutual need," instructing its audience in a willingness "to live as a soft body rather than an armed fortress"; *Therapy of Desire*, 498, 250, 275–77.

125. *DRN* 4.623–24, 4.620–21 (Loeb/Geer).

126. Indeed, the above-outlined theory of figuration also works in this way to sweeten Lucretius's unrelenting mortalism: through it, atomic attrition as a process that "withdraws [things] from our eyes" (*ex oculis vetustatem subducere nostris*) is simultaneously that which showers our eyes with visions. I mentioned above that in introducing the theory of simulacra, Lucretius compares them to—indeed, derives their existence from—more macroscopic exfoliations everywhere evident in the sensible world: as "when the slippery serpent casts off his vesture among the thorns (for we often see the brambles enriched with their flying spoils)" (4.60–62). The snake's discarded skin is a perfect figure for figuration in the Lucretian mode: sensible experience is taken to reliably express similar motions that are too small to see, so that the poet's gesture is not to create a fresh figure but to draw attention to this given form of disclosure: as if to say, look, there are figures for figuration caught everywhere on the bushes! But describing such husks as "flying spoils" (*spoliis volitantibus*) by which the brambles are "enriched" (*auctas*), Lucretius also casts transience as the vehicle of what richness and sweetness life offers. In their involuntary and inevitable passing, the items of the world accidentally regale each other in spoils: "for assuredly we see many things cast of particles with lavish bounty [*largiri multa*]" (4.72). Lucretius reminds the sentient readers of his poem, fearful of mortality, that their experience of sensuous life—of sensing and appearing—is co-extensive with their own and others' dissolution. Without this, there would be no sight, touch, pleasure, pain, or knowledge: nothing but an insensibility would differ little, after all, from death.

127. For instance, Peter Gay's classic argument in *Enlightenment: An Interpretation*.

128. Nonetheless, the philosophy's lingering association with the work of enlightenment is so powerful that even expert readers of *De rerum natura* frequently reverse Lucretius's negation of the power of bright rays. See, for instance, Monica Gale, "the power of *naturae species ratioque* to drive away our fears is compared to the power of the *radii solis* to drive away the darkness," confirming the "illuminating power of philosophy," *Myth and Poetry*, 203.

129. Lewis and Short, *Latin Dictionary*. "*Species*" gets lost in Creech's "those eternal Rules / Which from firm Premises true Reason draws / And a deep Insight into Nature's laws" (1.178–80) and John Mason Good's "These [terrors] alone / To Nature yield, and Reason" (1.165–66), whereas Lucy Hutchinson's "nature's contemplation" (1.151) catches Lucretius's emphasis on the visible look of things and a meaningful duplicity regarding to what or whom such looking belongs.

130. For deft and interesting glosses of "species" in suitably poetic contexts, see Goodman, *Georgic Modernity and British Romanticism*, 17–22; and Tiffany, *Toy Medium*, esp. 199–211.

131. See Priestman, *Romantic Atheism*, and again, Jonathan Sachs on the (no longer oxymoronic) topic of Romantic classicism in *Romantic Antiquity*.

132. Lauren Berlant, "Thinking about Feeling Historical," 6n17.

CHAPTER 1

1. Blake, *Jerusalem* 98.34–37, 9. Except where otherwise noted, Blake citations refer to *Blake's Illuminated Books*, ed. Bindman.

2. Herder, "On Image, Poetry, and Fable," 358–59 [*Werke in zehn Bänden*, 4: 635–36]. Erasmus Darwin, *The Temple of Nature*, e.g., 3.144n144, 342.

3. Blake, *First Book of Urizen* 18.50–21.1 Hence Erin M. Goss's sensitive poststructuralist assessment of Blakean geneses as "locat[ing] the body somewhere uncomfortably between the discursively constructed and the ontologically extant." But I do not think this is because, as Goss argues, "the body is revealed as a sign," an "imposition" marking "the failure of signs to refer to embodied and worldly experience" and a "figure" disclosing the way "embodiment and embodied experience escape all efforts of comprehension and naming"—for these claims in fact reinforce the difference between epistemology and ontology that Goss (and Blake) wish to challenge (*Revealing Bodies*, 89–91). Instead, the body in Blake is productive of and dependent upon figures and signs and therefore cannot be summoned to substantiate a distinction between being and meaning. Another way of putting the difference is that while Goss unfolds, through Blake, Judith Butler's famous admonition that the body "escapes linguistic grasp," I am indebted to Butler's sometimes forgotten corollary: when cultural theory speaks of linguistic "inscription" or imposition on the body, it summons "language" to act in a peculiarly disembodied way, to "proceed as if the body were not essential to its own operation." In fact, Butler argues, "language cannot escape the way it is implicated in bodily life." Butler, "How Can I Deny That These Hands and This Body Are Mine?," 4–5.

4. Blake, *Milton* 37.53. Denise Gigante's *Life: Organic Form and Romanticism* and Saree Makdisi's *William Blake and the Impossible History of the 1790s* represent this position. But see Stefani Engelstein's beautifully nuanced account of Blake's "immanently divine and mutable materiality," in which creative power is conceived as always inherently "bi- or even multidirectional" and perception as a form of "linking" that "influenc[es] the organizations of all involved." Engelstein and I share the view that Blake's "sweet Science" represents not a wholesale rejection of the empiricist project of knowing the world through the senses, but a sophistication of what that project will have to entail: "Every object is a product and part of a history between creator, observer, and world, and therefore constantly mutating." *Anxious Anatomy*, 106, 76–78.

5. Lamarck, *Philosophie zoologique*, 4. See Elizabeth A. Williams, *The Physical and the Moral.* The term *biologie* appeared among multiple authors between 1797 and 1802 (Roose, Burdach, Lamarck, and Treviranus) and vied with other names for a science specifically devoted to life; see Caron, "'Biology' in the Life Sciences."

6. Gigante, *Life*, 28, 126.

7. Blake, Milton 30.18–20; *Makdisi, William Blake and the Impossible History of the 1790s*, 266; Gigante, *Life*, 158, 276n9.

8. Blake, *Milton* 21(20).17.

9. In *Milton*, these earthly voices issue the following complaint against the ideal of totipotent creativity advanced by their "Fathers & Brothers" and "loud-voiced Bard[s]":

But the Emanations trembled exceedingly, nor could they
Live, because the life of Man was too exceeding unbounded
His joy became terrible to them they trembled & wept
Crying with one voice. Give us a habitation & a place
In which we may be hidden under the shadow of wings
For if we who are but for a time, & who pass away in winter
Behold these wonders of Eternity we shall consume
(*Milton* 30.21–27)

I return to this scene at this chapter's end, but suffice it for the moment to notice that their demand for shelter against "life . . . too exceeding unbounded" does not go unmet either: "there appeard a pleasant / Mild Shadow above: beneath: & on all sides round" (30.32–33). The soft enclosure Blake names "Beulah" relativizes and literally infantilizes pretenses at infinite and eternal life: "As the beloved infant in his mothers bosom [is] round incircled / With arms of love & pity & sweet compassion," so Eternity itself is *being held* (30.11–12).

10. Blake, *Jerusalem* 98.54. "[T]he Living Creatures of the Earth" express apparently legitimate concern about the sense, if any, in which they would survive admission into a "Life of Immortality" that would "Humanize" them "according to the wonders Divine / Of Human Imagination"; *Jerusalem* 99.4, 98.44, 98.31–32. I am indebted to Eric Santner's *On Creaturely Life* for my sense of their predicament. And compare W. J. T. Mitchell on the dialectic of boundary creation and destruction in Blake, for which the "Egg form'd World" is also a privileged example: "The act of creation, for instance, is a demonic act in so far as it encloses man in the 'Mundane Shell,' but from another point of view it is an act of mercy that prevents man from falling endlessly into the indefinite." *Blake's Composite Art*, 56.

11. *Jerusalem* 99.2–3.

12. *Milton* 2(b).38, 3(a).4.

13. *Milton* 45(44).15.

14. Thus as Simon Jarvis argues in a fresh dualist reading of Blake, "The natural body is a series of limiting illusions into which we fall"—chief among them the empiricist construction of the body as passively offering up raw sense impressions to be processed into knowledge by a separate mind or soul; "Blake's Spiritual Body," 13–32. Against this, Jarvis argues, Blake offers the precise equivalency of "Soul or Spiritual Body," positing "bodily experience [as] immediately and already a kind of knowledge," 25–26. But Jarvis, like George Gilpin, risks eliding the difference between classical empiricism and mechanics ("Bacon & Newton & Locke") and the newer sciences of life, a difference that Blake tends rather to exploit; see Gilpin, "Blake and the World's Body of Science," 35–56. Directing experimental scrutiny toward the unique properties of living matter, the life sciences had in fact shifted natural-philosophical focus from the inert bodies exemplary for classical physics to live bodies endowed with active and specific propensities. To represent a living body as something other than "any old kind of external object," then, Blake need not reject "the body natural" for "the body spiritual": he need only participate (albeit critically) in the pan-cultural attempt to do the same. Gigante is excellent on this point (*Life*, 2–3, 154); and see Richard Sha, who argues that contemporary medicine *supports* Blake's optimism about spiritual regeneration (*Perverse Romanticism*, 208–12) and Engelstein, *Anxious Anatomy*, 108.

15. Blake, "A Vision of the Last Judgment," *Complete Poetry and Prose*, 555.

16. But see Lily Gurton-Wachter's *Watchwords*, which argues that Blake's texts induce a kind of sweet, straying, divided attention that likewise challenges the poet's reputation for polemical vehemence; Gurton-Wachter's examination of the physiology of reading in Blake and Darwin deepens the case I try to make below for an experimental habitus that gives objects license to take shape. See, especially, 22–26, 29–58.

17. Blake, *First Book of Urizen* 16.6.

18. In *Blake's Agitation*, after all, Steven Goldsmith gorgeously proves that Blake offers the

canniest instruction in the allure and the limits of critical enthusiasm. Goldsmith casts Blake as pioneering a post-Revolutionary attitude of impassioned, yet passive, critical enthusiasm—an attitude that can serve as an alibi for the radical politics Blake is frequently presumed to possess, and that present-day turns to affect may less critically perpetuate. Goldsmith's discussion of the scientific and sentimental discourse of the heart is particularly relevant in our context, anticipating, from a different angle, the point that the body in Blake cannot be trusted as reservoir of resistance to discursive norms and forms. From Goldsmith's vantage, this mistrust arises less because the body co-authors such norms and forms than because Blake represents recourse to the unconventional authenticity of bodily feeling as itself conventional: the well-worn discourse of a veritable "pity industry." See *Blake's Agitation*, introduction and chapters 1, 3, and 4.

19. Blake's training and career as a commercial engraver go a long way toward resolving the mystery of his poems' startlingly technical medical lexicon and knowledge. In her pioneering investigation of Blake's relation to evolutionism and prescient expression of the biogenetic law in *The First Book of Urizen*, Carmen S. Kreiter first suggested John Hunter (1728–93) as a likely embryological informant among Blake's personal and professional acquaintances. See Kreiter, "Evolution and William Blake." John and his brother William (1718–83) were preeminent Scottish medical men at the late-century London vanguard of comparative anatomy, embryology, obstetrics, surgery, and vitalist physiology, as well as curators of an extremely important medical museum and library. The man-midwife William Hunter was famous for his lavish obstetrical atlas, *Anatomy of the Human Gravid Uterus Exhibited in Figures* (1774), while John Hunter's posthumous writings on the blood and gunshot wounds fueled a sensational debate in the eighteen-teens about the "vital principle" controlling animal organization (chapter 4 below). Blake himself engraved a commission for John in 1793; both Hunters patronized Blake's engraving master, James Basire; and Blake attended the Royal Academy during William's tenure there. (John, I cannot help noting in our context, proudly reported having "kept a flock of geese for more than fifteen years" in order to keep himself flush with mundane eggs for experimental vivisection. See Hunter, "Progress and Peculiarities of the Chick," and plate 73, in Owen, *Descriptive and Illustrated Catalogue*, viii–xvii, 204–5.) On Blake and the Hunters, see Stefani Engelstein's luminous exposition of Blake's engagement with William's anatomical aesthetics in *Anxious Anatomy*, 39–90. For a revelatory account of Blake's sophistication with regard to the collusion between physiology and sentimentalism in a Romantic-era "affective turn," see Goldsmith, *Blake's Agitation*. And see Tristanne Connolly's *William Blake and the Body*, 25–72, for psychosexual issues of embodiment and natality. It is also important to recall that Blake's publisher Joseph Johnson was a major force in London medical publishing and that Blake engraved illustrations for works including John Brown's *The Elements of Medicine* (1795), Erasmus Darwin's *The Botanic Garden* (1791) and *The Poetical Works* (1806), David Hartley's *Observations on Man* (1791), and Rees's *Cyclopaedia* (1820); see the reproductions in Essick, *Blake's Commercial Book Illustrations*.

20. See Gigante, *Life*, and Müller-Sievers, *Self-Generation*. For a definitive history of the controversy, see Roe, *Matter, Life, and Generation*. Also Roger, *Life Sciences in Eighteenth-Century French Thought*.

21. Gigante, *Life*, 6; Jacyna, "Immanence or Transcendence," 311–329; see also Desmond,

Politics of Evolution, on the cultural politics of London life science in the decades before (Charles) Darwin.

22. Important examples of the first tendency include Gigante, *Life*; Müller-Sievers, *Self-Generation*; Mensch, *Kant's Organicism*; and Pfau, "'All Is Leaf'"; for the second, see Armstrong, *Romantic Organicism*; Cunningham and Jardine, Introduction; Hamilton, "Romantic Life of the Self"; John H. Zammito's "Teleology Then and Now" and "Lenoir Thesis Revisited"; Roe, *Matter, Life, and Generation*; Reill, *Vitalizing Nature*; Richards, *Romantic Conception of Life*; and Gasking, *Investigations into Generation*. Robert Mitchell's *Experimental Life: Vitalism in Romantic Science and Literature* constitutes a sophisticated exception, especially where it begins to break with epistemological and aesthetic approaches to Romantic life philosophy inherited from Kant; e.g., 63–73 and 203.

23. In a trenchant interlude on Blake in *Science and Sensation in Romantic Poetry*, Noel Jackson anticipates my argument in stressing that, in Blake, historical time is coeval with that of the body, 92–93. In general, however, Foucault's understanding that it is strong-form organicism that inaugurates a properly historicist "history of nature" around 1800 (superseding ahistorical "natural history") prevails. (In *The Order of Things*, it is Cuvier's internalist organicism that shatters the spatial continuum of "classical" natural history, "reveal[ing] a historicity proper to life itself," 274–75.) Compare Wolf Lepenies's *Das Ende der Naturgeschichte*, Lyon and Sloan, *From Natural History to the History of Nature*, and Richards, *Romantic Conception of Life*, for methodologically diverse accounts that converge in identifying historicism in the natural sciences with organicist teleology.

24. Makdisi, *William Blake and the Impossible History of the 1790s*, 79–154. Makdisi sets a dazzling precedent for reading Blake as a *critic* of the rhetoric of self-organization in its biopolitical dimensions, yet for reasons advanced below, I depart from his conclusion that Blake offers a vitalist "life of affirmative and creative *power*. . . . of pure potential, of endless striving, of *élan vital*," as a clear alternative (318).

25. As Evelyn Fox Keller dubs the twentieth century in her book by the same name.

26. William Harvey, *Anatomical Exercitations concerning the Generation of Living Creatures* (1653), 2.

27. Blake, *First Book of Urizen* 10.5–6, 10.2–3.

28. On Blake's access to Harvey, see Kreiter, "Evolution and William Blake," 110–18, and note 19 above.

29. Harvey, *Anatomical Exercitations*, 3.

30. Ibid., 2.

31. Ibid., Preface, [ii, xi], 2; Blake, *Milton* 34.33.

32. Ibid., 222–23.

33. This paragraph's Harvey citations are drawn from *Anatomical Exercitations*, 222–28.

34. Roger, *Life Sciences in Eighteenth-Century French Thought*, 393 and 533–42. On epigenesis as "narrative," see also Reill, *Vitalizing Nature*, 8, and "Hermetic Imagination," 42. And compare Pfau, "'All Is Leaf'," on the Romantic revision of form and substance into progressive differentiation in time.

35. Harvey, *Anatomical Exercitations*, 204, 205, 197.

36. On preformism, see Clara Pinto-Correia's now classic work of intellectual history, *The*

Ovary of Eve: Egg and Sperm and Preformation, which rescued preformist theory from subsequent derision in the history of science by reconstructing its logical rigor and aesthetic elegance from an eighteenth-century point of view.

37. Erasmus Darwin, *Zoonomia*, 1: 502.

38. Blackmore, *Creation*, 4.289–90.

39. Ibid., 4.286–92.

40. Ibid., 4.292, 284–85.

41. Bonnet cited in Joseph A. Needham, *History of Embryology*, 213.

42. On Haller and his legacies, see Vila, *Enlightenment and Pathology*; Reill, *Vitalizing Nature*; Zammito, *Kant, Herder, and the Birth of Anthropology*, 221–53, 302–7, 327–32; Riskin, *Science in the Age of Sensibility*, 19–68; and Elizabeth A. Williams, *The Physical and the Moral* and *A Cultural History of Medical Vitalism*.

43. Henry, *Memoirs of Albert de Haller*, 63.

44. Haller, *First Lines of Physiology*, 207–8.

45. Haller, Vorrede, b3.

46. *Milton* 24(25).42, 25(26).31–33, my emphasis.

47. *Milton* 34.31.

48. *Milton* 23.60, 23.58–59. Engelstein points out that this combination of geography and physiology belongs to the genre of the anatomical "atlas," too; *Anxious Anatomy*, 84.

49. "Cambel" is a recurring character in Blake's books, a daughter of Albion whom S. Foster Damon identifies as "the Warring Female" associated with England's southern coastal counties. *Blake Dictionary*, 66.

50. *Jerusalem* 83.33–37.

51. My gratitude to Steve Goldsmith for his help with this formulation.

52. For instance, in Wai Chee Dimock's fine argument that Blake uses the flexible temporality of literary history to refute deterministic Newtonian time, "biological time" emerges reflexively as a synonym for the Newtonian kind—despite Blake's awareness that different biologies carried different theories of history with them to term; "Non-biological Clock," 153–77. Noel Jackson notices instead that Blake navigates between "biological determinism" and "naïve idealism" by representing historical, biological, and poetic time as coeval and subject to concerted (if improbable) transformation; *Science and Sensation in Romantic Poetry*, 99, 92–93.

53. Bourdieu, *Outline of a Theory of Practice*, 89.

54. Braunstein, "Le concept de milieu"; Roger, *Life Sciences in Eighteenth-Century French Thought*, 541–42; Canguilhem, "The Living and Its Milieu," 10–11.

55. The insight historicizes science, too, since natural-philosophical assessments of the shape of the earth and its place in the cosmos will "fluctuat[e]" according to the social formation of the sensoria that seek, empirically to observe them: "And sometimes the Earth shall roll in the Abyss & sometimes / Stand in the Center & sometimes stretch flat in the Expanse," *Jerusalem* 83.40–41.

56. *Milton* 34.33, 31.33, 37.53.

57. Goethe, "Die Absicht eingeleitet" (The Intention Introduced), MA 12: 16–17, and see chapter 3, below.

58. Noel Jackson, *Science and Sensation in Romantic Poetry*, 97.

59. Compare Goldsmith on the virtues of extreme activity and passivity, as well as active passivity and passive activity, in the context of Blake's Jesus; *Blake's Agitation*, 199–209, 243–46.

60. Kant, *Critique of the Power of Judgment*, §65 [245–46, ed. Guyer].

61. Ibid., §§65, 66 [244–45, 248, ed. Guyer].

62. Ibid., §65 [247, ed. Guyer]. On the way Kant's metaphysical commitment against hylozoism (shared with Blumenbach) competes with his attempt to curtail organicism to a merely "regulative" mode of judgment (which scruple Blumenbach and the Göttingen embryologists did not share), see Zammito, "Lenoir Thesis Revisited" and *Genesis of Kant's Critique of Judgment*, 189–227.

63. On the institutionalization of biology, see Nyhart, *Biology Takes Form*, 1–64. On a tradition of "teleo-mechanical" thinking obedient to Kant's delimitation of teleology to mere heuristic, regulative judgment, see Lenoir, *Strategy of Life*. However, Richards and Zammito have shown that Kant's reservations were (generatively) misunderstood and ignored in contemporary *Naturphilosophie*-inflected life science (Richards, "Kant and Blumenbach on the *Bildungstrieb*" and *Romantic Conception of Life*, 207–51; Zammito, "Lenoir Thesis" and "Teleology Then and Now"). Concurring with them that Kant did not supply the master research program for Romantic biology, I differ by examining Romantic alternatives to Kant that manifested *less*, rather than more, teleological organicism. On Coleridgean organicism, see Wylie, *Young Coleridge and the Philosophers of Nature*; Armstrong, *Romantic Organicism*; and Desmond, *Politics of Evolution*. Desmond persuasively casts the organicism Coleridge developed in the teens—disseminated through the College of Surgeons by his collaborator and protégé, Joseph Henry Green—as something of a calculated counterinsurgency campaign against the "leveling threat" of the materialistic, French, comparative morphology wielded by medical reformers agitating for democratization of the professional Royal Colleges in particular, the profession in general, and the society at large, 11–15, 262–74.

64. Zammito, "Lenoir Thesis," 129.

65. Roe, *Matter, Life, and Generation*, 151–52; Gasking, *Investigations into Generation*, 151; Keller "Organisms, Machines, and Thunderstorms"; Packham, *Eighteenth-Century Vitalism*, 208–16. And see Goethe, "Bildungstrieb," in *Zur Morphologie* 1.2, MA 12: 100–102; chapter 2, below.

66. Burwick, Introduction, ix.

67. Müller-Sievers, *Self-Generation*, 4.

68. Gigante, *Life*, 3, 126, 25.

69. Blake's description of the embryon "Orc" in *First Book of Urizen* 17.21–23.

70. Ibid., 3.6, 3.1.

71. Ibid., 3.3, 3.21, 4a.18, 4.16, 12.52.

72. Ibid., 21.25–26.

73. Coleridge to Thelwall, 31 December 1796; quoted in Wylie, *Young Coleridge*, 124.

74. *First Book of Urizen* 4a.19, 4a.38–40.

75. Ibid., 9.20, 4.38, 26.21, 14.13.

76. Ibid., 10.20, 4.36, 9.21, 9.32, 21.26.

77. Donna V. Jones, *Racial Discourses of Life Philosophy*, 4.

78. Gigante, *Life*, 106–54, particularly 107, 115. Compare Makdisi on "life as an infinitely

prolific number of re-makings, re-imaginings, re-becomings, the joyous life of pure potential,"
William Blake and the Impossible History of the 1790s, 318, and 313–24 generally.

79. *First Book of Urizen* 10.23–24, 4.28–29.

80. Ibid., 10.5, 14.13, 16.1; *Milton* 28.19. On this point, extensively, see John Tresch, *The Romantic Machine*; and see Goldsmith, "Strange Pulse," on the uncanny persistence of mechanism in the vitalist and sentimental discourse of the heart; *Blake's Agitation*, 226–61.

81. *First Book of Urizen* 3.2, 5.8, 9.27.

82. Ibid., 9.27, 4.3, 9.37–38.

83. Ibid., 23.10.

84. Ibid., 3.42–43.

85. Ibid., 4.8–11.

86. Ibid., 4.9, 21.4–6, 21.3.

87. Ibid., 7.12, 12.28–32; W. J. T. Mitchell, *Blake's Composite Art*, 152. Eyes, nostrils, and mouth crowd Urizen's foreshortened face, in contrast to his full-body organ of touch, the skin, which maintains extensive contact with the sea. In this passage, ocular dominance—seeing "by degrees"—defensively inhibits sensations that collapse distance through incorporation and contact. This alerts us to a strange feature of Blake's "sense catalogues," those recurrent scenes of sense-organ formation that are the poet's most distinctive and prescient descriptions of ideological imbrication (see Connolly, *William Blake and the Body*). In such catalogues, Blake tends to stop at four organs—the "shrunken eyes," "Ears in close volutions," "Nostrils bent down," mouth like a "Hungry Cavern" (*First Book of Urizen* 11–13). What of touch? In a rare, still indirect, reference to this elided sense, the mocking Fairy who dictates *Europe* suggests that through touch, "Man" could "himself pass out what time he please, but he will not" (K 3.5). W. J. T. Mitchell's brilliant taxonomy of the sense-linked forms that permeate Blake's pictures also stops at four: eye/orb, ear/spiral, tongue/arabesque, nose/arch. Tactility emerges instead in Mitchell's discussion of Blake's flat, elemental backgrounds: their refusal of perspectival depth undermines the safe distance between figure and ground, consciousness and its objects, and "restore[s] a tactile, synaesthetic quality to pictorial form," putting viewers into contact with "nonvisual sensations of heat, cold, wetness, dryness, hardness, and softness rather than the sensory alienation of visual distance," *Blake's Composite Art*, 59–60. In an intriguing parallel, Gernot Böhme's exploration of roads not taken in the construction of modern natural science leads to Aristotelian chemistry, which classified substances based on their palpable qualities—warm, cold, rough, soft—to the touching (and touched) naturalist, unapologetically rendering the sensuously experienced qualities of things as their objective characteristics. Haptic science, Böhme shows, presumes that humans experience bodily substances in their capacity *as* bodily substance, prompting his call for an "affected science" (*Betroffenenwissenschaft*) predicated on interaction, rather than on distanced and disinterested observation. *Alternativen der Wissenschaft*, 112–18, 20–21.

88. Oyama, *Evolution's Eye*, 126–27; Blake, *Four Zoas*, "Night the First," 4.43–45, *Complete Poetry and Prose*, 302.

89. Indeed, Roger casts this attention as a logical consequence of epigenesists' import of animal form from the realm of impervious first causes to the contingent domain of secondary ones: "epigenesis allowed and even required the study of everything that could be acting upon it, the study of heredity and environment [*milieu*]," *Life Sciences in Eighteenth-Century French Thought*, 534.

90. Darwin, *Zoonomia*, 39.4; 492.

91. Ibid., 39.4; 492–93.

92. Ibid., 3.1, 14.2; 14–17, 111–13.

93. Ibid., 2.2; 11.

94. Ibid., 2.2, 3.1, 14.1–2, 22.3, 39.6; 11, 14–17, 108–17, 253–57, 516–24. I explore this overly compressed suggestion in two articles under way as this book goes to press.

95. Lamarck and Étienne Geoffroy Saint-Hilaire were appointed to the newly created chairs in zoology at the Muséum in 1793; two years later, Cuvier joined to teach comparative anatomy; Appel, *Cuvier-Geoffroy Debate*, 11–39.

96. Lamarck, *Recherches sur l'organisation des corps vivans*, 58; compare Bichat, *Physiological Researches upon Life and Death*, 9–10. Pietro Corsi points out that Lamarck's *Recherches* garnered more contemporary attention than his now better-known *Philosophie zoologique* (1809); *Age of Lamarck*, xii.

97. Lamarck, *Recherches sur l'organisation des corps vivans*, 60.

98. Ibid., 76, 62, 76.

99. Ibid., 76, 70.

100. Ibid., 77.

101. Harvey, *Anatomical Exercitations*, 227.

102. Lamarck, *Recherches sur l'organisation des corps vivans*, 76. Representing embryogenesis as an induction of pressure and motion predicated on a power/susceptibility to be so moved, such epigenesis theories articulate something like the biological correlate of Judith Butler's double-aspect theory of subject formation: "A power *exerted* on a subject, subjection is nevertheless a power *assumed by* the subject," an involution of power marked by the signal trope of "turning back upon oneself" as the ungrounded moment of identity-formation; *Psychic Life of Power*, 11, 3. The physiological corollary of this psychic "life" might entail "becoming" as corporealization through incorporation of power.

103. Lamarck, *Philosophie zoologique*, 4; *Recherches sur l'organisation des corps vivans*, 44.

104. *Recherches sur l'organisation des corps vivans*, 89.

105. Heringman, *Romantic Rocks*, 196 (and on the geology of *Urizen* 94–107); Alan Bewell, "Cosmopolitan Nature," 120; Noel Jackson, "Rhyme and Reason"; Goodman, "'Uncertain Disease'"; Packham, *Eighteenth-Century Vitalism*, 147–74; and Priestman, *Poetry of Erasmus Darwin*.

106. Noel Jackson, "Rhyme and Reason."

107. Appel, *Cuvier-Geoffroy Debate*, 1–105.

108. Desmond, *Politics of Evolution*, 25–100, 236–75; see also Sharon Ruston, *Shelley and Vitality*, 24–73.

109. For a good introduction to the vagaries of Lamarck's reception, see La Vergata, "Il Lamarckismo," and the essays on "La postérité de Lamarck" collected in Laurent, *Jean-Baptiste Lamarck*.

110. "The spirit of animation has four different modes of action"—"or, in other words the animal sensorium possesses four different faculties, which are occasionally exerted"; in "their inactive state [they] are termed *irritability, sensibility, voluntarity*, and *associability*; in their active state . . . *irritation, sensation, volition* and *association*"; Darwin, *Zoonomia*, 5.1; 32.

111. Makdisi, *William Blake and the Impossible History of the 1790s*, 83, 266–67, 269, 281.

112. Canguilhem, "The Living and Its Milieu," 8 ["Le vivant et son milieu," 167].

113. Ibid.

114. Given the explicitly life-sustaining role of environing *milieux* in Lamarck's *Recherches*, I depart from Canguilhem's ultimate assessment of Lamarck's biology as a "naked vitalism" (12) that represents a desolated relation between organism and milieu.

115. Canguilhem, "The Living and Its Milieu," 26, 11 ["Le vivant et son milieu," 196, 172].

116. Ibid.,11 [172]. It is in this spirit that Richard Lewontin, in his classic essay "Gene, Organism and Environment" (1983), would observe that "we only recognize an 'environment' when we see the organism whose environment it is," and go on to research the processes of "niche construction" through which organisms make an environment out of select "bits and pieces of the external world" (63–64). Compare E. Darwin, *Temple of Nature*, 155–68.

117. Canguilhem, "The Living and Its Milieu,"27 ["Le vivant et son milieu," 197].

118. Blake, *Milton* 2(b).29.

119. Darwin, *Temple of Nature*, 3.294, n294; Blake, *Songs of Innocence and Experience*, 25.

120. Herder, "Fragments on Recent German Literature," 55; Blake, *Songs of Innocence and of Experience*, 48.

121. On Lamarck's self-conscious acknowledgment of circumstantial force in his own intellectual development, see Burkhardt, "Animal Behavior and Organic Mutability," 80.

122. On Blake's relation to Hunter, see note 19 above.

123. Blake, "An Island in the Moon," *Complete Poetry and Prose*, 454.

124. *Jerusalem* 22.20–24.

125. *Milton* 21(20).16–17. Laura Quinney justly assesses Antamon's action as "palliative care": we are "left longing," she writes, "recalled to our transcendental vocation but not raised to it." Thus her welcome reading moves quickly past Antamon in search of more "radical therapy," whereas I am interested in the persistence of this qualified and compromised dimension in Blake; *William Blake on Self and Soul*, 142–43.

126. *Milton* 27.3, 25(26).27.

127. *Milton* 27.3, 27.2, 27.10–11.

128. See Engelstein's kindred stress on Blake's "mutating" objects, grasped as "part of a history of interactions between creator, observer, and world"; *Anxious Anatomy*, 76.

129. *Milton* 27.8–20.

130. *Europe* 16(17).17–19.

131. Damon, *Blake Dictionary*, 24–25.

132. Damrosch, *Symbol and Truth in Blake's Myth*, 320.

133. On the "intimate connection" between engraving and anatomy's "exploration of hidden physical or material topographies" with chemical and probing instruments, see Connolly, *William Blake and the Body*, 32.

134. "Con-science," in Engelstein's terms: science predicated upon "encouraging the *flexible* interaction of bodies no longer envisioned as discrete individuals"; *Anxious Anatomy*, 107–12.

135. *Milton* 31.2–3.

136. Blake, *Descriptive Catalogue*, in *Complete Poetry and Prose*, 550.

137. Nelson Hilton's *Literal Imagination: Blake's Vision of Words* follows the visual and verbal polysemy of Blakean "lines" for the length of two generative chapters, from "Chains of Being" to "Fibres of Being," 56–100.

138. *Milton* 30.22.

139. Harvey, *Anatomical Exercitations*, 230.

140. This section draws heavily on two dozen seminal perspectives collected in Oyama, Griffiths, and Gray, *Cycles of Contingency*; on the analyses of the epigenetic turn collected in Van Speybroeck, Van de Vijver, and de Waele, *From Epigenesis to Epigenetics*; and on Scott F. Gilbert and David Epel's textbook, *Ecological Developmental Biology: Integrating Epigenetics, Medicine and Evolution*. For the framing of epigenetics as a renewal of epigenesis against modern genetic preformation, see especially Oyama, Griffiths, and Gray's introduction to *Cycles of Contingency*, 4. And compare Pfau's examination of contemporary "evo-devo" as an echo of Goethean life science (albeit under the sign of organicism); "'All Is Leaf,'" 26–29.

141. Jacob, *Logic of Life*, 2.

142. In confirmation of Blake's admonition that theories of generation are theories of secular history, Jacob's lead historiographical metaphor for the "history of heredity" he undertook to write in *The Logic of Life: A History of Heredity*, is preformism's quintessential topos: "a series of organizations fitted into one another like nests of boxes or Russian dolls" (16). Each successive epoch in the life sciences, Jacob argues, has broken open one container to reveal "another structure of a higher order," from the early modern focus on visible surfaces, to the late Enlightenment turn to internal organization, to the nineteenth-century cell theory, to the twentieth century's nucleic acid molecule (16–17, 207). On the subtleties of genetic preformationism, see Oyama, Griffiths, and Gray, Introduction, 4.

143. Oyama, Griffiths, and Gray, Introduction, 4.

144. Oyama, *Evolution's Eye*, 47; James Greisemer, "What Is 'Epi' about Epigenetics?," 98–99.

145. Jacob, *Logic of Life*, 16, 207.

146. Ibid., 3. See Gilbert and Epel, *Ecological Developmental Biology*, appendix C.

147. Dawkins, *Selfish Gene*, 21.

148. Keller, *Century of the Gene*, 7.

149. Van Speybroeck, Van de Vijver, and de Waele, *From Epigenesis to Epigenetics*, 2–3.

150. "The Lamarckian Dimension" is the subtitle of geneticist-theorists Eva Jablonka and Marion Lamb's groundbreaking *Epigenetic Inheritance and Evolution*; see also their short piece with E. Avital, "'Lamarckian' Mechanisms in Darwinian Evolution"; and Steele, Lindley, and Blanden, *Lamarck's Signature*, xvii–25.

151. Gilbert and Epel, *Ecological Developmental Biology*; Jablonka, "Systems of Inheritance"; Nijhout, "Ontogeny of Phenotypes"; and Klopfer, "Parental Care and Development."

152. See Laland, Odling-Smee, and Feldman, "Niche Construction," 125, and Lewontin, "Gene, Organism, and Environment," 63–65; and compare Erasmus Darwin, *The Temple of Nature, or, The Origin of Society* 1.255–68.

153. Called "experimental teratology" in the nineteenth century, this field was arguably inaugurated by Lamarck's protégé at the Muséum, Geoffroy Saint-Hilaire (1772–1884). Geoffroy focused Lamarckian zoology on development by systematically manipulating the environmen-

tal conditions surrounding mundane eggs to study their effects on animal organization; see Hall, *Evolutionary Developmental Biology*, 60.

154. Gottlieb, "Developmental Psychobiological Systems View," 42–51.

155. Oyama, *Evolution's Eye*, 88.

156. Oyama, "Terms in Tension," 185, 187–88.

157. Oyama, Griffiths, and Gray, Introduction, 3–4; Oyama, *Evolution's Eye*, 70. While DST invokes "self-organization," the term does not imply an autotelic capacity separating life from the rest of material nature. As Bruce H. Weber and David J. Depew explain, at stake is rather the thermodynamic, "prebiological fact" that some physical and chemical systems are "autocatalytic, chemically dissipative systems before they are anything else"; DST thus operates against the "autonomy of biology" standpoint. See their "Developmental Systems, Darwinian Evolution, and the Unity of Science," 242–44, 246; and Oyama, "Terms in Tension."

158. Jablonka and Lamb gloss their Lamarck-positive alternative to gene-centered neo-Darwinism at the outset of *Evolution in Four Dimensions*: "there is more to heredity than genes; some hereditary variations are nonrandom in origin; some acquired information is inherited; evolutionary change can result from instruction as well as selection" (5). The text explores the interaction between genetic, epigenetic, behavioral, and symbolic/linguistic systems in biological heredity.

159. For example, Zimmer, "How Microbes Defend and Define Us."

CHAPTER 2

1. Foucault, *Order of Things*, 160.

2. Jacob, *Logic of Life*, 87. The quote concerning death comes from Lamarck's *Philosophie zoologique*, 1:106.

3. Both Foucault and Jacob are drawing heavily on Georges Canguilhem; see, for instance, "Du singulier et de la singularité en epistemologie biologique" (1962). More recent histories moderate the rhetoric and nuance the dating but generally affirm an epochal turn toward "life" as a specific problem for knowledge (an "explanatory scandal," in Charles Wolfe's phrase) between the two centuries. See, for instance, Roe, "Life Sciences"; Duchesneau, *La physiologie des lumières*; and Müller-Wille and Rheinberger, *Cultural History of Heredity*.

4. Studies of Goethean morphology frequently focus on the early *Metamorphosis of Plants* essay and Goethe's archetypal notions of *[Ur]Typus* and *Ur-phänomen*, and tend to explain Goethean science (like and alongside *Naturphilosophien*) as a response to problems posed in Kantian philosophy—above all, as a means of synthesizing Kantian ruptures into higher unities and totalities that overcome the limits the critical philosopher placed on human knowing. See the following valuable studies, most treated in greater depth in chapter 3's discussion of Goethe's Kant-engagement: Richards, *Romantic Conception of Life*; Pfau, "'All Is Leaf'"; Förster "Die Bedeutung" and *Twenty-Five Years of Philosophy*; Jonas Maatsch's Introduction to *Morphologie und Moderne*, 1–15; and Wellmon's "Goethe's Morphology of Knowledge." In this chapter, I instead take the journals' temporal and generic disjunctures seriously as connected to a changed argument about metamorphosis hazarded therein; see Eva Geulen's like-minded pursuit of the possibility that the journals' "apparent lack of form may be another

form of form" ("Serialization in Goethe's Morphology," 58). This approach, I think, receives powerful corroboration from Andrew Piper's *Dreaming in Books*, which argues from the perspective of Romantic media history that Goethe's late-life publishing practices were redefining the novel "as something material, processual, and spatially dispersed," 25, 1–51. Safia Azzouni's *Kunst als praktische Wissenschaft*, on collective and objective authorship in these writings, and Stefan Blechschmidt's *Goethe's Lebendige Archiv* also contribute to an increasing understanding of Goethe's constructivist, rather than expressivist, textual forms. In this they build on Stephen Koranyi's *Autobiographik und Wissenschaft im Denken Goethes*.

5. Goethe, "Das Unternehmen wird entschuldigt" (The Undertaking Is Excused), *Zur Morphologie* 1.1, MA 12: 12.

6. See Jocelyn Holland's insightful comparison of Goethe's prose and elegiac versions of plant metamorphosis, *German Romanticism and Science*, 19–55, and Dorothea von Mücke's excellent examination of *Zur Morphologie* as a commentary upon the emergent polarization between literary and scientific author-functions. Mücke's conclusion that the journal's experimentation with genre reveals that scientific ideas are "highly tolerant of paraphrasing, reformulation and experimentation with different formats," while literary ideas are not, seems to summon an opposition that is in fact undone by the experimentation her article so well chronicles. See Mücke, "Goethe's Metamorphosis," 38, 45. Chad Wellmon's exploration of Goethe's resistance to specializing nomenclature ultimately reads it as advocacy of a "transcendental Science"; "Goethe's Morphology of Knowledge."

7. As chapter 1 reflected, from William Harvey's *De Generatione Animalium* (1651) to the works Goethe takes up in *On Morphology* (Wolff, Blumenbach, etc.), the sciences of life emergent in the very long eighteenth century were above all sciences of *generation*. It is telling that the word *gerontology* was not coined until the early twentieth century.

8. Geulen, "Serialization in Goethe's Morphology," 58. While preparing this manuscript for press, I had the pleasure of reading a pre-print of *Aus dem Leben der Form: Goethes Morphologie und die Nager*, Geulen's monograph on Goethe's radical theorization and practice of form in the morphology serials: among many other virtues, this text offers trenchant and eloquent corroboration of a rigorously antivitalist, antiteleological, nonholistic approach to the formal problem of life in Goethean morphology, particularly as argued through the journal's generic and theoretical commitment to provisional accumulations, collections, series, and strategically equivocal ordering and reasoning.

9. Goethe, "Die Absicht eingeleitet," MA 12: 14, 16.

10. On Goethe's ecological thinking in relation to recent ecocritical challenges to the purportedly Romantic metaphysics of "nature," see Dalia Nassar's "Romantic Empiricism after the 'End of Nature,'" as well as the *Goethe Yearbook* special issue, "Goethe and Environmentalism," that Nassar has co-edited with Luke Fischer. See also Breidbach, *Goethes Metamorphosenlehre*.

11. Koselleck, *Futures Past*, 241–48.

12. Here I am indebted to Theresa Kelley's pioneering attention to "wayward exfoliating contingency" in Romantic botany as a mode of figural and factual resistance to (its own) taxonomic and totalizing impulses. See "Romantic Nature Bites Back," 202, and *Clandestine Marriage* (on which more below).

13. Goethe to Karl August, 1826, cited in R. H. Stephenson, "Cultural Theory of Weimar Classicism," 150, 155–58.

14. Goethe to Knebel, 14 February 1821, Goethe, *Werke*, 4: 34.

15. Lepenies, *Das Ende der Naturgeschichte*, esp. 97–114; Crary, *Techniques of the Observer*, 69–70; Daston and Galison, *Objectivity*, 55–113.

16. Reill, *Vitalizing Nature*, 182. I join Reill in seeking to recover antisystematic revisions of mechanist natural philosophy that should not be confused with the "totalizing solutions of nineteenth-century scientism, positivism, and romantic *Naturphilosophie*" (182). Reill argues that such solutions should be understood as concerted rejections of late-Enlightenment, mediating vitalisms, and not their culmination (4, 14, 199–201). Goethe's texts sit uneasily within even Reill's revised periodization, however, and part of the aim of this chapter is to recognize the continued availability and strategic recovery of the less absolute approaches into the first decades of the nineteenth century (174). Reill's research on the blurring and lost boundary between life and death in the late eighteenth century has special resonance here (171–82).

17. On the relative lateness and richness of Lucretius's reception in Germany, see Hugh Barr Nisbet's "Lucretius in Eighteenth-Century Germany" and "Herder und Lucrez." Regine Otto's "Lukrez bleibt immer in seiner Art der Einzige" documents Goethe and Knebel's correspondence about the translation, including strategies for introducing such a notorious materialist to a hostile public. In a review of Knebel's translation published in *Über Kunst und Altertum*, Goethe announced his (never fully realized) intention to write a thorough treatment of Lucretius as "Man, Roman, Natural Philosopher, and Poet," and Nisbet shows how Herder, Schiller, Goethe, Knebel, Steffens, and Schelling all appear to have entertained the hope of writing a modern didactic epic on the model of *De rerum Natura* around 1800. In her commentary on Goethe's relation to the genre of didactic poetry, Dorothea Kuhn (superb editor and curator of Goethe's complete scientific writings) suggests that Goethe's late poem cycle *Gott und Welt*, which included poems published in *Zur Morphologie*, might be a late offshoot of his earlier neo-Lucretian aspiration to write a "Roman über das Weltall." See Kuhn, "Natur und Kunst," 474–75, and chapter 3, below.

18. For the historical emergence of Epicurean philosophy in a time of crisis in Athens, see, for instance, Howard Jones, *Epicurean Tradition*; and see Hardie on *De rerum natura*'s attunement to the complexity of its context and neglected importance as a model for a kind of cultural and political historicism for which the Augustan poets are better known; *Lucretian Receptions*, 13–18.

19. Goethe to Charlotte v. Stein, 21 June 1785, *Die Schriften zur Naturwissenschaft*, LA 2.9a: 321. Margrit Wyder's sensitive and thorough survey of Goethean science is among the only studies I could discover that attended to Goethe's early microscopy and its sources. Wyder argues a transition from an early Goethe keen, with Herder and Knebel, to point out continuities between mineral crystallization and vegetable and animal generation, toward one who insists on distinctions between the three realms. I think that as late as the morphological papers bodies of all kinds are (again?) built from a contingently determined but ontologically flexible material minim (*Punkt*). See *Goethes Naturmodell*, esp. chapters 6 and 8.

20. Herder to Knebel, *Goethes Gespräche*, 1: 362; LA 2.9a: p. 319.

21. Von Gleichen-Rußworm, *Abhandlung über die Saamen- und Infusions-Thierchen* (1778), 72–73. John Turberville Needham, *Summary of Some Late Observations*, 26.

22. On the role of rediscovered Lucretian atomism in early-modern experimental philosophy, see Catherine Wilson, *Epicureanism at the Origins of Modernity,* chapter 2; Richard Kroll's *Material Word*; and Picciotto, *Labors of Innocence,* especially 211–24.

23. Goethe, "Infusions-Tiere," LA 1.10: 25–40; 10 May, 38.

24. Ibid., e.g., 14 April, 28–31.

25. Ibid.

26. Ibid.

27. A recurrent signal of Lucretian poetic science; see chapter 4, below.

28. Blumenbach first publishes on the term in a 1780 article for the *Göttingisches Magazin der Wissenschaften und Litteratur,* and again in the popular *Über den Bildungstrieb und das Zeugungsgeschäfte* (1781). See Kant, *Critique of the Power of Judgment,* "Second Part: Critique of the Teleological Power of Judgment," §81 [290–93, ed. Guyer] and the following section of this chapter.

29. See, however, Wyder, *Goethes Naturmodell,* and Stephenson, *Goethe's Conception of Knowledge and Science,* for Goethe's keen awareness of the risk of sweeping particular differences away by philosophical generalization. In her discussion of Goethe's *Naturlehre* essay, which publicly refutes Knebel's private work on "ice-flowers," Wyder sees Goethe transforming the gradated *scala naturae* into a nonhierarchical topography of hills and valleys, yet one that relinquishes its earlier interest in transitional forms in favor of epitomes of each type of being and its specific mode of becoming. *Goethe's Naturmodell,* 203, 206–7.

30. Geoffroy expresses optimism that the climate is changing, since the Academy has just awarded an honorable mention to a paper that applies "physico-chemical researches to the study of animal organization." "Mémoire sur la théorie physiologique désignée sous le nom de VITALISME." By this time we are indeed at the threshold of the *next* "revolution" in biology—cell theory—that will shift attention back from organic wholes to their quasi-independent *parts.* See Duchesneau, *Genèse de la théorie cellulaire.*

31. Blumenbach, *Über den Bildungstrieb,* 79–80.

32. Ibid., 80–81.

33. Compare Bichat's 1800 definition of life as a "permanent principle of reaction" against being "influenced incessantly by inorganic bodies," *Physiological Researches upon Life and Death,* 9–10.

34. Mensch, *Kant's Organicism,* 8–9. On the question of teleological thinking as heuristically valid in German biology (Blumenbach, Treviranus, Kielmeyer, Meckel, von Baer, Johannes Müller, Carl Bergman, Rudolph Leuckart), see Timothy Lenoir's *The Strategy of Life: Teleology and Mechanics in Nineteenth-Century German Biology* and the opposition it provokes. Lenoir argues for the consistency of a Kantian tradition that accepts teleology as a (merely) regulative/heuristic presumption and *not,* as in *Naturphilosophie,* the substance of a living universe that culminates in self-consciousness. But his valuable study also makes clear the centrality of an utterly unique, autotelic causality to living beings in both traditions. And, as Zammito has persuasively shown, Kant's followers were not nearly so circumspect in their methods or conclusions. See Zammito, "Lenoir Thesis Revisited" and "Teleology Then and Now," and Robert J. Richards, "Kant and Blumenbach on the *Bildungstrieb,*" as well as the more extensive discussion and notes in chapter 1.

35. Mensch, *Kant's Organicism,* 12–14.

36. Goethe, "Bildungstrieb," MA 12: 100–102. On Wolff, see Roe, *Matter, Life and Generation.*

37. Goethe, "Bildungstrieb," MA 12: 101.

38. Ibid.

39. Ibid. Helmut Müller-Sievers's *Self-Generation: Biology, Philosophy, and Literature around 1800* (1997) gave the groundbreaking critique of epigenetic, autopoetic ideology and its collusion with the Romantic literary absolute, and mischievously experimented in defending preformism as the available alternative view. And Jocelyn Holland's *German Romanticism and Science: The Procreative Poetics of Goethe, Novalis, and Ritter* argues against the stark polarization between epigenetic and preformist positions frequently attributed to the period, attending to important subtleties in the "procreative poetics" of her focal authors.

40. Goethe, "Bildungstrieb," MA 12: 101–102.

41. Goethe, "Versuch einer Witterungslehre" (1825), LA 1.11: 245, first published in 1833 in the *Ausgabe Letzter Hand.* I am indebted to Wolf von Engelhardt's *Goethe im Gespräch mit der Erde* for pointing to these late, unpublished essays on *Knochen-* and *Vergleichungslehre* as important documents of Goethe's engagement with the Kantian idea of organism. But to describe Goethe's project here as a "modification" of Kant's idea of organic purposiveness and his hypothetical "synthetic-contemplative" understanding (following Eckart Förster's interpretation) seems to underestimate the gravity of Goethe's divergence from the basic premises of Kantian biological thinking. Whereas for Kant, the study of living beings permits us to think of their organization as (internally) purposive—not because this is objectively true, but because it is subjectively legitimate as a regulative maxim for our understanding—Goethe quite unequivocally enjoins researchers to discard the heuristic of "determinations and purposes" in favor of "relationships and connections." Compare Geulen, *Aus dem Leben der Form,* 18.

42. In *The Romantic Conception of Life,* Robert J. Richards redeems the seriousness and influence of Goethean morphology in preparing the ground for Darwinian evolutionism: "one might even say, without distortion, that evolutionary theory was Goethean morphology running on geological time" (407). Breidbach aptly glosses the kind of co-determination at stake in language that draws Goethe into agreement with the positions voiced in the previous chapter by Erasmus Darwin and Lamarck: "form is first constituted in an environment and absolutely according to that environment's resources . . . a consequence of interactions that bind themselves into the different life forms" (*Goethes Metamorphosenlehre,* 240–42). From our present perspective the ecological scope and tenor of this view, in its conviction that organized beings cannot be adequately thought without a global view of their interdependency and co-determination, are unmistakable:

> The whole plant kingdom, for instance, would appear to us as a tremendous sea, which is just as necessary to the conditional existence of the insects, as the world ocean and the rivers are to the conditional existence of the fish, and we would see that a tremendous number of living creatures are born and nourished in this plant-ocean; indeed, we would ultimately view the whole animal world again merely as a great element, in which one species, if it does not arise from, then certainly sustains itself in and through, the other. ("2. Abschnitt, Versuch einer allgemeinen Vergleichungslehre," LA 1.10: 122)

43. "Abiogenesis" and "heterogenesis" are post-facto terms that help to clarify dimensions of the historical controversy. See Farley, *Spontaneous Generation Controversy*.

44. Marx, "Economic and Philosophic Manuscripts of 1844," *MER*, 91.

45. John Turberville Needham, *Summary of some late Observations*, 646; Goethe, "Infusions-Tiere," LA 1.10: 28; Von Gleichen-Rußworm, *Abhandlung über die Saamen- und Infusions-Thierchen*, 76.

46. Marx and Engels, *German Ideology*, *MER*, 171.

47. Marx, "Economic and Philosophic Manuscripts of 1844," *MER*, 90, 115.

48. Priestley, "Observations and Experiments Relating to Equivocal, or Spontaneous Generation," 129, 120, emphasis added.

49. Ibid., 128, 125, 120.

50. John Turberville Needham, *Summary of some late Observations*, 658–59, 652, 654.

51. Althusser, "Underground Current," *Philosophy of the Encounter*, 163–207.

52. Ibid., 168–69.

53. Kramnick, *Actions and Objects*, 78.

54. Farley, *Spontaneous Generation Controversy*.

55. In *Pandora's Hope*, Latour takes the initial self-evidence of spontaneous generation, and its subsequent displacement by Pasteur's bacteriological theory, as a case study in why science studies "should be able to talk calmly about [the] *relative existence*" of natural/scientific objects. "An entity gains in reality," Latour explains, "if it is associated with many others that are viewed as collaborating with it": working technicians, textbooks, schools, experiments, belief structures, industries, etc. "We never say 'it exists' or 'it does not exist,'" Latour concludes, "but 'this is the collective history that is enveloped by the expression spontaneous generation, or [alternatively] germs carried by the air'"; *Pandora's Hope*, 153, 156, 159. As he put it in another context: "there might exist many metaphysical shades between full causality and sheer inexistence" (*Reassembling the Social*, 72). And see his more detailed study of the inversely proportional credibility of germs and spontaneous generation in *The Pasteurization of France*.

56. Von Gleichen-Rußworm, *Abhandlung über die Saamen- und Infusions-Thierchen*, 72–73.

57. See Jacques Lezra's superb discussion of the customary dimension of Lucretian atomic theory in *Unspeakable Subjects*, 1–34.

58. Goethe, "Die Absicht eingeleitet," MA 12: 15–16.

59. *DRN* 1.243, 5.830–31 (Loeb/Geer); Goethe, "Die Absicht eingeleitet," MA 12: 12–13.

60. Goethe, "Infusions-Tiere," LA 1.10: 31–32, 16 April.

61. Goethe, "Die Absicht eingeleitet," MA 12: 14.

62. Ibid., 13. On this point, see also Grave, "'Beweglich und bildsam.'"

63. See "Bedeutende Fördernis durch ein einziges geistreiches Wort" (Meaningful Progress through a Single Witty Word), MA 12: 309 (chapter 3, below).

64. I allude to the essays "Fate of the Manuscript" and the "Fate of the Publication" that accompanied Goethe's republication of his 1790 *Metamorphosis of Plants* treatise in *On Morphology*'s first issue (1817). On establishing precedence in scientific periodicals and their burgeoning numbers around 1800, see Lepenies, *Das Ende der Naturgeschichte*, 97–114; on Goethe in particular, Mücke's "Goethe's Metamorphosis" and Koranyi, *Autobiographik und Wissenschaft*, 115–32.

65. Goethe, "Verstäubung, Verdunstung, Vertropfung," MA 12: 212–23. Hereafter cited in text and in my translation.

66. Becker et al., MA, 12: 999. See Breidbach's brief but illuminating treatment in *Goethe's Metamorphosenlehre*, 244–48.

67. Goethe, "Versuch," MA 12: 30–31.

68. Ibid., 30.

69. Ibid., 31 and Kelley, *Clandestine Marriage*, 234. Kelley perceptively diagnoses these contingent metamorphoses as "hover[ing]" at "the edges of the *Metamorphosis* essay," where death is also implicit in Goethe's choice of an annual as his exemplary plant; see Kelley, "Restless Romantic Plants," 190, and *Clandestine Marriage*, 235. In the "Verstäubung, Verdunstung, Vertropfung" essay, these issues occupy center stage.

70. Citing this very essay, *Das Deutsche Wörterbuch* gives the Latin *dissipatio* for *Verstäubung*: "auflösung, zersetzung in die bestandtheile." Schelver's potentially "heretical" subversion has been neutralized for readers of English by the term *Verstäubung*'s translation as "pollination" in Goethe, *Goethe's Botanical Writings*, trans. Mueller.

71. The reviewed book was Schelver's son-in-law August Henschel's *Von der Sexualität der Pflanzen*; *Isis*'s editor Lorenz Oken was likely the review's author. See *Isis* 10, 667. I am indebted to Kuhn's characteristically fine editorial notes for this lead; see LA 2.10a: 825–31.

72. Yet see Kelley, *Clandestine Marriage*, 229–33, for gender-bending of the sexual system already on evidence in Goethe's 1790 *Metamorphosis* essay, as well as for a kind of tactical restraint on figuration through which Goethe, Kelley argues, eschewed analogies that humanize plant anatomy.

73. On Schelver's specific contribution to Romantic *Naturphilosophie*, see Bach, "Für wen das hier gesagte nicht gesagt ist."

74. The title of a table in the third issue of *On Natural Science in General*—"Eye, receptive and reactive [*Auge, empfänglich und gegenwirkend*]"—glosses receptivity as an epistemic virtue, while the "Meaningful Progress" essay (see chapter 3 below) more extensively celebrates the morphologist's susceptibility to transfiguration by the object under view: "Every new object, well seen, opens a new organ in us" (*Jeder neue Gegenstand, wohl beschaut, schließt ein neues Organ in uns auf*), MA12:556, 306.

75. Kelley, *Clandestine Marriage*, 5. For Kelley, who titles her study after Linnaean *Cryptogamia*, the "resilient hiddenness" that finally attains acknowledgment in this class is in fact synecdoche for the ineffable materiality and difference that characterize plants in general: an ability "to resist taxonomic order and invite figuration" that constitutes their ongoing allure in Romantic botanical writing (3, 5). For a difference on the relation between matter, figure, and resistance, see my Introduction, note 66, and the biosemiotic vantages explored in chapter 3.

76. Goethe's colleague Johann Wolfgang Döbereiner performed the chemical analyses; see "Verstäubung, Verdunstung, Vertropfung," 221–22.

77. The parenthetical takes inspiration from Barbara Johnson's *Persons and Things* and Sandra Macpherson's *Harm's Way* on the surprising risks and pleasures of personal thingfulness, an issue unfolded more carefully in the second part of chapter 3. Compare Catherine E. Rigby's ecocritical exploration of the link between Romanticism and biosemiotics, particularly her observation that Goethe's perspective differs critically from the more familiarly Romantic "book" or "language of Nature" topos. At stake instead is "the possibility that other-than-

human entities are engaged in their own communicative exchanges, regardless of what we make of them" (34). Rigby, "Art, Nature, and the Poesy of Plants," 23–44. More on biosemiosis in chapter 3 below.

78. See Robert Mitchell's resonant essay on "mutual atmospheres" in Romantic botany, which centers on the particular allure of *Cryptogamia* (such as mushrooms); Goethe's essay mischievously threatens to relegate *all* pollen to this catch-all class for reproductive misfits. Mitchell, "Cryptogamia," 615–35.

79. Jesper Hoffmeyer explains this way: "The semiosphere is a sphere like the atmosphere, hydrosphere, or biosphere. It permeates these spheres from their innermost to outermost reaches and consists of communication: sound, scent, movement, colors forms, electrical fields, various waves, chemical signals, touch, and so forth—in short, the signs of life." *Biosemiotics*, 5.

80. With "Elasticity" Goethe intends the "property of spontaneous expansion," rather than the now familiar sense of springy shape-retention (*OED*, no *Deutsche Wörterbuch* listing).

81. Bishop, "One Art," *Geography III.* Gerard Passannante thought of Elizabeth Bishop, too, in his very different neo-Lucretian context. See "Reading for Pleasure," 89.

82. Compare Piper, *Dreaming in Books*, note 4, above.

83. Goethe, "Das Unternehmen wird entschuldigt" (The Undertaking Is Excused), MA 12: 12.

84. Schwartz, *After Jena*, 15–40.

CHAPTER 3

1. Compare Dalia Nassar's delineation of the unacknowledged strain of "Romantic empiricism" in the context of German philosophy: "What differentiates this strand of romanticism is an interest in nature as a phenomenon to be observed (and not merely theorized or systematically constructed), and a concern with investigating and transforming our capacity to observe and grasp the natural world." "Romantic Empiricism after the 'End of Nature,'" 300.

2. "Tender empiricism" (*zarte Empirie*), from *Sämtliche Werke nach Epochen seines Schaffens*, MA 17: 532; "objectively active," from "Meaningful Progress by Way of a Single Witty Word" ("Bedeutende Fördernis durch ein einziges geistreiches Wort") (1823), MA 12: 306–9.

3. *Von der Natur der Dinge*, trans. Karl Ludwig von Knebel, 1.167. On the Knebel translation, see chapter 2, above. Compare Lucretius, "in luminis oras exoritur" (with variations "exit," "effert," "tollunt," "profudit," "eregit") at 1.22–23, 1.170, 1.179, 2.577, 5.224–25, 5.1455.

4. "Morphologie," LA 1.10: 128.

5. Interpreting Goethe's "Versuch als Vermittler," Nassar similarly argues not that idea and experience are in unproblematic unity in Goethe, but that the gap between them is spatial and temporal, rather than absolute; "Romantic Empiricism after the 'End of Nature,'" 302.

6. See Rigby "Art, Nature, and the Poesy of Plants," 23–44.

7. Danesi and Sebeok, *Forms of Meaning*, 5–6. Jesper Hoffmeyer's *Biosemiotics: An Examination into the Signs of Life and the Life of Signs* makes especially clear the affinities between this emergent discipline and the Romantic perspectives recovered in the present study because it begins by stating frankly that "processes of sign and meaning cannot, as is often assumed, become criteria for distinguishing between the domain of nature and culture." Rather, "semiosis and interpretative processes are essential components in the dynamics of natural systems" (4, xiv). As in our case, the perspective has to be carefully distinguished from vitalism, meta-

physics, reductionism, and anthropomorphism (5–15) and has been especially difficult to pursue under the terms of the neo-Darwinist "Modern Synthesis" and its purely digital/symbolic sense of semiosis as "coding" (52–109).

8. Goethe, "Höheres und Höchstes," in *Selected Verse, With Plain Prose Translations of Each Poem*, 268–9.

9. Rigby, "Art, Nature, and the Poesy of Plants," 37; Rigby also gives a good overview of the discipline's genealogy through Jakob von Uexküll (1864–1944), Thomas A. Sebeok (1920–2001), Charles Sanders Peirce (1839–1914), and, more recently, Hoffmeyer and Claus Emmeche at 25–27, including suggestions that the resemblance to Romantic attitudes toward nature may be familial, rather than accidental. The ecological ethos Rigby puts forward in this essay is inspired by Schelling's Spinozistic conception of the natural world as a "creatively self-organizing and dynamically self-transforming" "meta-organism . . . of which human consciousness too was integrally a part"—that is, strongly organicist in a way I think it is important to question (30–31). Compare her Heideggerian perspective in "Earth, World, Text." And see Hoffmeyer's comments on choosing *bio* over *physio*semiosis at *Biosemiotics*, 5n3.

10. Goethe's manuscript draft reads "Es spricht re/Sache sich selbst durch sich laut aus," Goethe-Schiller Archiv 78/134 (Reimar Bestand). The same words have been used recently to summarize indexical signification: "things speak themselves; they are not spoken"; Doane "Indexicality," 3.

11. Goethe, *Zur Farbenlehre*, MA 10: 485. Thirteen lines of Knebel's translation of *De rerum natura* also close the section on "Intentional Colors" in Goethe's historical treatment of the seventeenth century in the *Farbenlehre*, MA 10: 657.

12. *DRN* 4.

13. Borrowing exquisitely accurate terms from Walt Whitman's poem "Eidólons," which hails them as the "The ostant evanescent," *Leaves of Grass*, 4, line 21.

14. "Die Absicht eingeleitet," MA 12: 16–17.

15. Höre nun, wie sich dieselben so leicht und flüchtig bewegen,
Unaufhörlich entfließend, und stets abgleitend den Körpern.
Immer ein äußerstes quillt empor in Fülle von Dingen,
Welches sie von sich schießen . . .
(4.143–46)

16. Kant, *Critique of the Power of Judgment* §66 [249, ed. Guyer].

17. Bichat, *Physiological Researches upon Life and Death*, 9–10.

18. Thanks to Steve Goldsmith for helping me to specify that formulation. Monica Gale writes of Lucretian science that "there are no separate forces of creation and destruction . . . simply atoms, aggregating and disaggregating in space, in accordance with one natural law." "Lucretius and Previous Poetic Traditions," 63.

19. I am indebted to Anne-Lise François's graceful and mind-opening account of nature's "open secrecy" in Goethe, one among numerous forms of noninsistent disclosure that literature registers as there for the taking—or not taking. *Open Secrets*, 1–65. A single verse on nature's "open secret" (*öffentlich Geheimnis*) connects Goethe's radical validation of appearance (*Schein*) to the kind of form we have been tracking through his morphology. The poem "Epirrhema" disabuses readers of their suspicion of appearances by instructing them in

the real plurality of the objects under view in the observation of nature (*im Naturbetrachten*): "Freuet euch des wahren Scheins,/Euch des ernsten Spieles:/Kein Lebendiges ist ein Eins,/ Immer ists ein Vieles." (Rejoice in the true illusion, in the serious play: No living thing is a One, always it's a Many.) *Selected Verse*, 273.

20. De Man, *Resistance to Theory*, 11. Hereafter cited in text.

21. Even at this preliminary level, the Lucretian mode of connecting "reference with phenomenalism" runs athwart the disciplinary system de Man sketches: in *De rerum natura* it is first and foremost the *rhetorical*, rather than the logical, dimension of the poem and of things—tropes, figures, simulacra, images—that works to connect or confuse linguistic and phenomenal realities. Lucretius was in fact communicating a philosophical system perhaps as eager to undermine the supremacy of logic/dialectic as de Man: Epicureanism was notable (or notorious) in the ancient world for minimizing the role of logic in favor of the two other traditional pillars of philosophy, ethics and physics. Asmis, *Epicurus' Scientific Method*, 19, 30.

22. See Cynthia Chase's eloquent exposition in *Decomposing Figures*, 3–5.

23. See Roman Jakobson, "Closing Statement: Linguistics and Poetics."

24. De Man, "The Rhetoric of Temporality," *Blindness and Insight*, 187–228. Cited in text hereafter.

25. See Marc Redfield's virtuosic *The Politics of Aesthetics: Nationalism, Gender, Romanticism* for elaborations of the ethics of allegory in this vein. My account here of Goethe's against-the-grain science of *Bildung* may be thought of as a counterpart to Redfield's account of the *Bildungsroman*, which sets up *Bildung* as "*the* narrative" of aesthetics, the ideological program that pretends to integrate "a particular 'I' into . . . the universal subjectivity of humanity." Redfield, *Phantom Formations*, 10, 38.

26. De Man, "Phenomenality and Materiality in Kant," *Aesthetic Ideology*, 89–90.

27. Deleuze, "Simulacrum and Ancient Philosophy," 267.

28. Dorothea Kuhn suggests that Goethe's late poem cycle *Gott und Welt*, which included "Dauer im Wechsel" and multiple poems published in *Zur Morphologie*, can be seen as a partial realization of his earlier neo-Lucretian aspiration to write a "Roman über das Weltall." See Kuhn, "Natur und Kunst," 474–75.

29. A para-text sharpens this connection: Goethe enclosed "Dauer im Wechsel" in a thank-you note to the psychologist J. C. Reil, describing it as an "attempt to express poetically" what Reil had written on page 58 of his recent book. Page 58 concerns the atomic undoing of the "I":

> We regard the person in question from the present moment backwards to the first dark point of our existence . . . Still, this I, that persists so stubbornly in our consciousness, is, in reality, a highly changeable thing. The elder believes he is still what he was eighty years ago. But he is not the same anymore. Not a single atom out of all of them that he was, eighty years ago, is still there. Time has, with every forward step, nibbled at his soul and body, and has more than once completely redone him, has developed moral and physical perfections in him and then destroyed them again.

C. Reil, *Rhapsodien über die Anwendung der psychischen Kurmethode auf Geisteszerrüttungen*, 58, from William Stephen Davis, "Subjectivity and Exteriority in Goethe's 'Dauer im Wechsel,'" 454, my trans. Davis notes that the poem was in fact probably completed in 1801, before Goethe received Reil's book in August of 1803.

30. In Luke's prose translation the "let" carries from line 33's imperative *Laß*. Goethe, *Selected Verse*, 195–96.

31. Lines 17–32. Here is the complete poem:

Hielte diesen frühen Segen,
Ach, nur Eine Stunde fest!
Aber vollen Blütenregen
Schüttelt schon der laue West.
Soll ich mich des Grünen freuen,
Dem ich Schatten erst verdankt?
Bald wird Sturm auch das zerstreuen
Wenn es falb im Herbst geschwankt.

Willst du nach den Früchten greifen,
Eilig nimm dein Teil davon!
Diese fangen an zu reifen,
Und die andern Keimen schon;
Gleich mit jedem Regengusse,
Ändert sich dein holdes Tal,
Ach, und im demselben Flusse
Schwimmst du nicht zum zweitenmal.

Du nun selbst! Was felsenfeste
Sich vor dir hervorgetan,
Mauern siehst du, siehst Paläste
Stets mit andern Augen an.
Weggeschwunden ist die Lippe,
Die im Kusse sonst genas,
Jener Fuß, der an der Klippe
Sich mit Gemsenfreche maß.

Jene Hand, die gern und milde
Sich bewegte wohlzutun,
Das gegliederte Gebilde,
Alles ist ein andres nun.
Und was sich an jener Stelle
Nun mit deinem Namen nennt,
Kamm herbei wie eine Welle,
Und so eilts zum Element.

Laß den Anfang mit dem Ende
Sich in Eins zusammenziehn!
Schneller als die Gegenstände
Selber dich vorüberfliehen!

Danke, dass die Gunst der Musen
Unvergängliches verheißt,
Den Gehalt in deinem Busen
Und die Form in deinem Geist.

[Ah, if this early bounty of blossoms could but last one hour! But the warm west wind already shakes them and they rain down abundantly. Shall I take pleasure in the green leaves in whose shade but lately I stood? Soon they will tremble sere in autumn and a storm will scatter them too.

If you wish to grasp the fruit, make haste to take your share of it! for some of it is beginning to ripen, and some is already germinating. Swiftly, with every shower of rain, your sweet valley changes: and alas, in one and the same river you cannot swim a second time.

And you yourself! At all those things that seemed to stand forth before you firm as rock, at those walls and palaces, you look with ever-changing eyes. That lip has wasted away whose pain was once healed by kisses, that foot which once boldly leapt like a mountain goat to the precipice's challenge.

That hand, that articulated structure, once so gladly and generously extended to give happiness—all these things are different now. And that which in their place now calls itself by your name came hither like a wave, and is hastening likewise to mingle with the elements.

Then let the beginning and the end contract into a single point, and let yourself speed by even more swiftly than these objects! Give thanks that the favour of the Muses promises an imperishable thing: the meaning in your heart, and the form in your mind.]

Goethe, *Selected Verse*, 195–96.

32. It is also relevant, in the context of such structures, that Goethe once described the chronic unfinishedness of his biographical/-logical periodicals this way: "Unfortunately there is no shortage of additional links to attach, since life does not cease to enjamb." Cited in Koranyi, *Autobiographik und Wissenschaft im Denken Goethes*, 144.

33. *Deutsches Wörterbuch* (DWB).

34. Peirce, "Nomenclature and Divisions of Triadic Relations," *Essential Peirce*, 2: 291.

35. Krauss, "Notes on the Index," 59–64, 80.

36. Peirce, "What Is a Sign?" *Essential Peirce*, 9.

37. Ibid.

38. Compare *DRN* 1.265–328.

39. Doane, "Indexicality," 3.

40. Marjorie Levinson's neo-Spinozist, or "Renaissance" reading of Wordsworth's "I wandered lonely as a cloud" manages to "[present] space or ground as its hero, not over-against figure, but as continuously engendering and absorbing figure." See "Of Being Numerous," 643. See chapter 5 below on figure/ground relations in Shelley's materialist allegory.

41. Krauss, "Notes on the Index" (2), 64. Here is a revealing, slant echo with Jonathan Culler's seminal work on apostrophe as the signal trope of lyric. Culler distills the lyric (as opposed

to narrative) force in poetry to that which substitutes the fictive *now* of its own discourse for the empirical time of reference. In addressing absent, inanimate, and insensate entities, lyric poems evince a flagrantly fictional, but not thereby ineffective, power to summon objects and others into their discourse, "as elements of the event which the poem is attempting to *be*." At stake in the "brazenness" and "strangeness" of lyric discourse, Culler argues—in what could scarcely be a better description of the uncanny "now" of Goethe's *gegliederte Gebilde*—is "the uncalculable force of an event." Culler, "Apostrophe," *Pursuit of Signs*, 149, 152–53.

But the differences are also instructive: whereas lyric apostrophe is said to eschew extrinsic, empirical time and reference in favor of the productive event of its own discourse—"the time of enunciation"—"Dauer im Wechsel" attempts to stage the overwhelm of its own speaking by "another now." Rather than enfold deixis into *lyric* apostrophe (classically epitomized in the expressions "O" and "Ach"), we might take indexical *this*'s as signs that a different kind of nonnarrative eventfulness is under way: a *didactic*, indexical poetics that works by suppressing the vocative efficacy of its own enunciation in hopes of collaborating with something, someone else. In this sense it is important that here the poem's interpellation of its reader as the "Du" in question ("What, in that place / calls itself by *your* name") becomes irresistible, because we can add, to Culler's sense that in such moments we "los[e] our empirical lives" to the fictive event, Goethe's emphatic sense that the poetic event requires another set of hands and eyes as its support (159).

42. On *elementa* as an "abecedarian coinage," see Kennedy, *Rethinking Reality*, 86. Kennedy carefully reads *De rerum natura* for insights into the debates between realism and constructivism, "discovery" and "invention," in postpositivist science studies. And yet the study is inconclusive, above all because Kennedy fails to let the text query the presumption, shared by realists and constructivists alike, that representation, where it occurs, is an exclusive product of human knowing. *Rethinking Reality* thus depicts an aporia between a kind of realism insistent upon a preexisting reality invulnerable to our representations and a constructivism that construes the "real" as our effect. *De rerum natura* in fact presents an alternative that neither position can see: that things (*res*) themselves play a part in the making of representations of them. Though it must be kept in mind that the making and reading of "text" is only one among many partial metaphors for such participations, see Joanna Picciotto for a fitting characterization of the atomized book of nature as an "open text" that is at once visible "testimony of its own production" and susceptible to "editorial intervention." *Labors of Innocence*, 218–219.

43. neve putes aeterna penes residere potesse
    corpora prima quod in summis fluitare videmus
    rebus et interdum nasci subitoque perire.
    (*DRN* 2.1010–12)

44. Serres, *Birth of Physics*, 149–150.

45. Ibid., 140.

46. Auerbach, *Scenes from the Drama of European Literature*, 17. Fortunately, we are now in a position to modify Auerbach's disappointed conclusion that "Lucretius' most original creation"—his vision of atomic motion as "a dance of figures"—was "without influence."

47. Lezra, *Unspeakable Subjects*, 5.

48. See Marković, *Rhetoric of Explanation*: "Lucretius only uses carefully studied analogies, similitudes, and metaphors, which reveal *causal links* between various natural phenomena" (96). And Garani, *Empedocles REDIVIVUS*, esp. 18–25.

49. "[T]he various sounds of the tongue nature drove them to utter, and convenience moulded the names for things" (*At varios linguae sonitus natura subegit/mittere, et utilitas expressit nomina rerum*). *DRN* 5.1057, 5.1028–29.

50. Unseld, *Goethe and His Publishers*, 212–91; Piper, *Dreaming in Books*, 19–52.

51. Althusser, *Reading Capital*, 186–89.

52. Barbara Johnson, "Using People: Kant with Winnicott," *Persons and Things*, 94. Compare Bywater, "Goethe: A Science Which Does Not Eat the Other," 298.

53. Barbara Johnson, "Using People: Kant with Winnicott," *Persons and Things*, 95.

54. Daston and Galison, *Objectivity*, 17.

55. Haraway, "Situated Knowledges," 583.

56. Daston and Galison, *Objectivity*, 242.

57. Ibid., 247.

58. The critical literature matches the idealist philosophical climate from which Goethe increasingly attempted to dissociate his work (whether Kantian-transcendental, Jena Romantic, or Schelling-style *naturphilosophisch*) in giving disproportionate weight to the ideational pole—especially the notion of *Typus*, which decreased significantly in centrality in Goethe's later epistemology. Robert J. Richards's valuable study exemplifies the trend in its claim that Goethe progresses toward an "absolute ideal-realism," *Romantic Conception of Life*, 408; see also engagements with Thomas Pfau, Frederick Amrine, Eckart Förster, below. And see Eva Geulen, "Metamorphosen der Metamorphose: Goethe, Cassirer, Blumenberg," on the critical virtue of extricating Goethe's thinking from the sign of pantheism and "All-Natur" through which it continues to be received; 205–7.

59. In the 1844 Manuscripts, Marx diagnoses idealist phenomenology's antagonism against "*objectivity* as such" as the hyperbolic outgrowth of its need to deny the object-dimension of human being in particular, its status as an acting and suffering body among bodies. Marx points out that "objectivity" in general has been collateral damage in a weightier contest between naturalist and theological accounts of the human. For where "human being" is defined as "*spiritual* being" against "natural, corporeal, sensuous, objective being"—against "*suffering*, conditioned and limited creature[s], like animals and plants"—the aspiring human is strongly motivated to repudiate what Marx, too, calls "his *objective* activity." Here Marx's alternative to Hegelian phenomenology, like Goethe's to the Kantian kind, is the open avowal of "*objective* activity" in a double sense: first, to understand as essential, rather than illusory, his work on, in, and among "real, sensuous objects"; but second, to accept that such objective efficacy entails susceptibility to being the object of another's action. A being can "establish" objects, Marx insists, only "*because* he is established by objects—because at bottom he is *nature*. . . . To be objective, natural and sensuous . . . or oneself to be object, nature and sense for a third party, is one and the same thing." Or, more pithily, "As soon as I have an object, this object has me for an object." Marx, "Economic and Philosophic Manuscripts of 1844," *MER*, 113–16.

60. Goethe, "Anschauende Urteilskraft" (Intuitive Judgment), MA 12: 98.

61. Eckart Förster's classic essay on the subject, "Die Bedeutung von §§ 76, 77 der 'Kritik der

Urteilskraft' für die Entwicklung der nachkantischen Philosophie [Teil I]," enfolds Goethe's scientific method within the Kantian architectonic, as a kind of real belief in the category of "Intuitive Judgment" (*Anschauenden Urteilskraft*) that Kant had put forward as hypothetically possible, if unattainable for human, discursive understanding. Though Förster's essay gives a lovely account of many aspects of Goethean looking (see, in particular, the gloss on Goethe's attempts to construct experiments in series, connected in recollection), the Kantian lens transforms this refined empiricism into an idealism that seeks to transcend contingency and particularity toward "that generality that manifests itself in countless spatial-temporal variations and shapes, each of which empirically—that is, in a limited and imperfect way—represents the idea" (186). See also the thesis's long-awaited elaboration, especially vis-à-vis Goethe's Spinoza engagement, in Förster's *Twenty-Five Years of Philosophy*.

62. Eva Geulen "Serialization in Goethe's Morphology," 65. Please see Geulen's forthcoming *Aus dem Leben der Form* for the full and forceful elaboration of this much-needed argument.

63. Goethe, "Anschauende Urteilskraft," MA 12: 98.

64. See for instance Kant, *Critique of the Power of Judgment*, §28 [147, ed. Guyer].

65. Compare again Jennifer Mensch's *Kant's Organicism* and Helmut Müller-Sievers, *Self-Generation* for just how thoroughly the organicist logic of autotelic self-generation underwrites Kant's theory of cognition and rational freedom.

66. Goethe, "Das Unternehmen Wird Entschuldigt" (The Undertaking Is Excused), MA 12: 11. The German reads: "Wenn der zur lebhaften Beobachtung aufgeforderte Mensch mit der Natur einen Kampf zu bestehen anfängt, so fühlt er zuerst einen ungeheuren Trieb, die Gegenstände sich zu unterwerfen. Es dauert aber nicht lange, so dringen sie dergestalt gewaltig auf ihn ein, daß er wohl fühlt wie sehr er Ursache hat auch ihre Macht anzuerkennen und ihre Einwirkung zu verehren."

67. Ibid.

68. For these rival logics of epigenesis, see chapter 1 above.

69. "I will therefore take the liberty of representing my former experiences and observations [*Erfahrungen und Bemerkungen*], and the ways of thinking that sprung from them, historically in these pages [*in diesen Blättern geschichtlich darzulegen*]." Goethe, "Meaningful Progress," MA 12: 309.

70. See LA 2.10a: 904–9.

71. Goethe, "Meaningful Progress," MA 12: 306–9.

72. Goethe, *Gespräche mit Eckermann*, 1 February 1827.

73. Goethe, "Meaningful Progress," MA 12: 306. My translation here and throughout, although Bertha Mueller gives a rendering in *Goethe's Botanical Writings*, 235–37. Mueller's translation must be treated with some caution; for instance, it renders the key phrase "Jeder neue Gegenstand, wohl beschaut, schließt ein neues Organ in uns auf" as "Each new subject, well observed, opens up within us a new vehicle of thought" (MA 12: 306 [Mueller, 235]). Goethe in fact writes: "Each new *object*, well observed, opens up a new *organ* in us."

74. Coleridge observes: "The oddity of finding anything new in bringing the results of meditation to bear on *the notices of Observation*, and vice versâ . . . amuses me! What in the name of common sense is the meaning of the word, Contemplation [*Anschauung*], but this so quaintly called gegenständliche Thatigkeit [*sic*]"? (*Marginalia*, 2: 1025). Coleridge here misses

the opportunity Goethe finds in Heinroth's phrase for rendering thinking as saturated by "the notices of Observation," rather than as the prior "results of meditation" brought to bear on the present news. Glossing Heinroth again as the attempt "to unite activ[e] and passive Seeing in reference to the same Object" (1025), Coleridge also reveals that he thinks the force of Heinroth's (unoriginal) argument is to add more subjective *activity* to thinking, missing the issue to which Goethe is in fact sensitive: that it is precisely the *passivity* of Contemplation, the possibility of receptivity to realities freely "given," that is under threat of extinction in a post-Kantian climate. (An argument with profound affinities to Hannah Arendt's in *The Human Condition*. See Robert Mitchell, "Romanticism and the Experience of Experiment.")

75. Goethe, "Meaningful Progress," MA 12: 306.

76. As Frederick Amrine has written on the topic of Goethe's philosophy of science, "there is a real sense in which one *becomes what one perceives*"; or as Thomas Pfau puts it, "knowledge for Goethe is above all a sharing in the structure of appearance by way of sustained observation." This complication is fortunate, rather than disabling, for knowledge: "For Goethe," Amrine continues, "the conventional method of isolating phenomena is tantamount to wearing blinders . . . having already abstracted from the phenomena, one can no longer *develop with them*." Unlike Pfau and Amrine, however Goethe remains attuned to the deflationary senses of partiality within this participatory form of knowledge: to the belated, accidental, and incomplete aspects of mutually transfigurative perception—and the heterogenous form of periodical representation such empiricism elicits—rather than to its testimony to "nature's over-arching unity as a self-originating (epigenetic) and self-organizing totality of metamorphosis" (Pfau). See Amrine, "Metamorphosis of the Scientist," 40 and Pfau, "'All Is Leaf,'" 8–9.

77. Compare Joseph Vogl's emphasis on the *Ununterschiedbarkeitszone*, the zone of indistinguishability, between Subject and Object in a Goethean empiricism sensitive to the "ecstatic effects of the phenomenon." Yet I would counter the anamnestic temporality and inward trajectory Vogl ascribes to this possibility—"a path that leads from the realm of the unfolded world of appearance to the unborn life of things and beings," for which "each affection is auto-affection, each relation, self-relation"—with an emphatically late style. Joining defenselessness to wry sophistication, Goethe's attitude is closer to what Nietzsche called the "second, dangerous innocence" that emerges after lengthy and arduous trial: "[F]rom such abysses, from such severe sickness, also from the sickness of severe suspicion, one returns *newborn*, having shed one's skin, more ticklish and malicious [*gehäutet, kitzliger, boshafter*], with a more delicate taste for joy, with a tenderer tongue [*zarteren Zunge*] for all good things, with merrier senses, with a second dangerous innocence in joy, more child-like and yet a hundred times subtler [*raffinierter*] than one has ever been before" (*Gay Science*, 37–38 [*Die fröhliche Wissenschaft*, 10–11]). It is interesting that Nietzsche, a great admirer of what he called the "Goethean glance," soon after commends the capacity to honor skins, surfaces, and the visible that Goethe, as we have seen, positioned as the keystone of morphology: "What is required . . . is to stop courageously at the surface, the fold, the skin, to adore appearance, to believe in forms, tones, words, in the whole Olympus of appearance." See Vogl's "Bemerkung über Goethe's Empiricismus," 117, 121, 123. And Paul Bishop and R. H. Stephenson, *Friedrich Nietzsche and Weimar Classicism*.

78. This understanding of "objectivity" in the late journals converges with Stephenson and Bishop's work on Goethe and Schiller's classicism as against Jena Romanticism's self-reflexive

absolutes. Such classicism, they argue, consists in "a precise love" for "what is other . . . *as being an other being*. . . without the *arrière pensée* of 'thinking makes it so.'" The formulation is borrowed from Adrian Stokes; see Stephenson, "Cultural Theory of Weimar Classicism," 155–58.

79. Heinroth, *Lehrbuch der Anthropologie*, 105, 107, 110.

80. Ibid., 111.

81. Compare Breithaupt, *Jenseits der Bilder*, 16.

82. Hennig, "Goethes Begriff 'Zart,'" 79, 83.

83. Ibid., 78, 81, 92. The six-page list of natural objects to which Goethe applied the term ranges from "noodle-dish" and "Galvonometer" to "gypsum crystals"; it includes wave, cloud, light and shadow forms, a "leaf-fiber," and, of course, a "point" (89).

84. Ibid., 101–2.

85. Goethe, LA 2.10a: 112, emphasis added.

86. Thus Breithaupt has legitimately discerned in Goethean epistemology a "politics of perception" aspiring precisely toward "an escape from the dominance of the aesthetic": that is, an attempt to render contingent and negotiable the established conventions of seeing within any given epoch, *Jenseits der Bilder*, 9–20.

87. Amrine, "Metamorphosis of the Scientist," 36, 38.

88. As its title makes plain, Henri Bortoft's *The Wholeness of Nature: Goethe's Way toward a Science of Conscious Participation in Nature* sees Goethe's science as exemplifying "the principle of organic wholeness" (ix). The same can be said of many of the phenomenological approaches in *Goethe's Way of Science: A Phenomenology of Nature*, ed. Seamon and Zajonc, and of the *Janus-Head* issue dedicated to Goethe's "Delicate Empiricism," many of which are expressly indebted to Bortoft. Pfau's erudite reading of Goethe's morphological writings as temporalized differentiation links them splendidly to an Ovidian concept of general tropism and the "world as a welter of profoundly and vividly *related* things" ("All Is Leaf," 13). Yet he, too, ultimately positions Goethean morphology within a general Romantic "organicism" consonant with Hegelian temporality, Blumenbach's notion of an autopoetic *Bildungstrieb*, and Coleridgean observations on biological and textual bodies; see esp. 11, 18, 25, 30. As I argued at the close of the last chapter, Goethe links the release of the emphatically belated and incomplete *Morphology* periodicals to a fresh consciousness of the historical impossibility "of an all-encompassing One" in which biological and biographical forms might participate (Pfau, 12). As Goethe puts it in a melancholy "Zwischenrede" (Interlude): "Produced from alternating points of view . . . written down at different times, [these essays] will never again flourish as a unity." But as we have seen, this biographical form ("pieces of a human life," the "Zwischenrede" calls it) correlates to a revisionary biology that has ceased to require unity of life, and maybe even of its flourishing. See Goethe, "Zwischenrede," MA 12: 93, and compare Geulen, *Aus dem Leben der Form*, 69.

89. Amrine, "Metamorphosis of the Scientist," 41.

90. Among the many earlier works subject to reinflection and revision in Goethe's late periodical project was his great theory of optics and colors, the *Farbenlehre* of 1810. The *On Natural Science in General* portion of the late journal project immediately reopened the topic, and in the fourth issue a new set of investigations, essays, and schemata appear under the heading *Chromatik* (MA 12: 553–609). The series begins with a table entitled "Auge, empfänglich und

gegenwirkend" (Eye, receptive and reactive), and among the thirty-one short pieces that follow is one that confronts the link between life, materiality, and visibility in explicitly Lucretian terms: "27. Der Ausdruck Trüb" (The Expression Turbid), 603–6. "Turbidity" (*die Trübe*) is for Goethe the indispensable medium of sight and, he insists, the (notably clouded) "pure concept, upon which the whole *Color Theory* rests." The essay shares the morphological essays' permissive interest in life as a tissue in uneven progress toward the visible, and links *Trübe* to the French and English word "trouble."

91. "Der Ausdruck Trüb," MA 12: 604.

92. Ibid., 605–6, citing "what the Romans called *rarus* (Lucret. II. 106)."

93. Serres, *Birth of Physics*.

94. Goethe, "Attempt at a Meteorology," LA 1.11: 245.

CHAPTER 4

1. Montaigne, *Complete Essays*, 808 (translation slightly modified) [*Essais*, 3: 266].

2. Ibid., 809 [268].

3. Ibid., 809 [267].

4. Ibid., 796 [252].

5. Ibid., 796 [256].

6. Ibid., 809 [267].

7. Shelley, *Defence of Poetry*, §37, *Shelley's Poetry and Prose*, 530. Shelley's journals and Mary Shelley's reading list record reading Montaigne in September–November of 1816; a letter of 1818 also has him expecting the *Essais* from the binder (*Letters* 1: 480 and 2: 591). But more to the point is Montaigne and Shelley's mutual interest in Lucretian corporeality: Lucretius is Montaigne's second-most-cited poet, after Horace, and Shelley is known to have read *De rerum natura* in school and reread it in 1810, 1816, 1819, and 1820. See Ford, "Lucretius in Early Modern France," and Turner, "Shelley and Lucretius."

8. Paul de Man, "Shelley Disfigured," *Rhetoric of Romanticism*, 93–123.

9. De Man, "Autobiography as De-Facement," *Rhetoric of Romanticism*, 70. Turner seems to have first recognized the allusion, now a standard note in the Norton *Shelley*.

10. The scene invites us, as Marjorie Levinson has been doing with precision and imagination, to suspend a still regnant "presumption of representation" grounded in "the philosophically founding distinction between subjects and objects," a challenge that recovering the "pre- or non-Cartesian paradigms" operative within Romantic writing can help present-day readers meet. Such "weak" Romanticisms, in their nonmodern nonparticipation in that basic dialectical subject/object agon (and its self/other, nature/culture entailments), Levinson argues, resonate with renewed value as postclassical sciences and postmodern humanities grapple with the ecological crisis that may at last render that dialectic not just theoretically unconvincing but practically useless. "Of Being Numerous," 649, "Pre- and Post-Dialectical Materialisms," 115, 116, 118; see also her "Romantic Poetry: The State of the Art."

11. To my knowledge, only Paul Hamilton's "A French Connection: The Shelleys' Materialism" attends specifically (and splendidly) to the fact that what Shelley retrieves from *De rerum natura* for *The Triumph of Life* is Lucretius's materialist theory of signs. *Metaromanticism*, 139–55. Three attempts to grasp Shelley's whole oeuvre under the sign of "Lucretianism"

in various broad senses are noteworthy. Jerrold Hogle has Shelley's Lucretianism contribute to an overarching poststructuralist and psychoanalytic theory of the oeuvre as powered by a subliminal logic of "radical transference": the incessant, processual, "centerless displacement of figural counterparts by one another," rather than the adherence to a single philosophical system (skeptical, idealist, skeptical idealist) by which scholars long sought to explain Shelley's work (*Shelley's Process*, 10). The systematicity of *Shelley's Process* itself, however, puts the insight under strain: enfolding Shelley's whole oeuvre into a developmental narrative that culminates in *The Triumph*'s purported reworking of all prior transgressions, the book's ethos of relentless transformation identifies change with emancipation in a way *The Triumph* does not, and also too easily avows linguistic transpositions as "inherently iconoclastic and revolutionary" (14). For Hugh Roberts in *Shelley and the Chaos of History*, meanwhile, the Lucretian *clinamen* leads from Shelley's Enlightenment sources through Michel Serres and into contemporary chaos theory, affording to Shelley a theory of chaotic creativity that navigates between recuperative historicism and deconstructive skepticism. In *Shelley's Intellectual System and Its Epicurean Background*, Michael Vicario proposes another resolution of the persistent skepticism vs. idealism crux, arguing that the poet espouses a consistent "Platonic atomism" in the tradition of early modern Christianizing adaptations of Epicurean philosophy. Yet the part of Epicurean philosophy that *The Triumph* stages in slow motion was devised against the Platonic theory of forms. See Asmis, *Epicurus' Scientific Method*, 30, and Deleuze, "Simulacrum and Ancient Philosophy." This is not to say that Shelley was never a "Platonist" or "Platonic atomist"—but not here.

12. Coleridge, "Hints Towards the Formation of a More Comprehensive Theory of Life," *Coleridge's Poetry and Prose*, 597.

13. Hadot, *Veil of Isis*, 209. Among scholars of English Romanticism, Martin Priestman has done the most to restore this "second British Lucretian moment," roughly 1790 to 1820, through the lens of the nonconformists and freethinkers who translated, edited, embraced, and imitated the poem as a "scourge of religious orthodoxy" during a period of war with France that put their progressive rationalist ideology "under ever more furious attack as godless, pro-French and unpatriotic." Priestman, "Lucretius in Romantic and Victorian Britain," 289; and see his *Romantic Atheism*. Continuing to track, in particular, Romantic-era interest in the theory of figuration that requires and generates turbidity (rather than pure light) as the indispensible medium of knowledge, this chapter proceeds by casting Romantic neo-Lucretianism as a revisionary shadowing of that "bullish, confident," high-Enlightenment strain ("Lucretius in Romantic and Victorian Britain," 298). Sensitized to the violence that enlightening can do, this revisionary attitude instead looks to Lucretius to represent the astonishing complexity of history as collective action and passion.

14. Noel Jackson's "Rhyme and Reason" remains extremely insightful on the forthright instrumentalism of Darwin's neo-Lucretian didactics and the aestheticist reaction they provoked from some quarters; David Duff, however, fills in the other side of the antididactic story, including the ongoing strength and interest of Darwin's genre and the persistence of "didactic" by other names. See my chapter 5, Duff's *Romanticism and the Uses of* Genre, 111–12, as well as Priestman's important new monograph on Darwin, *The Poetry of Erasmus Darwin: Enlightened Spaces, Romantic Times*, which maps the astonishing extent of his literary network and ends with important suggestions about unlikely Romantic aftereffects. On freethinking and

antiquarian circles and their works and motives, see Priestman's *Romantic Atheism* and Heringman, *Sciences of Antiquity*.

15. Heringman's study of a different category of Romantic natural material—rocks—offers a key precedent for recognizing there a form and understanding of materiality "located precisely between the two materialities recently competing for the objects of Romanticism, that of the letter, and that of history." *Romantic Rocks*, 20.

16. "Sensitive Plant," line 69; *Prometheus Unbound* 3.4, lines 65, 67; I am thinking especially of stanzas 4–5 in "Ode to the West Wind," and in "Mont Blanc" of lines 37–48 (*Shelley's Poetry and Prose*, 288, 265, 300–301, 98). In some ways this chapter follows through on Shelley's rhetorical question about the similarity between botanical and bibliographic leavings in the "Ode to the West Wind," line 58: "What if my leaves are falling like [the forest's] own!" (punctuated as an exclamation, not a question). And on Frances Ferguson's revelation that "Mont Blanc" keeps positing epistemological questions in a "love-language" that admits the "inevitability of any human's seeing things in terms of relation." In light of Goethe's work on epistemic tenderness, we might read Ferguson's formulation as an insight into the material condition of seeing rather than as a unilaterally human inability to "tak[e] the material as merely material." See "What the Mountain Said," *Solitude and the Sublime*, 207, 211. This would be the option Shelley models in the second stanza of "Mont Blanc" in particular, where the expected opposition between active and passive aspects or accounts of mental process is instead depicted as a difference between modalities of reception: "my human mind, which passively / Now renders and receives fast influencings" (lines 37–38). Where "render" makes of the mind's giving a giving *back*, relationality ("unremitting interchange" [line 39]) is not mind's posit but its ontological condition, given by and shared with the rest of the "universe of things around" in a way that makes thought "wild" for the thinking subject (lines 39, 41).

17. A phrase from Shelley's fragmentary 1819 essay "On Life"—"We live on, and in living we lose the apprehension of life"—whose resonances Ross Wilson eloquently unfolds to monograph length in *Shelley and the Apprehension of Life*. (And see Shelley, "On Life," *Shelley's Poetry and Prose*, 506.) I appreciate Wilson's formulation that "Poetry and 'life' are, in Shelley's work, mutually disclosing," but would dissent from several key aspects of its rationale: the purported fixity and precision of scientific approaches to life, the animism he concludes subtends the mutual disclosure of poetry and life in Shelley, and the human exceptionalism with which he constrains qualitative nature to its apprehension "in *us*." See *Shelley and the Apprehension of Life*, 5–7, and compare 148, 161 for these positions' extension to *The Triumph of Life*, in particular.

18. See Meeker, *Voluptuous Philosophy*, which chronicles the flourishing and decline of neo-Lucretianism in eighteenth-century France. Meeker argues that over the course of the eighteenth century matter was purged of its figurative impact (and vulnerability to figuration) and became the inert object of a type of scientific inquiry—a development that simultaneously engendered literature as a supremely figurative domain autonomous from (but also impotent with regard to) material effects.

19. Coleridge, *Collected Letters*, 4: 761. For direct influence, we can speak mostly (but extensively) of Goethe's *Faust 1*, portions of which Shelley translated; see Burwick, *Damnation of Newton*, and Webb, *Violet in the Crucible*, chapters 4–5.

20. See especially Berman, *Social Change and Scientific Organization*, chapters 3–4, for the Regency production of professional science as "a general instrument for creating a smoothly functioning social order" (123) and the rising class of professionals who embodied it, in contrast to the regnant late eighteenth-century combination of aristocratic patronage and agricultural entrepreneurialism. For the ideology of service that eventually made scientific wage work into a motivated "vocation," see Duman, "Creation and Diffusion"; for a comparative perspective on the emergence of professional scientific research and its institutional supports, see Joseph Ben-David, "Profession of Science" and *Scientist's Role*. In "De-centring the 'Big Picture,'" Andrew Cunningham and Perry Williams argue that this early nineteenth-century moment deserves at least as much attention as the "Scientific Revolution": it was not until then that practitioners began to conceive of their activity as "science" (as opposed to natural philosophy), to posit its origins in the seventeenth century, and to have the option of pursuing it as a career. Cunningham and Williams stress the desacralizing theme of this transformation: "'science' was the new collective name of the new secular disciplines for studying the natural world as a secular object," for extracting its resources, "and for acquiring knowledge in a secular sense for material and social improvement" (424). On life science in the German university context, see Nyhart, *Biology Takes Form*. These accounts may, however, underestimate the way in which seventeenth-century experimentalism already understood itself as a vocation, and a profoundly corporate and collective one, albeit sometimes virtually so; see Picciotto, *Labors of Innocence*.

21. Shelley, *Defence of Poetry*, §39, *Shelley's Poetry and Prose*, 531. In the interest of keeping comparative momentum, I postpone until chapter 5 discussion of some specifically English literary-historical backgrounds for Shelley's neo-Lucretian experiments.

22. Goodman, *Georgic Modernity and British Romanticism*, 1–16; and see James Chandler's seminal case for English culture around 1819 as marked by a pressing, self-conscious preoccupation with "contemporaneity" that prefigures and still conditions rival modalities of critical historicism, including through the very idea of a "case" (*England in 1819*, esp. 94–154). Salient to both is Reinhart Koselleck's diagnosis of the heightened consciousness, after 1800, "of the noncontemporaneities which exist within chronologically uniform time," and of the obstacles to grasping one's own historical moment during the time of its occurrence; *Futures Past*, 246– 48. I take up the audible resonances with Derrida's *Specters of Marx* in the Coda.

23. Peacock, "Four Ages of Poetry," 24.

24. Ibid., 22, 24.

25. Abrams, *Natural Supernaturalism*, 441.

26. Chandler, *England in 1819*, 105–14, 23–46, 483–554. On the production of a "disembodied" Shelley through the posthumous editing of his works, see Fraistat, "Shelley Left and Right," and see Andrew Bennett's "Shelley and Posterity" and Karen Weisman's "Shelley's Ineffable Quotidian" on transience.

27. Rajan, *Supplement of Reading*, 351. De Man's influence over what Romanticists and other rhetorical readers mean by "figuration" and "materiality" is difficult to overstate; see especially Arac, *Critical Genealogies*; Wang, *Fantastic Modernity*, 37–68; and the essay collections *Material Events: Paul de Man and the Afterlife of Theory*, ed. Tom Cohen, and *Legacies of Paul de Man*, ed. Marc Redfield.

28. De Man, "Shelley Disfigured," *Rhetoric of Romanticism*, 116–17.

29. Ibid., 122.

30. Ibid., 120, 116, 122.

31. Ibid., 120.

32. Ibid. Carefully read, "Shelley Disfigured" instructed Romanticists at once in the most exclusively linguistic understanding of figuration and the most antifigural understanding of nonlinguistic "actual"/"material" events. In a final moment, language and material events are cast as very closely related indeed—but only once each has been recast as violently destructive of all relation. De Man prepares the initial polarity by construing figures as agents of cognition and signification, exemplified, above all, by the poem's dazzling and devastating "shape all light" (line 352). When this figure tramples "the gazer's mind/ . . . /into the dust of death," she reveals, for de Man, that figuration makes meaning not by mediation (whether aesthetic, sensuous, historical) between language and its others, but by the pure, violent, positing power of language considered "in itself" (lines 386, 388; *Rhetoric of Romanticism*, 116). Only a symmetrically "unrelated power" could interrupt that of "cognition and figuration" so conceived, and it is in such a capacity that de Man summons up the materiality of "actual events" in his essay. But in choosing Shelley's "defaced body" to exemplify material eventfulness, de Man has chosen an incident so brutal that it enables him to say of "actual events" exactly what he has said without them—that is, what he has said of "language considered by and in itself" (120, 116). Both exert, against the figural "illusion" of relation, a relationless power that "warns us that nothing . . . ever happens in relation . . . to anything that precedes, follows, or exists elsewhere" (122). As Jonathan Arac asks, "why choose to figure history in the rigor of a corpse?" Arac's answer: de Man "staged a history so horrid as to scare us back into the text." *Critical Genealogies*, 107.

33. Kaufman, "Aura, Still," 48; Theodor Adorno, "On Lyric Poetry and Society," 37–54. Kaufman's brilliant defenses of auratic, lyric autonomy by way of post-Kantian aesthetics converge with my reading in their end point, for in Kaufman "the new" that lyric's formal contentlessness discloses is defined not as radical vacancy, but as the "previously obscured aspects of the social" or "all that is emergent in the social" (51, 48). But neo-Lucretian materialism seeks frankly to visualize these liminal and emergent contents without passing through the paradoxical moment of lyric withdrawal (one reason why lyric, as I suggest in chapter 5, may not be the genre through which to understand Romanticism's Lucretian moments). Also at stake is a touch-based account of sense and figuration that renders the Kantian aesthetic virtue of impartial, or "non-partisan" (Kaufman), contemplation quite literally impossible. See also Kaufman's "Legislators of the Post-Everything World."

34. Pyle, "Kindling and Ash." In such a reading, Pyle sensitively and unapologetically deploys what Simon Jarvis has since diagnosed as a materialism of "perfected disenchantment," which invokes "matter" as an all-purpose limit to ideological mystification, particularly pretenses at reference, knowledge, and sense (*Wordsworth's Philosophic Song*, 78). The specific atomist minima Shelley revives in the poem operate differently: in various concrescences, they constitute percepts, concepts, and images, as well as physiological bodies and poetic texts; they also work to "populate" rather than vacate the scene of perception. See, however, Pyle's more expansive reading of *The Triumph*'s "materialist poetics" in *The Ideology of Imagination*, where they open onto a protostructuralist "knowledge of the laws" of historical domination (128)—a point I try to make (by different means) in chapter 5.

35. *Triumph of Life*, lines 352, 386, 388; de Man, "Shelley Disfigured," *Rhetoric of Romanticism*, 115.

36. De Man, "Shelley Disfigured," *Rhetoric of Romanticism*, 94.

37. I allude to Jonathan Culler's classic essay on "Apostrophe" in *The Pursuit of Signs*, 142–43.

38. John Milton calls death a "Grim Feature" at *Paradise Lost* 10.279. See Monique Allewaert on the privileged role of synecdoche in materialist figuration, "Toward a Materialist Figuration," 68.

39. Wang, *Fantastic Modernity*, 55.

40. See Jacyna, "Immanence or Transcendence," and Packham, *Eighteenth-Century Vitalism*. In *The Politics of Evolution*, Adrian Desmond characterizes philosophical organicism in the English context as something of a calculated counterinsurgency campaign against the "leveling threat" of materialistic French comparative morphology (especially in the hands of a new cadre of non-Oxbridge-educated medical men agitating for reform of the profession).

41. William Lawrence and John Abernethy's dispute concerned whether this special power should be conceived as immanent to organization or as a divine "superaddition." For the Shelleys' relationship with Lawrence, Percy's early immersion in London's fractious medical community, and an excellent summary of the vitality controversy, see Sharon Ruston's *Shelley and Vitality*; also Marilyn Butler's seminal introduction to her edition of Mary Shelley's *Frankenstein*.

42. Guyer, "Biopoetics, no pag."

43. Blumenbach, *Über den Bildungstrieb*, 89.

44. See Lawrence's introduction of the term, from Treviranus, in the second and third of his *Lectures on Physiology, Zoology, and the Natural History of Man*, 52, 58. On his transmission of continental science and that science's connotations in London in the teens, see Desmond, *Politics of Evolution*.

45. Johann Friedrich Blumenbach, *A Short System of Comparative Anatomy*, and Lawrence's dedication to the *Lectures on Physiology, Zoology, and the Natural History of Man*, iii–iv.

46. Lawrence, *Lectures on Physiology, Zoology, and the Natural History of Man*, 67, 6.

47. Ibid., 59, 66, 68.

48. Ibid., 60.

49. Jacyna, "Immanence or Transcendence," 311–29.

50. See Robert Mitchell's distinction between prescriptive "theoretical" and open-ended "experimental" vitalisms in *Experimental Life* (4–10), and nuanced approaches to the history of literary vitalism in Packham, *Eighteenth-Century Vitalism*; Holland, *German Romanticism and Science*; and Reill, *Vitalizing Nature*.

51. Lawrence, *Lectures on Physiology, Zoology, and the Natural History of Man*, 86–87.

52. [George d'Oyley] quoted in Mary Shelley, *Frankenstein*, appendix C, 249.

53. Lawrence, *Lectures on Physiology, Zoology, and the Natural History of Man*, 95.

54. See Esposito, *Bíos*, chapter 4.

55. Compare Roberts on an anti-organicist "option" informing Shelley's work, which he credits to Diderot and connects to the Gothic tradition. *Shelley and the Chaos of History*, 239.

56. Hamilton aptly expresses Rousseau's "error" this way: "This is the error which allows life to be felt as an intolerable imposition from without and our best self to be conceived of as a consciousness anterior to all physical circumstance." *Metaromanticism*, 154.

57. Goethe, "Verstäubung, Verdunstung, Vertropfung," 210 (see chapter 2 above).

58. Wang, *Fantastic Modernity*, 37–68. Taking Napoleon's Arc de Triomphe to exemplify the takeover of French Revolutionary vocabulary by an imperial discourse, Wang argues that *The Triumph* collapses revolutionary and reactionary victory monuments in order to "dramatiz[e] what happens when a revolutionary discourse forgets its own rhetoricity," 63.

59. Compare Pyle, *Ideology of Imagination*, 120–25, on the spectators' false choice and its connection to Walter Benjamin. And see Robert Kaufman's reconstruction of a venerable tradition of "Left German Shelleyanism" stretching from Marx and Engels to Benjamin, Bertolt Brecht, and beyond; "Aura, Still" and "Intervention and Commitment Forever!"

60. Terada, "Looking at the Stars Forever," 279; Faflak reads *The Triumph*'s pageantry as keyed to the "nearly psychotic balance between maddening rigidity and barely controlled madness" characterizing Regency social life. "Difficult Education," 75, 73.

61. Shelley's sonnet, "Lift not the painted veil," line 1, *Shelley's Poetry and Prose*, 327.

62. Two valuable works that have lately taken up Shelley's relation to the new biology, Gigante's *Life* and Ruston's *Shelley and Vitality*, together devote fewer than five pages to the only one of Shelley's poems named expressly for their subject. But the early-century vitalist perspectives that Gigante and Ruston reconstruct to evaluate the poem justify, and indeed prescribe, this outcome, since *The Triumph of Life*, I would argue, is an experiment in refusing such vitalist epistemology. Ruston writes about the poem in terms that presuppose that a poem preoccupied with death and materialism cannot have taken the vitality controversy as its subject, and that, even within such a poem, Shelley's "vital" poetry must be working against such morbid preoccupations: "Even in this dark and pessimistic poem, where the materialist imagery used serves only to portray the world as one of death, vitality is possible through poetry," 180. The poem's intense interest in incidental, metonymic generativity is also at odds with Gigante's reading of Shelley's biopoetics as symbolic, vitalist, and not very distinct from Coleridge's (*Life*, 155–207). Ross Wilson's "Poetry as Reanimation" is sensitive to the value of "evanescence" in Shelley but ultimately affirms the frequent judgment that where "life" is represented as "erosion," at issue is "death-in-life" and therefore "no life," 127–29. Robert Mitchell's *Experimental Life* does not take up the poem, but its discussions of Shelley on sensation, vital media, and "mutual atmospheres" nonetheless indirectly illuminate *The Triumph*.

63. Ross Wilson, *Shelley and the Apprehension of Life*, 5–7. But see 146, where Wilson concedes to Bichat some sympathetic sophistication on this score, albeit as an exception to scientific theory still construed as "a fixed set of relations around a conceptual core."

64. Lawrence, *Lectures on Physiology, Zoology, and the Natural History of Man*, 62.

65. Gigante, *Life*, 206.

66. See Barbara Johnson, *Persons and Things*, 4.

67. See Kramnick, *Actions and Objects*, 61–99; and Kennedy, *Rethinking Reality*. This technicality, however, does not prevent *De rerum natura* from being put to vitalist uses, including in recent new materialist revivals. See Steven Goldsmith's far-reaching critique of this trend in "Almost Gone" (important to this book's Coda).

68. Coleridge, "Hints," *Coleridge's Poetry and Prose*, 597; Shelley, *Triumph of Life*, line 466, *Shelley's Poetry and Prose*, 497.

69. In a kindred vein, albeit one that stays within the self-referential frame of an allegory of

reading, Rajan questions de Man's choice of a figural representative in *The Triumph*, pointing toward alternative figures for figuration in the poem that "[allow] form to emerge only through disfiguration." The poem, she observes, "cannot simply be reduced to its most traumatic images"; its depictions of figuration are "too complex to be summed up in the term 'effacement.'" *Supplement of Reading*, 327–28.

70. *DRN* 2.112–24.

71. On the movement from poetic image to causal link in this analogy and its centrality to Lucretius's scientifically correct poetics, see Marković, *Rhetoric of Explanation*, 93, 96.

72. Shelley, *Defence of Poetry*, §40, *Shelley's Poetry and Prose*, 532.

73. See Robert Mitchell, *Experimental Life*, 144–89, on the contamination between vital and communicative senses of "media" in the period.

74. Berlant, "Thinking about Feeling Historical," 5.

75. Shelley to Hogg, 6 July 1817, quoted in Turner, "Shelley and Lucretius," 269.

76. *DRN* 4.49–50, trans. Good. See Vicario, *Shelley's Intellectual System*, chapter 4, on Good's translation in the Shelleys' context.

77. One modality of Romantic distance-taking on the Enlightenment's Lucretius (see Coda).

78. Because coming into being as anything other than an individual atom occurs in the plural, by composition—atoms are, philosophically and etymologically, the only true "individuals" (indivisibles) in the universe—"to become" means, in *De rerum natura*, at once and immediately, "to begin to come undone." As Monica Gale observes, Lucretian science brooks "no separate forces of creation and destruction"; "Lucretius and Previous Poetic Traditions," 63.

79. William Hazlitt, quoted in Khalip, *Anonymous Life*, 117; Benjamin, *Arcades Project*, 370 [J81, 6], for which reference I am grateful to Kaufman's "Intervention and Commitment Forever!"

80. Keach, *Shelley's Style*, 73–78 and 44–45; Chandler, *England in 1819*, 554. Shelley and Blake, Keach observes, are more apt than Wordsworth and Coleridge to defy "the usual idea of metaphorical or figurative imagery," which "takes for granted a dualism of mind and matter" (45).

81. Bewell, *Romanticism and Colonial Disease*, 209, 218. Turner suggested the same, wondering lightly if Shelley's "tendency to attribute a kind of solidity to things of the mind" might not be "a stylistic habit caught from early and frequent study of a poet for whom . . . even thoughts and dreams are caused by material *simulacra* floating through the air." "Shelley and Lucretius," 282.

82. Shelley, *Defence of Poetry*, §48, *Shelley's Poetry and Prose*, 535. See, in addition to Noel Jackson's "Rhyme and Reason" and the sociologies noted above, two excellent new accounts of Romantic poetry and modern disciplinarity, Robin Valenza's *Literature, Language, and the Rise of the Intellectual Disciplines in Britain* and Jon Klancher's *Transfiguring the Arts and Sciences*. See also Raymond Williams, *Culture and Society*, 30–48, and the trenchant introductions to Noel Jackson, *Science and Sensation in Romantic Poetry*, and Maureen McLane, *Romanticism and the Human Sciences*.

83. Peacock, "Four Ages of Poetry," 11; *DRN* 2.1050–51 (2.1062, trans. Good). See chapter 3's discussion of this line.

84. Shelley, *Defence of Poetry*, §37, *Shelley's Poetry and Prose*, 530.

85. Peacock, "Four Ages of Poetry," 12; Shelley, *Defence of Poetry*, §44, *Shelley's Poetry and Prose*, 534. Eighteenth-century neo-Lucretianism, Meeker argues in the context of her study of its role in the French Enlightenment, "is the materialism that must be forgotten for the Kantian movement into maturity to begin," *Voluptuous Philosophy*, 13. Yet as we are beginning to see, this movement was neither as total nor as totally Kantian as it has seemed.

86. Lawrence, *Lectures on Physiology, Zoology, and the Natural History of Man*, 72.

87. This reading thus touches, by different ontological and epistemological reasoning and with different consequences, an insight about the relation between philosophy (by extension, natural philosophy and science) and literature that Cynthia Chase unfolds from de Manian disfiguration: "if it is impossible finally to locate a *proper* statement, an utterance proper to its speaker, a meaning coincident with intentions not qualified by the possibility of quotation, of an implied 'as if'—then the fictional status of enunciation in literary texts does not deprive them of the sharpest possible significance," *Decomposing Figures*, 4–6. In the alternative materialist reading developed here, literary fictions also attain the "sharpest possible significance": less because the figural aspect of language (as rhetorical, performative "force") *afflicts* the possibility of reference in any kind of discourse, making literary signification no less reliable than other kinds, but rather because the figural aspect of language (as rhetorical affection, perhaps, and in description) grants one possibility of reference by way of impropriety-in-common between language and its others.

88. Anne-Lise François notices that utterances about "the weather (most often rain, sometimes snow)" are paradigmatic examples of constative utterance: "the double object and subject—the subject without agency or action without an agent"; see her "Unspeakable Weather," 147–48.

89. The parallel to the "rain of resemblance" Levinson has disclosed in her neo-Spinozan reading of Wordsworth's daffodil poem is astonishing: here, too, "the self dissipates" into a "scene brought forth by the living air in which the blur of the self has melted" ("Of Being Numerous," 650). But the differences are also instructive: whereas the Spinozan plenist epiphany will move toward totality, timelessness, and continuous space, the Lucretian atomist picture remains constitutively shot through with the spatial and temporal discrepancies secured in the notion of "void." This would be one difference between the "living air" that suffuses "Tintern Abbey" and *The Triumph*'s "living storm."

90. Goldsmith, *Unbuilding Jerusalem*, 221, and chapter 4, "Apocalypse and Politics: Percy Bysshe Shelley's 1819."

91. *Prometheus Unbound* 2.4, 5–6, *Shelley's Poetry and Prose*, 246.

92. *Queen Mab* 8.116, 8.41, 8.12, 9.65–66, *Shelley's Poetry and Prose*, 63, 61, 67.

93. *Queen Mab* 4.153, *Prometheus Unbound* 3.4, 65, *Shelley's Poetry and Prose* 39, 265.

94. Hamilton, *Metaromanticism*, 146. Hamilton argues by way of La Mettrie and Lucretius that the final vision reveals "the sensuous production of all experience, a material production that gives the lie to skepticism and religious transcendentalism alike," in a like-minded interpretation that recognizes the possibility of a "materialist" reading of *The Triumph of Life* that sees the poem's end as "climactic rather than just interrupted" (152). Whereas Hamilton takes Shelley's depiction of rapid aging as a lapse in Lucretian "authentic creativity" from outgoing self-expression into "a depleting and repressive search for truth," I think *De rerum natura*'s account of simulacra production has led Shelley to depict "self-expression" as inevitably co-

incident with depletion (153–54). That is, "decay, degeneracy, the extinction of higher vision, mechanical subordination, and so on," are for me not just "demonizations" of this form of materialism, but irredeemable features of whatever consolation it provides (154).

95. An enthusiastic spokesman for Priestley's pneumatic enterprise, Walker "was aligning the campaign for aerial improvement with that against the corruption of an unrepresentative parliament and the established church." Golinski, *British Weather and the Climate of Enlightenment*, 165; see also his *Science as Public Culture*, chapter 4. Walker went on to publish a monograph exclusively devoted to the health effects of such all-pervasive air, as had Shelley's friend Thomas Forster in 1817: see Adam Walker, *A System of Familiar Philosophy in Twelve Lectures, Being the Course Usually Read by Mr. A. Walker* . . . (London, 1799) and *Analysis of A Course of Lectures in Natural and Experimental Philosophy*, as well as Thomas Forster, *Observations on the Casual and Periodical Influence of Particular States of the Atmosphere on Human Health and Diseases* (London, 1817). On Walker and Shelley's relationship, see Cameron, *Young Shelley*, 8, 80, 294; and Ruston, *Shelley and Vitality*, 33.

96. Favret, *War at a Distance*, 120.

97. Khalip, *Anonymous Life*, 14, 11.

98. Erasmus Darwin, *The Temple of Nature*, 1.247–50.

99. Priestley, "Observations and Experiments Relating to Equivocal, or Spontaneous Generation." Kant, *Critique of the Power of Judgment* §72 [263, ed. Guyer].

100. Khalip, *Anonymous Life*, 182–83, 185 (quoting Adorno, my emphasis). *De rerum natura*, Roberts astutely observes, tells "a history of errors becoming essence" that his *Shelley and the Chaos of History* spins out into a generative exploration of the middle space between teleological totality and sheer contingency in psychic, historical, artistic, and evolutionary dimensions. See, especially, *Shelley and the Chaos of History*, chapter 8. Using chaos theory to conclude that each death and each loss "becomes, through reiteration, a turbulent, negentropic 'gain'" (286), though, Roberts makes good on a kind of inevitable change to which Lucretius and Shelley do not ascribe value, positive or negative, and also undercuts the reality of specific, irrevocable loss that forms the difficult core and incessant refrain of Lucretian philosophical consolation (see Coda). More troubling is Roberts's characterization of personal openness to chance and change and to living "as fully as possible within the moments" (404, 407) as amounting to the "new *politics* of poetry" that his expansive study claims to seek—even as it has so very little to say about Shelley's concertedly political and topical poems.

101. François, *Open Secrets*, 21.

102. Shelley, *The Mask of Anarchy*, lines 147, 149, 151, *Shelley's Poetry and Prose*, 320.

103. On cleaving in this sense, I am indebted to Kevis Goodman; see her "Romantic Poetry and the Science of Nostalgia."

104. In light of Monique Allewaert's study of the kinds of personhood elaborated in colonial ecologies, moreover, we can discern in this "parahuman" figure something other than a patently essentializing association between the feminine, the animal, and the colonial subject under the sign of savage nature. Such heterogeneous personations might instead express the actual possibility of "tenuous and revolutionary alliance between tropical elemental forces and subaltern persons" that Allewaert attributes to persons' strategic exploitation of their intimacy with exploited ground: practices of personhood that contested majoritarian models of enlightened agency by making "disaggregation and dispersal" ontologically primary; (*Ariel's*

Ecology, 3). See, in this connection, Bewell's reading of *The Triumph*'s death-dance imagery as referencing the Asiatic cholera pandemic of 1817, Hastings's march through India, and their nostalgic representation in Thomas Medwin's *Pindarees*. The poem's "Rousseau," Bewell shows, is depicted as a colonial invalid: as specifically exposed, that is, to the kind of "clime" that threatened personally "disorganized and disorganizing" (*Ariel's Ecology*, 7) effects in the colonial imagination. Bewell, *Romanticism and Colonial Disease*, 209, 239.

105. Wordsworth, Preface (1802), Wordsworth and Coleridge, *Lyrical Ballads*, 258–60.

106. Wordsworth wielded the word *atmosphere* quite technically in the Preface. As Jayne Elizabeth Lewis shows, for eighteenth-century writers of fiction and of pneumatic chemistry alike, "atmosphere *was* appearance": the name for what happens when air *displays* itself through the absorption of "the Steams, Effluvia, and all the Abrasions of Bodies on the Surface of the Earth." In both novels and experimental philosophy, representing "atmosphere" amounted to "the making apparent, if not strictly visible, of an otherwise invisible medium." In terms of our inquiry, this would be another inescapable instance of the fidelity of figural techniques to the physical phenomena they sought to describe. Compare Lewis on the way "air grants writing that presents itself as essentially, not accidentally figurative—a claim to literal truth." Lewis, *Air's Appearance*, 3–4.

107. Wordsworth, Preface (1802), 259; Shelley, *Defence of Poetry*, §36, *Shelley's Poetry and Prose*, 530. On Wordsworth and the evolution of the intellectual labor market from the "gentleman amateur" model toward that of the "occupational professional," see Catherine E. Ross's, "Professional Rivalry," which explores the increasing competition between Wordsworth and Humphry Davy in response to these changes as motivated by initial "functional and philosophical *similarities* between poets and scientists" (35).

108. Hence the *Defence*'s famous conclusion that poetry's ongoing relation with specializing forms of "scientific and oeconomical knowledge" has to be worked out in the sphere of *legislation*; see chapter 5, below. Shelley, *Defence of Poetry*, §§37, 48 *Shelley's Poetry and Prose*, 530, 535. Shelley's focus on "legislation" picks up on the increasing involvement of science in government, an "administrative revolution" entailing a new "legal ideology of science" and conjoined to transformations in the profession of law (for example, the bureaucratization of medicine into public health); see Berman, *Social Change and Scientific Organization*, 107, 114. Berman identifies the Regency consolidation of this new sense of "science" in answer to ruling classes' experience of political insecurity and social siege ("widespread crime, disease, and social dislocation"): previously a leisure activity, rational entertainment, disinterested humanitarian or interested commercial venture, "science" is by 1820 a "tool by which to construct an ordered society," wielded by a class of professionals who believed and marketed the idea that "stability could be arranged or managed through technical know-how" (106, 109). In this sense *The Triumph* stands in the crosshairs between early and late Foucauldian preoccupations with "life" in the period—as an epistemic crux that reconfigures the organization of knowledge and a sphere that would elicit the elaboration of new, biopolitical technologies of governmental power.

109. Shelley, "The Sensitive Plant," line 69, in *Shelley's Poetry and Prose*, 288. See Robert Mitchell's insightful exposition in *Experimental Life*, which recognizes in such atmospheres an "intimacy," "collectivity," and ontological dependency that the Kantian category of aesthetic experience is ill-suited to grasp; 190–217, esp. 201–9. In *Clandestine Marriage*, 210–215, moreover, Kelley connects "The Sensitive Plant" to *The Triumph of Life* by way of "an earthbound"

form of uncompromisingly *"embodied life"* that humans are learning, in each case, from "the damaged life of plants": it is, after all, as an "old root" that "what was once Rousseau" teaches, despite himself, a lesson about human life "over and against romantic aspirations for forms of life beyond death" (215).

110. Goodman, *Georgic Modernity and British Romanticism*, 3. "What, once again, is physics?" asks Michel Serres in his monograph on Lucretius: "It is the science of relation." *Birth of Physics*, 123.

111. Benjamin, "On the Concept of History," thesis 2, 390; "Über den Begriff der Geschichte," 693–94.

112. An "atmospheric index," as Tobias Menely has put it in a searching and kindred essay on the "smoggy intermingling of human affairs and atmospheric phenomena" in Cowper's poetry, that serves as "an archive of the other times that remain." Menely, "'Present Obfuscation,'" 488–89. Menely's essay (which I regret not reading before the article version of this chapter went into print) sees Benjamin as invested in a "baseline of climatic continuity" that can be relied upon as history's counterpoint. Thus Benjamin trades nostalgically in an "Enlightenment division of knowledge [that] separates weather from history," the unwonted precipitate of which is an "unreadable human atmosphere" to which *The Task*, for instance, attests (480, 489–90). Whereas Menely stresses such atmospheres as obscuring the possibility of intelligible atmospheric signification—an older possibility of reading divine portents in the skies—in this chapter and the next, I am more interested in the way eighteenth- and nineteenth-century poems treat such apparitions as indexically significant in a mundane sense that is not stymied by the intermixture of nature and history (and indeed presumes and expects this "intermingling"). Despite the differences, there is something remarkable in our having both alighted on Benjamin's second thesis for taking in the natural and historical complexity of post-Enlightenment atmospheres.

113. Benjamin, "On the Concept of History," theses 17 and 2, 396 and 389; "Über den Begriff der Geschichte," 703, 693.

114. Benjamin, "On the Concept of History," thesis 2, 389–90; "Über den Begriff der Geschichte," 693.

115. Benjamin, "On the Concept of History," theses 17 and 16, 396; "Über den Begriff der Geschichte," 702.

116. Benjamin, "On the Concept of History," thesis 2, 390; "Über den Begriff der Geschichte," 694.

117. Benjamin, concerned throughout his writing with the question of how history imbues memory, explains in *Berliner Chronik* that his childhood images "belong" to the later decades of the nineteenth century: "not in the manner of general images, but of images that, according to the teaching of Epicurus, constantly detach themselves from things and determine our perception of them." Childhood, he continues, "having no preconceived opinions, has none about life. It is as dearly attached . . . to the realm of the dead, where it juts into that of the living, as to life itself." *Reflections*, 28–29.

118. Mill, "What Is Poetry?," 13.

119. Althusser, "Underground Current," *Philosophy of the Encounter*, 168–69. And recall, in this context, Keach's brilliant observations on *The Triumph*'s rhymes as "arbitrations of the arbitrary." *Shelley's Style*, 191.

CHAPTER 5

1. See especially Hogle, *Shelley's Process*; Priestman, *Romantic Atheism*; Roberts, *Shelley and the Chaos of History*; and Vicario, *Shelley's Intellectual System* (see chapter 4 above).

2. Shelley, *The Mask of Anarchy, Written on the Occasion of the Massacre at Manchester*, in *Shelley's Poetry and Prose*, 316–26. Line numbers are cited in text.

3. Important readings of *The Mask*'s aesthetics and politics from Susan Wolfson, Marc Redfield, Anne Janowitz, Steven E. Jones, Forest Pyle, Theresa Kelley, Steven Goldsmith, Joel Faflak, Kir Kuiken, and Matthew Borushko are engaged below, though this list is by no means complete.

4. Two generic relocations different from mine provide very refreshing views on this crux: Steven E. Jones's reading of *The Mask of Anarchy* in the technical terms of satire, and Anne Janowitz's in terms of the tradition of "communitarian lyric" that her groundbreaking study restores. I draw frequently on each throughout. See Jones, *Shelley's Satire: Violence, Exhortation, and Authority*, and Janowitz, *Lyric and Labour in the Romantic Tradition*.

5. Compare Lily Gurton-Wachter's powerful exposition of the "unexpected intersections of national security, perception, and natural history" in Charlotte Smith's "Beachy Head," and Kevis Goodman's analysis of the geological ground in that poem as an image of history's absent cause. Gurton-Wachter, *Watchwords*, 113, 110–39; Goodman, "Conjectures on Beachy Head." My thinking on the historical intelligence of natural history in *The Mask of Anarchy* is inextricably indebted to these projects; more on each below.

6. There is a much reiterated thesis that the earlier modern epistemic formation called "natural history" in fact possessed no capacity to treat nature historically. What is remarkable about the original contexts of this thesis is their near-total equation of "history" with a notion of organic "development." In Michel Foucault's *The Order of Things,* natural history's spatial vision of nature as a vast, visually available, and temporally synchronous plane of similarities and differences has to be shattered by the new biologies—by their invention of "life" as an inscrutable, internal relation in defiance of lateral relation among kinds—for "historicity" to be "introduced into nature." The organic being "wraps itself in its own existence, breaks off its taxonomic links of adjacency, tears itself free from the vast, tyrannical plan of continuities"— only thus is it enabled to enter the "great temporal current" of evolutionary time that Foucault has in mind as "a 'history' of nature" that deserves the name. See *The Order of Things*, 274–76.

Bringing the Foucauldian thesis to German sociology of science, Wolf Lepenies argued "the end of natural history" around 1800 under the pressure of vast new quantities of empirical information that could be systematized only through temporalizing disciplines. Like Foucault, he concluded that the idea of a "history of nature" was previously "not thinkable": classical natural history used the term *history* "without the thought of development [*Entwicklung*]," intending merely "the description of those bodies belonging to the kingdoms of nature" (*End of Natural History*, 30). But did the absence of history conceived *as development* really amount to the total absence of historical thinking in prior forms of natural history? What forms of natural historicity does this argument marginalize in its heroic, organicist aspiration to think the enabling conditions of evolutionary deep time? Compare Goodman, "Conjectures on Beachy Head," 992, on the way Foucault's thesis has overshadowed Charlotte Smith's form of dialectical and geohistorical intelligence.

7. Cowper, "Yardley Oak," lines 45–48.

8. Natural history, as James Secord rightly notices, persists to this day under the same name and others despite the fact that "it seems to have come to an end so often" ("Crisis of Nature," 449). Didactic poetry, as classicist historians of the genre Robert M. Schuler and John G. Fitch remark, has no more striking feature than "exuberant" persistence despite a millennium's worth of critical attempts "to denigrate, formalize, or confine the genre"; see "Theory and Context of the Didactic Poem," 2. On the Romantic version of didactic disparagement, see Noel Jackson's "Rhyme and Reason: Erasmus Darwin's Romanticism," which argues that neo-Lucretian didactic poetry was driven to extinction on account of its "instrumental aesthetic logic," which openly wielded poetic pleasure as a means to radical political ends (173, 177). As for natural history, Heringman, ed., *Romantic Science: The Literary Forms of Natural History*, collects a series of case studies in this capacious category's Romantic-era survivals; see Heringman's introduction for the methodological challenges natural history poses for present-day historicist criticism and its role in precipitating disciplinary divisions and fostering ongoing collaborations. And see Jardine, Secord, and Spary, eds., *Cultures of Natural History*, for essays on the discipline's material, social, and literary practices and functions from the sixteenth century to the present.

9. Virginia Jackson, *Dickinson's Misery*, 6.

10. Shelley to Hunt, 1 May 1820, *Letters*, 2: 191.

11. Susan Wolfson's chapter on *The Mask* in *Formal Charges*, Goldsmith's in *Unbuilding Jerusalem*, and Redfield's in *The Politics of Aesthetics* are particularly astute on each of these dilemmas, respectively.

12. Wolfson, *Formal Charges*, 198.

13. Keach, "Rise like Lions?"

14. See Archibald Prentice's *Historical Sketches and Personal Recollections of Manchester,* (1851), citing the published results of the citizens' committee that he and other residents established to "arriv[e] at an approximation to the extent of death and calamity inflicted" through "a careful and rigid inquiry . . . made for many successive weeks," 166–67. The committee also raised and distributed funds for the wounded: "They were disabled from work; not daring to apply for parish relief; not even daring to ask for surgical aid, lest, in the arbitrary spirit of the time, their acknowledgment that they had received their wounds on St. Peter's Field might send them to prison—perhaps to the scaffold," 166–67.

15. Sidmouth; cited here is his letter's reprinting in the Manchester serial, *Peter-Loo Massacre, containing a faithful narrative of events which preceded, accompanied and followed the fatal sixteenth of August, 1819, edited by an observer*; no. 2, 17.

16. *Peter-Loo Massacre*, no. 1, 10.

17. Chandler, *England in 1819*, 83, my emphasis.

18. Ibid., 82–83.

19. Steven E. Jones, *Shelley's Satire*, 102.

20. Kelley, *Reinventing Allegory*, 147, 96.

21. Redfield, *Politics of Aesthetics*, 159–60. More to the point is Walter Benjamin's praise of "Shelley's *grasp* of allegory"—he is writing of "Peter Bell the Third"—for the way it "incorporate[s] into itself the most immediate realities. . . . bailiffs, parliamentarians, stock-jobbers," and the like. I owe the connection to Robert Kaufman, yet would not read it as clearly

supporting "the auratic distance linked to lyric poetry and aesthetic autonomy": it seems instead that Benjamin is stressing proximity to the quotidian hubbub of politics and commerce. See Benjamin, *Arcades Project*, 370 [J81, 6], and Kaufman, "Intervention and Commitment Forever!," 10.

22. *The Medusa; or, Penny Politician* 1, no. 1 (20 February 1819): 1–2.

23. Anne Janowitz, *Lyric and Labour in the Romantic Tradition*, 102.

24. Shelley to Hunt, 3 November 1819: "we hear that a troop of the enraged master manufacturers are let loose with sharpened swords upon a multitude of their starving dependents & in spite of the remonstrances of the regular troops that they ride over them & massacre without distinction of sex and age, & cut off women's breasts & dash the heads of infants against stones." *Letters*, 2: 119.

25. Though often mistaken for a mode of giving body to transcendent, spiritual meaning, remarks Walter Benjamin in his study of the Baroque allegorical mode, the real function of allegorical personification is "to give the concrete a more imposing form by getting it up as a person"—to "see everything in terms of figures (not souls)." *Origin of German Tragic Drama*, 187.

26. Teskey, *Allegory and Violence*, 25.

27. Ibid., 23, 30, 181.

28. Ibid., 170.

29. Ibid., 16.

30. Kuiken, "Shelley's 'Mask of Anarchy,'" 95–106. And see his *Imagined Sovereignties*, which contextualizes this reading within a broader interpretation of the relationship between imagination and radical futurity in Shelley through the figure of metalepsis, 169–210.

31. Kuiken, "Shelley's 'Mask of Anarchy,'" 97.

32. Redfield, *Politics of Aesthetics*, 155.

33. Goldsmith, *Unbuilding Jerusalem*, 244–45, 257.

34. Borushko, "Violence and Nonviolence in Shelley's *Mask of Anarchy*," 105–6.

35. Ibid., 106.

36. Steven E. Jones, *Shelley's Satire*, 112.

37. Shelleyans will recognize recycled textual matter here, namely, *Laon and Cythna*'s governing emblem of the struggle between an imperial eagle and a popular snake. This emblem is also given as a backlit atmospheric spectacle that confuses wings and scales: "For in the air do I behold indeed / An Eagle and a Serpent wreathed in fight / . . . / Feather and scale inextricably blended" (1.8.3–4, 1.9.3, *Works*, 1: 112). The image is explained: "Such is this conflict—when mankind doth strive / With its oppressors in a strife of blood / . . . / The Snake and Eagle meet—the world's foundations tremble!" (1.33.1–2, 9, *Works*, 1: 120).

38. The "steam of gore" formulation appears in a draft manuscript version of the poem, quoted in Steven E. Jones, *Shelley's Satire*, 110. Jones links the vaporous Shape to the materials of popular graphic satires and optical illusions—"transparencies," graphic medleys, dioramas, and pantomimes—finely articulating what it would be tempting to call the "print-culture" dimension of Shelley's materialism, did Jones not connect print media so fluidly to visual and plastic mediation, 94–123.

39. Marx and Engels, *German Ideology*, *MER*, 154.

40. Shelley, *Philosophical View of Reform*, 47.

41. Shelley, Preface to *Prometheus Unbound, Shelley's Poetry and Prose*, 208.

42. Wordsworth, 1802 Preface, Wordsworth and Coleridge, *Lyrical Ballads*, 250.

43. See Kelley, *Reinventing Allegory*, 96.

44. Jerrold Hogle also recognizes, in the Shape, the softening presence of "Lucretian Venus" amid "the militance of the Freedom-figure mythologized in the French Revolution," "this demi-goddess neither looks to nor offers herself as a self-sufficient model for mimesis." *Shelley's Process*, 137.

45. *DRN* 2.23, trans. Good.

46. Shelley, *Defence of Poetry*, §48, *Shelley's Poetry and Prose*, 535.

47. Steven E. Jones, *Shelley's Satire*, 97–98, 112–15.

48. Lucretius teaches that the particles of mind are the finest and most sensitive of all, moved even by the light touch of the subtlest ambient simulacra: "light, innum'rous semblances of things / . . . / with ease / Pierce they the porous body, reach, recluse, / Th' attenuate mind, and stimulate the sense" (*DRN* 4.740, 4.746–48, 4.107–9, trans. Good). See also *DRN* 3.417–33.

49. See Kelley's hint about "a strange semiotic cross between iconicity and indexicality" in Romantic materialist allegory, and her suggestions that allegory needs to be understood in synecdochal and metonymic terms. *Reinventing Allegory*, 100, 47, 78.

50. Teskey, *Allegory and Violence*, 24.

51. In her ongoing and unsurpassed investigation of the historical production of the affectively "intimate publics," Lauren Berlant carefully marks the risk of "the displacement of politics to the realm of feeling," which at once opens a more aesthetically and pedagogically tractable dimension of politics and revives "the difficulty of inducing structural transformation out of shifts in collective feeling." *Cruel Optimism*, xii. *The Mask of Anarchy* is about this difficulty.

52. See Hörmann, *Writing the Revolution*, 91–148, for a recent evaluation especially attuned to the difference between radicalism in the political sphere and Marx's notion of social revolution.

53. Compare Jarvis, *Wordsworth's Philosophic Song*, 84–107.

54. *German Ideology, MER*, 155–56. That Shelley viewed "the multitude" coming into self-conscious visibility, victimization in 1819 as a historically new class, whose immiseration differed in kind and degree from prior poverty, is even more explicit in the draft prose treatise meant to accompany the popular songs. Since the previous English revolutionary period in the seventeenth century, explains *A Philosophical View of Reform*, "A *fourth class* [has] appeared in the nation . . . . a denomination which *had not existed before*" and has "no constitutional presence in the state": their "labor produces the materials of subsistence and enjoyment to those who claim for themselves a superfluity of these materials"; in fact, behind the monarchy, aristocracy, and government, "the power which has [really] increased is the power of the rich," making of monarchy merely "the string which ties the robber's bundle" (50–52). On early nineteenth-century theorizations of the "proletariat," see Hörmann, *Writing the Revolution*, 93.

55. *German Ideology, MER*, 155–56.

56. Foucault, "Society Must Be Defended," 239–40.

57. Nixon, *Slow Violence and the Environmentalism of the Poor*, 2.

58. Ibid., 2–3. In our context, Nixon's emphasis on "attrition" as an important tempo of violence suggests the aptitude of the Lucretian materialist imaginary to its perception and representation.

59. Hope's "father Time" emblematized their biopolitical predicament, supervising a pattern of premature mortality instead of generational cyclicity: "He has had child after child / And the dust of death is piled / Over every one" (lines 94–96).

60. Benjamin, *Origin of German Tragic Drama*, 167, 185, 182. The most complete recent study of Benjaminian natural history, Beatrice Hanssen's *Walter Benjamin's Other History: Of Stones, Animals, Human Beings, and Angels*, concludes that Benjamin's use of the term *Naturgeschichte* was generally pejorative, "essentially express[ing] the dehistoricization of history that baroque drama put on display" (50–53). But Hanssen perhaps too readily perpetuates Lepenies's account of natural history as "the atemporal, ahistorical conception of nature typical of the natural sciences and common before the advent of evolution theory" (50–53). In the case of Benjamin, a "*spatial*" approach ought not to be so easily dismissed, given the centrality of a historical materialist's practice of "constellation" that draws temporally dispersed elements into tense simultaneity. When Benjamin writes of a form of allegorical artistry that "recognized history" in nature's "eternal transience," he delineates, I would argue, a legitimately historical point of view from which the face of nature is not ahistorical by virtue of being spatial (179 [*Ursprung des deutschen Trauerspiels*, 355]).

In natural decay, Benjamin suggests, people, their effects, and their manufactures are caught becoming air and landscape, decomposing into the matter of the world: "in the process of decay, and in it alone, the events of history shrivel up and become absorbed in the setting" (179 [355]). From the point of view Benjamin calls "Baroque," then, as from the point of view this project has been investigating as neo-Lucretian, there are no properly raw materials for the composition of natural, verbal, or manufactured bodies—every particle has its history. And artistry, Benjamin explains, is less about expressive creativity than about the arrangement of historically saturated found materials. The allegorist, in Benjamin's account, is a skilled manipulator of these rich natural-historical materials rather than a creator who produces fresh truth from the depths of his interiority. On the allegorical stage, he writes in a difficult but crucial formulation, "*Auf dem Antlitz der Natur steht 'Geschichte' in der Zeichenschrift der Vergängnis*" (On the countenance of nature stands 'History' in the characters of transience). Baroque artistry, Benjamin suggests, performs an "allegorical physiognomy" of this "nature-history," a discursive interpretation that discloses the history in nature's face. What Benjamin calls the "*Zeichenschrift* of transience" manifest on the face of nature—where *Zeichen* (sign, mark, drawing, figure) attaches to *Schrift* (script) broad pictorial and plastic possibilities—is taken as an invitation to tell history in the partial, restrictively human alphabet that Benjamin marks off with quotation marks: "'History.'" *Origin*, 177–78 [353], translation modified.

61. Goodman, *Georgic Modernity and British Romanticism*; Favret, *War at a Distance* (especially "A Georgics of the Sky," 131–37). And see Menely on the haze in Cowper's georgic: "the ontic status of the haze is indeterminate: as sign or context, physical or cognitive condition, natural, divine, or human artifact"; "'Present Obfuscation,'" 485.

62. Goodman, *Georgic Modernity and British Romanticism*, 5–8, 105.

63. Ibid., 3–4, 12, 90, 98, 110, 126.

64. Ibid., 3. In this connection, Philip Hardie interrogates the prima facie contrast between *De rerum natura* and Virgil and Ovid's political and historical turns and investments, arguing for *De rerum natura* as "a major model for all exercises in culture history in the Augustan poets." See, in particular, Hardie's discussion of key topoi in Virgilian historicism as Lucretian adaptations. *Lucretian Receptions*, 13, 44–46.

65. This is a bit different from a formal mimesis of contents that leaves the gap between representation and its referent intact: from, for instance, the "similarity" Goodman uncovers between the geologically composite character of "the rock formation called Beachy Head" and the verbally composite character of "the poetic formation also called Beachy Head," wherein Smith's heterogeneity of literary genres, sources, and citations mirrors, in the realm of words, the mixed texture of the rock in the realm of things (without, however, allowing *these* two domains to have mingled or com-posed). See "Conjectures on Beachy Head," 995–96.

66. Thomson, "Summer," lines 1052–65. See Bewell, *Romanticism and Colonial Disease*, on the not merely imaginary threat of corporeal disintegration posed by colonial climates, and Allewaert, *Ariel's Ecology*, on such climates' potentially anticolonial collaborations with "parahuman" inhabitants. In a gesture that prefigures the consequential tumult of near and distant happenings in Shelley's *The Triumph of Life*, Thomson concludes that "the wide enlivening air is full of fate" ("Summer," line 1083); the passage is itself multiply indebted to the depiction of the airborne pestilence that culminates in the plague of Athens in *DRN* 6.1090ff.

67. Thomson, *The Seasons*, "Spring," line 858. See Goodman's "Conjectures on Beachy Head," alert to the poetics of "parting the elements" of composite forms in Charlotte Smith, rather than those elements' affective disruption of poetic mediation (1000). Given the legacy of Marxist structuralism and cultural studies that Goodman brings brilliantly to bear on Smith's geological poetics, I am prompted to think that we are beginning to face up to a hitherto unfashionable didactic and scientific dimension of materialist poetry in the Lucretian tradition that gives this poetry something in common with Marxist ethical criticism itself (as laid out, for instance, in Jameson's *The Political Unconscious*), rather than with the individual aesthetic forms it sought to grasp, symptomatically, as "imaginary or formal 'solutions' to unresolvable social contradictions" (Jameson, 79).

68. Darwin, *Temple of Nature*, 4.450.

69. Cowper, "Yardley Oak," lines 105–7, 111, 118, 44–48, 137–43. For the fuller reading Cowper's poem deserves (it is also an elegiac nationalist allegory for the "pulverised . . . superstructure" and tenacious "foundations" of the British body politic), see Tim Fulford, "Britannia's Heart of Oak" and "Cowper, Wordsworth, Clare."

70. Contrast Gurton-Wachter's superb reading of "Beachy Head," for instance, for which Smith's concerted avoidance of polemic and conjecture is crucial to the poem's nonviolence, to an ethos of just noticing that resists enlisting nature or history in *Polemos*, war. See *Watchwords*, 131–37. What follows instead risks, with Shelley, the irony of a polemic on nonviolence. Goodman, too, is interested in the way Smith eschews conjecture—the Enlightenment genre of conjectural history—in favor of a "literally conjectural poetics" styled after the composite character of the terrain; "Conjectures on Beachy Head," 994.

71. See Redfield, *Politics of Aesthetics*, 157.

72. A phrasing that makes clear a path not taken that would relate this phrase to Abel's

spilled blood and consider this materialist view of the earth, as Coleridge hints in the *Biographia Literaria*, as a secular corollary of the "Dread Book of Judgment," retaining a material record of every deed and trespass (114).

73. I borrow "affordances" in this sense from Caroline Levine, who herself imports it from design theory "to describe the potential uses or actions latent in materials and designs." Levine, *Forms: Whole, Rhythm, Hierarchy, Network*, 6.

74. Milton's description of the face of his explicitly Lucretian *Chaos* in *Paradise Lost* is perhaps the most efficient gloss on the hard philosophical compromise that predicates decay for composite forms, but grants permanence to their constitutive particles. *Chaos* is described as an "Anarch old / With faultring speech and visage incompos'd" (2.998–99). The adjective *incomposed* means, on the one hand, lacking composure, "discomposed," such that this aged face allegorically betrays the coming-undone that befalls each extant thing, each particulate arrangement, in (Lucretian) nature. But *incomposed* also means *un*compounded, simple, without components, so that in confronting this "old / . . . visage incompos'd," Satan is looking a Lucretian atom, in all its ancient and featureless permanence, in the face. What the line captures about Lucretian materialism is a view of physical being that sees present, apparently unitary objects as decadent forms that will always instantiate at least two kinds of age: the age of the object that presents itself and the diverse and interminable "punctual" (Cowper) histories of its component parts. Old things, as in Cowper's "Yardley Oak," are especially exemplary for this lesson. In fraying, "faultring," decomposing, a purported *in*dividual reveals its divisible status as contexture of multiple beings. The significant-yet-nonidentical likeness between the aging faces of things and their aged constituents is yet another way in which nonhuman and human bodies evince a rhetorical and figurative dimension, their tendency to trope.

75. Erasmus Darwin, *Temple of Nature* 4. 419–24.

76. See Gurton-Wachter on a Romantic poetic tradition that cultivates "a *double attention* strengthened rather than diminished by its division," not least by way of Erasmus Darwin's physiology of reading; *Watchwords*, 33, and chapter 1, 33–58.

77. On Darwinian personification and its discontents, see Noel Jackson, "Rhyme and Reason," and Packham, *Eighteenth-Century Vitalism*, chapter 5. (Apropos of Darwin, see Priestman's *Poetry of Erasmus Darwin* for an extraordinary rehabilitation of the enjoyments of the poem's spatial imaginary on every order of magnitude.)

78. De Quincey, "The Works of Alexander Pope," *Works*, 16: 339.

79. Wolfson, *Formal Charges*, 200, 196.

80. Ibid., 196. The poem's first lines: "As I lay asleep in Italy / There came a voice from over the Sea / And with great power it forth led me / To walk in the visions of Poesy."

81. Redfield, *Politics of Aesthetics*, 155–57.

82. Ibid., 159.

83. Pyle, "Frail Spells," 7, quoted in Redfield, *Politics of Aesthetics*, 151; but see Pyle's Marxian engagements in *Ideology of Imagination*.

84. Kaufman's trenchant essays on Shelley's lyric poetry argue, in an Adornian idiom, that Shelley's lyrics respond to the dilemma that "thought determined by society—by society's own concepts of itself . . .—can never give a satisfactory picture of that society" with scrupulously negative aesthetics whose contentlessness leaves space for something genuinely new to be con-

ceived. And he traces Frankfurt School aesthetics' intellectual debts to Shelley's poetry. See "Intervention and Commitment Forever!" ¶1.

85. See Kuiken, *Imagined Sovereignties*, 169–210, for an overview of and new contribution to these future-oriented arguments.

86. Mill, "What Is Poetry?" (1833) (later combined with "The Two Kinds of Poetry" and republished as "Poetry and Its Varieties"), 12–13.

87. See Virginia Jackson, *Dickinson's Misery*, 129–33.

88. Dalzell, *Criticism of Didactic Poetry*, 25, 70.

89. Ibid., 44, 25–26, 33, 8.

90. Ibid., 34.

91. I am thinking of how in "Tintern Abbey" and "Frost at Midnight," exemplary poems for M. H. Abrams's 1965 essay "Structure and Style in the Greater Romantic Lyric," what seems to be a solitary discourse on the relation between self and scene takes a turn toward instructing a less experienced person (who is neither self, nor audience, nor evidently absent, inanimate, or dead like the best exemplars of the counterfactual force of apostrophe). And I quote Maureen McLane, on Wordsworth's "We Are Seven"; *Romanticism and the Human Sciences*, 59.

92. Dalzell, *Criticism of Didactic Poetry*, 46; I refer to the speaking situations of Darwin's *The Temple of Nature* and Shelley's *Queen Mab*, respectively.

93. Duff, *Romanticism and the Uses of Genre*, chapter 3, "(Anti-)Didacticism"; Shelley, Preface to *Prometheus Unbound*, *Shelley's Poetry and Prose*, 209.

94. Ibid.

95. Ibid. Stephen C. Behrendt glosses Shelley's "coordinated assault" on multiple audiences in "forms ranging from the carefully reasoned philosophical prose of *A Philosophical View of Reform* to the broad dramatic satire of *Swellfoot the Tyrant*, from sensational Gothic romance to powerful stage tragedy, from the highly charged polemic of the exoteric political poems to the sublime forms and 'beautiful idealisms' of *Prometheus Unbound.*" *Shelley and His Audiences*, 165.

96. As Chandler showed, at stake was the making of a new, nationalist literary canon through a decades-long repudiation of the neoclassical, francophilic, artificial splendor of the eighteenth-century didactic and satirical verse forms otherwise known as "Pope." The "Pope Controversy," as *controversy*, is one proof that antididacticism is but half the story; Duff, moreover, points out that "didactic" was an insult Romantics frequently leveled at each other (not just their predecessors—here Wordsworth is a particular lightning rod) and summons an impressive quantity of evidence to show that Erasmus Darwin's didactic epics catalyzed a revival in early nineteenth-century Britain unequaled since the time of Pope himself. See Chandler, "Pope Controversy," 481–509; and Duff, *Romanticism and the Uses of Genre*, 111–12, 114.

97. *OED* cites Hannah More's 1799 usage as the first instance of this kind: "1b. Of a teaching method, teacher, etc.: that conveys knowledge or information by formal means such as lectures and textbooks, rote learning etc. Freq. contrasted (often unfavourably) with teaching methods encouraging greater involvement or creativity on the part of those being taught." But note that the broader, older neutral sense (1a), "Esp. of literature: intended to instruct; having instruction as a primary or ulterior purpose," adduces a 1780 example that already slights "The dry, didactic character of the Georgics."

98. Duff, *Romanticism and the Uses of Genre*, 108, 115.

99. De Quincey, "Notes on Gilfillan's Gallery of Literary Portraits" [John Keats], *Works*, 15: 300–301, 301n.

100. Mill, "What Is Poetry?," 12–13.

101. De Quincey, "The Works of Alexander Pope," *Works*, 16: 336.

102. Ibid., 337, 336.

103. Ibid., 359.

104. Ibid., 361.

105. Ibid., 363.

106. Ibid., 359–60.

107. McGrath, *Poetics of Unremembered Acts*, 3–4.

108. Coleridge to Wordsworth, September 1799: "I am anxiously eager to have you steadily employed on 'The Recluse' . . . I wish you would write a poem, in blank verse, addressed to those, who, in consequence of the complete failure of the French Revolution, have thrown up all hopes of amelioration of mankind, and are sinking into an almost epicurean selfishness, disguising the same under the soft titles of domestic attachment and contempt for visionary *philosophes*. It would do great good, and might form a Part of 'The Recluse.'" Coleridge, *Collected Letters*, 1: 527.

109. For an overview of the pantheism controversy surrounding Spinoza, see Frederick C. Beiser, *Fate of Reason*, 44–92; also Zammito, *Genesis of Kant's Critique of Judgment*, 228–47; Marjorie Levinson's restoration of this "submerged philosophical context" and its estranging power in Wordsworthian lyric in "Of Being Numerous" and "A Motion and a Spirit"; and, with relation to Shelley and historical imagination, Chandler, *England in 1819*, 51–93 and 545–49.

110. A line by Coleridge in Southey's *Joan of Arc, an Epic Poem* (1796), book 2, line 37, later reworked to open Coleridge's "The Destiny of Nations" in *Sibylline Leaves* (1817). In *The Fall of Robespierre: Contexts*, edited by Daniel E. White with Sarah Copland and Stephen Osadetz, Romantic Circles Electronic Editions, March 2008, https://www.rc.umd.edu/editions/robespierre/joan.html

111. *DRN* 2.991–1001.

112. To return to the longer citation, De Quincey also suppresses the fact that physical hearts were lately quite notorious for continuing to pulsate separate from their animal systems.

113. "et elucet natura profundi," 2.1050–51 (2.1062, trans. Good).

114. De Quincey, "Notes on Gilfillan's Gallery of Literary Portraits" [John Keats], *Works*, 15: 300, my emphasis.

115. Ibid., 301, 301n.

116. De Quincey, "The Works of Alexander Pope," *Works*, 16: 336.

117. De Quincey, "Notes on Gilfillan's Gallery of Literary Portraits" [John Keats], *Works*, 15: 301.

118. "the powerful nature disdained the skill . . . and triumphed by means of mere precipitation, of volume and of headlong fury." Ibid.

119. Ibid.

120. Ibid.

121. See Coda.

122. See Goldsmith, "Almost Gone."

123. Shelley figures prominently in Jacques Khalip's theory of the second-generation Romantic ethos of anonymous self-dispossession, against coercive identifications of sympathy and solidarity; see *Anonymous Life*, 1–24, and 97–103. See, among many luminous moments, Khalip's commentary on what one Shelley letter calls "*self*—that burr that will stick to one," 118. On Shelleyan poetry as/and the "Spirit of the Age," see Chandler's magisterial argument to that effect in *England in 1819*, in particular the reading of the "Ode to the West Wind" that discovers a Shelley "led by the events of post-Revolution history to construct an account whereby he and post-Revolutionary history make each other," 554.

124. Shelley, *Defence of Poetry*, §21, *Shelley's Poetry and Prose*, 523.

125. Despite Shelley's famous Hellenism, Jonathan Sachs argues, "contemporary Britain as it actually exists is commonly understood through Rome" in Shelley's oeuvre. *Romantic Antiquity*, 12.

126. Wolfson, *Formal Charges*, 202.

127. See Redfield's very similar reading, *Politics of Aesthetics*, 158.

128. Ibid., 159.

129. This sense increases when one reads the eyewitness memoir and newspaper accounts from which Shelley has drawn very closely, even in moments that seem most flagrantly fictional or poetic. Archibald Prentice (*Historical Sketches and Personal Recollections of Manchester*), for instance, attests to sounds much like Shelley's "Let the tyrants pour around / With a quick and startling sound / Like the loosening of a sea," and even like the poem's famous image of the slaughter as "Eloquent, oracular; / A volcano heard afar."

130. Lindstrom, *Romantic Fiat*, 180.

131. The full third thesis reads: "The materialist doctrine that men are products of circumstances and upbringing, and that, therefore, changed men are products of other circumstances and changed upbringing, forgets that it is men who change circumstances and that it is essential to educate the educator himself. Hence, this doctrine necessarily arrives at dividing society into two parts, one of which is superior to society. The coincidence of the changing of circumstances and of human activity can be conceived and rationally understood only as revolutionising practice." Karl Marx, "Theses on Feuerbach," *MER*, 144.

132. Freire, *Pedagogy of the Oppressed*, 39, 54.

133. Ibid., 99, 106–7.

134. Ibid., 101, 70.

135. Ibid., 73.

136. Janowitz, *Lyric and Labour in the Romantic Tradition*, 105.

137. Wolfson, *Formal Charges*, 199, citing Thomas R. Edwards, *Imagination and Power: A Study of Poetry on Public Themes*.

138. See Janowitz, *Lyric and Labour in the Romantic Tradition*, 62–112; Hörmann, *Writing the Revolution*, 125–29.

139. "E. J. B. [Edward James Blandford]," "A REAL DREAM; *Or, Another Hint for Mr. Bull!*," *Medusa*, 17 April 1819, lines 56, 60, 61.

140. *Medusa*, 1 May 1819.

141. Janowitz, *Lyric and Labour in the Romantic Tradition*, 102.

142. Keach, *Arbitrary Power*, 147. For instance, from the fact that the poem's opening masquerade does leave "Anarchy" "dead" and "Hope" walking "ankle-deep in blood." *Mask of Anarchy*, lines 131, 127.

143. Keach, *Arbitrary Power*, 148; and see Wolfson, *Formal Charges*, 197.

144. Goldsmith, *Unbuilding Jerusalem*, 15, 252.

145. Ibid., 252.

146. "Julianus Probus," "A VIEW OF THE FUTURE STATE OF THE COUNTRY AND ARRIVAL OF THE DAY OF RECKONING," *Medusa*, 24 April 1819.

147. Goldsmith, *Unbuilding Jerusalem*, 252–53.

148. Butler, *Notes toward a Performative Theory of Assembly*, 8.

149. Ibid., 158–61.

150. Ibid., 18, 11.

151. Ibid., 75, 18.

152. Teskey, *Allegory and Violence*, 24.

153. Shelley, *A Philosophical View of Reform*, 77.

154. Borushko in fact casts "the aesthetic *as* nonviolence" (my emphasis), its only pure expression: "the poem resists the barbarism of late 1819 by staking out its autonomy from the coercive methods of practical politics, from the self-defeating violence of revolutionary politics—from, in short, the entire economy of instrumental rationality." "Aesthetics of Nonviolence," ¶¶ 3, 12.

155. Richard B. Gregg, *Power of Nonviolence*, 44–45. And at worst, the oration's unflinching commitment to nonviolence seems to verge on a sadomasochistic will to witness the sacrifice of another unarmed multitude:

"Let the fixed bayonet
Gleam with sharp desire to wet
It's bright point in English blood
Looking keen as one for food.

"Let the horsemen's scimitars
Wheel and flash, like sphereless stars
Thirsting to eclipse their burning
In a sea of death and mourning.

"Stand ye calm and resolute,
Like a forest close and mute,
With folded arms and looks which are
Weapons of unvanquished war,

. . .

"And if then the tyrants dare
Let them ride among you there,
Slash, and stab, and maim, and hew,—
What they like, that let them do.

"With folded arms and steady eyes,
And little fear, and less surprise
Look upon them as they slay
Till their rage has died away."
    (lines 311–22, 340–47)

156. Here is the full text of the stanzas in question:

"Let the Laws of your own land
Good or ill, between ye stand
Hand to hand, and foot to foot,
Arbiters of the dispute,

"The old laws of England—they
Whose reverend heads with age are grey,
Children of a wiser day;
And whose solemn voice must be
Thine own echo—Liberty!"
    (lines 327–35)

157. Shelley, *Philosophical View of Reform*, 80.

158. Ibid. It is something of a critical commonplace that Shelley's 1819 poems and *A Philosophical View of Reform* are ambivalent on the question of revolutionary violence. This judgment seems to me slightly but crucially misstated. These writings instead evince an unambivalent preference for something rigorously theorized as nonviolent revolution *and* an unambivalent awareness that such a question is not the author's to decide. What *A Philosophical View of Reform* calls "the power of decision" rests with a body that we saw the text identify as "A fourth class": a "denomination which had not existed before," has "no constitutional representation in the state," and differs from the *tiers état* that was the subject of the French Revolution (50–51). *A Philosophical View of Reform* does not waver in affirming the right of the working class to violent "insurrection" as "the legitimate expression of that misery" to which they are subjected: "the laboring classes, when they cannot get food for their labour, are impelled to take it by force" (58). In addition, just as *The Mask* recommends to assembled persons the persuasive power of a posture of overwhelming force held in ethical abeyance, *A Philosophical View of Reform* rhetorically wields the persuasive power of the as-yet-withheld right of revolution.

Change that "bears the aspect of revolution" is not what the author hopes for, but neither, he affirms twice in the essay, "can he hesitate under what banner to array his person and his power" should it come to pass: "the history of the world teaches us not to hesitate an instant in the decision" to choose "the mischiefs of uprising [and] popular violence" over "the mischiefs of tyrannical and fraudulent forms of government" (49, 74, 50, 48). But again, the "power of decision" does not reside with him and may be "already past" (74). If the decision *were* his, and its moment *not* past, he would accede to a course of incremental reform. But with the moment for reform past and the decision in no individual hands, Shelley also expresses his actual and present hope: a nonviolent revolution in which the existing government would stand down before an "unwarlike display of the irresistible number and union of the people" (81). If the exploited majority are "impelled by a uniform enthusiasm and animated by a distinct and powerful apprehension of their object—and full confidence in their undoubted power—the struggle is merely nominal." The empowered minority "perceive the approaches . . . of an irresistible force" and "divest themselves of their usurped distinctions; the public tranquility is not disturbed by the revolution" (76). Judgment on this proposal—as starry or clear-eyed, bad or good faith—probably depends on whether one believes, as Shelley did, in the necessity of nonviolence as the only way to outmaneuver the "secret sympathy between destruction and power, between monarchy and war"; and, what is more difficult, whether one is willing to accept the violence on the part of the state and the casualties on the part of the demonstrators—"let them hack, and maim, and hew"—that principled "non-violence" expects, tolerates, and puts to use (82).

159. See Coda on this feature of *De rerum natura.*

CODA

1. "*Die Sinnlichkeit des Menschen ist also die verkörperte Zeit, die existierende Reflexion der Sinnenwelt in sich.*" Marx, "The Difference between the Democritean and Epicurean Philosophy of Nature" (*Differenz der demokritischen und epikureischen Naturphilosophie nebst einem Anhang*), hereafter "Dissertation," *MECW*, 1: 64 [*MEGA*, 1.1: 50].

2. The caveat most concisely and famously formulated in *The Eighteenth Brumaire of Louis Bonaparte* (see below), but already at work in *The German Ideology*; see *MER*, 165, for example.

3. Marx, "Dissertation," *MECW*, 1: 65.

4. *DRN* 1.459–63, Loeb translation slightly modified.

5. Marx, *Eighteenth Brumaire of Louis Bonaparte, MER*, 595, translation modified: "Die Menschen machen ihre eigene Geschichte, aber sie machen sie nicht aus freien Stücken, nicht unter selbstgewählten, sondern unter unmittelbar vorgefundenen, gegebene und überlieferten Umständen," *MEGA*, 8: 115.

6. Althusser, "'1844 Manuscripts' of Karl Marx," *For Marx*, 159. For Althusser, the 1844 Manuscripts represent "the Marx *furthest from Marx*" above all because of their apparently uncritical *humanism*, because their lynchpin notion of "alienated labour" seems sanctioned by an "*essence of Man*" taken for prior, universal, and permanent. See also "Marxism and Humanism," *For Marx*, 226–27.

7. Althusser, "Introduction: Today," *For Marx*, 33. In the essays of the early 1960s gathered in *For Marx*, Althusser posited "an unequivocal 'epistemological break'" in 1845, separating "the 'ideological' period before and the scientific period after," with *The German Ideology* and the *Theses on Feuerbach* amounting to "Works of the Break." See *For Marx*, "Introduction: Today," "For My English Readers," and "On the Young Marx," 34–35, 12–13, 49–86.

8. Central to Marx's (ungenerous) confrontation with Hegel in the 1844 Manuscripts is the polemical revaluation of the "*suffering*, conditioned, and limited" human-as-animal ("species being") and human-as-object as something other than errors and liabilities to be overcome (*MER*, 115). Vindicating this dimension of human being, Marx uses the same words Goethe did—"*objective* activity"—likewise equating the sensuous ability to experience/examine objects to the ability "oneself to be object," that is, to *passionate* susceptibility to another's action rather than to dispassionate judgment: "As soon as I have an object, this object has me for an object . . . . To be sensuous is to *suffer* [*Sinnlich sein ist leidend sein*]. Man as an objective, sen-

suous being is therefore a *suffering* being . . . a *passionate* being. Passion is the essential force of man energetically bent on its object" (115–16). As for Blake, Shelley, and Goethe, making dependency and affection ontologically primary in this way means accepting for the human animal an *"inorganic* body" (75), in technical contradistinction to the word that, as the reader knows too well by now, designated the living body in its causal self-sufficiency and autonomy. Thus Marx writes in the Manuscripts of "Nature" as "man's *inorganic body*. . . . with which he must remain in continuous intercourse if he is not to die. That man's physical and spiritual life is linked to nature means simply that nature is linked to itself, for man is a part of nature" (75).

9. See Althusser, "Marxism and Humanism," *For Marx*, 224, 228.

10. *MER*, 90–91. See Foster, *Marx's Ecology*; Schmidt, *Concept of Nature in Marx*—and, more stridently, John L. Stanley's *Mainlining Marx*. Seeking to correct the "anti-biological phobia" and asceticism of the Western Marxist tradition, Sebastiano Timparano radically re-appropriates the naturalism typed as "vulgar" as a means of reorienting materialism toward "man's biological frailty," "the element of passivity in experience," and the "dependence on na-ture [that] persists amid our social life." He offers a *pessimistic* solidarity in limitations on sheer self-determination for which he credits Lucretius, Giacomo Leopardi, and the eighteenth- and nineteenth-century hedonistic materialisms Marx and Engels may have too hastily dismissed as "bourgeois." *On Materialism*, 214, 208, 18, 34, 45, 49, 67.

11. Marx, "Preface to the First German Edition of *Capital*," *MER*, 297.

12. Marx, *German Ideology*, *MER*, 165, 157 [*MEGA*, 3: 40, 3: 30].

13. Owing to the neglect of "the production of life, both of one's own labour and of fresh life in procreation" in history as previously conceived, Marx and Engels argue in *The German Ide-ology*, "the real production of life seems to be primeval history, while the truly historical appears to be separated from ordinary life, something extra-superterrestrial [*Extra-Überweltliche*]. With this the relation of man to nature is excluded from history and hence the antithesis of nature and history is created." See *German Ideology*, *MER*, 157, 165 [*MEGA*, 3:30, 3:40].

14. Levinson, "Pre- and Post-Dialectical Materialisms," 112–13.

15. Something closer to this version of Nature—at once more powerful and more vulner-able than its Newtonian image as "law"—might be the figure that is revisiting theorists of the anthropocene and the post-humanities as they transgress that taboo: "Quite a Trickster, this Gaia," comments Bruno Latour, in "Waiting for Gaia," 9. And compare Donna Haraway on "Coyote or Trickster" as an alternative to the "mother/matter/mutter" complex for addressing the earth and other bodies in their capacity as "material-semiotic-actor"; "Situated Knowl-edges," 593–96.

16. Compare the fresh emphasis on questions of signification in Serenella Iovino and Serpil Oppermann's introductory essay to the volume *Material Ecocriticism* (2014), "Stories Come to Matter," 1–17, for instance, to the (important) reaction against "a cultural turn that privi-leges language, discourse, culture and values" expressed in Diana Coole and Samantha Frost's introduction to the first anthology of new materialist approaches just two years earlier, *New Materialisms: Ontology, Agency and Politics*, 3. Monique Allewaert's *Ariel's Ecology* is perhaps the fullest realization of this possibility.

17. Marx, *German Ideology*, *MER*, 156 [*MEGA*, 3: 28]; and see *MECW*, vol. 1, and *MEGA*, vol. 1.2, for the poems.

18. Althusser's editors note that Marx's dissertation was at one phase acknowledged in a

footnote, though it was left out of later versions of the work. There Althusser referred to Marx's thesis as "a splendid piece of nonsense, which the thought of his 'youth' made inevitable: an interpretation of the 'clinamen' as freedom." *Philosophy of the Encounter*, 206n50.

19. Althusser, "Underground Current," *Philosophy of the Encounter*, 190.

20. Ibid., 167–68, 195–96. "That is," Althusser continues in terms examined in chapter 2 above, "instead of thinking contingency as a modality of necessity, or an exception to it, we must think necessity as the becoming-necessary of the encounter of contingencies," 193–94.

21. Marx, "Theses on Feuerbach," *MER*, 143.

22. Marx, "Dissertation," *MECW*, 1: 65 [*MEGA*, 1.1:50], trans.mod.

23. Ibid.

24. Ibid.

25. Marx, "Economic and Philosophic Manuscripts of 1844," *MER*, 90; *German Ideology*, *MER*, 154, 170–71.

26. "Dissertation," *MECW*, 1: 65. On Goethe and Haraway, see chapter 3, above.

27. "Dissertation, *MECW*, 1: 65.

28. Marx and Engels, *The Holy Family* (*Die Heilige Familie*) 6.3.3, "Critical Battle against French Materialism" (*Kritische Schlacht gegen die französische Revolution*), *MEGA*, 2: 136. Designating materialism the *"natural-born* son of Great Britain" (as opposed to metaphysical France), Marx and Engels praise Bacon, in his new scientific inductive method and his admiration of classical atomism, as "the real progenitor of *English materialism* and all *modern experimental science.*" In Baconian experimentalism, Marx and Engels observe, matter is afforded movement, and not in a *"mechanical* and *mathematical"* way. Groping though the words "drive" (*Trieb*), "living spirit" (*Lebensgeist*), and "tension" (*Spannkraft*) to describe the way this movement is conceived, they at last land on an expression from Jakob Böhme that opens the strange dimension of material passion (not just action) I have been arguing for. Material movement is conceived "als *Qual*—um den Ausdruck Jakob Böhmes zu gebrauchen—der Materie" (as an agony of matter, to borrow Jakob Böhme's expression).

29. See Introduction.

30. "Economic and Philosophic Manuscripts of 1844," *MER*, 87.

31. *German Ideology*, MER, 170–71.

32. "Economic and Philosophic Manuscripts of 1844," 89.

33. William Blake, *Milton*, 27.29–30; Lamarck, *Recherches sur l'organisation des corps vivans*, 44. As Noel Jackson has pointed out, Marx's observation relies on the Romantic period's "foundational understanding of the sensorium as a historically mutable object," *Science and Sensation*, 17–18.

34. Derrida, *Specters of Marx*, xix.

35. Marx, *Eighteenth Brumaire of Louis Bonaparte*, *MER* 595.

36. Goldsmith, "Almost Gone," 417.

37. Ibid. "Consolation" is anathema to critique for many. But it might be important to flag a presumption linking materialists as different as Paul de Man and Fredric Jameson, that the most ethical materialist criticism will be the criticism that *hurts most*. If it is naively anthropocentric and covertly violent, as Lucretius and many others have taught, to believe that material processes were designed *for* our use and enjoyment, or progress naturally toward their im-

provement, perhaps it is equally so to simply invert the premise and presume that the unvarnished truth of the matter would give *no pleasure?*

38. John Dryden, "The Latter Part of the Third Book of Lucretius, against the fear of death" [What has this Bugbear death to frighten man] (1685), lines 38, 37. Furthermore:

> Nay, though our atoms should revolve by chance,
> And matter leap into the former dance;
> Though time our life and motion could restore,
> And make our bodies what they were before,
> What gain would all this bustle bring?
> The new-made man would be another thing;
> When once an interrupting pause is made,
> That individual being is decayed.
> (lines 19–26)

39. "Sunt igitur solida primordia simplicitate," *DRN* 1.548.

40. Deleuze and Parnet, *Dialogues II*, 15. Compare "The Simulacrum and Ancient Philosophy," 363n30. Many thanks to Alley Edlebi for the more extensive reference.

41. The attitude shares something of the bereaved yet militant optimism Anahid Nersessian calls "adjustment": "the ability to receive loss as an ontologically positive entity integral to the material makeup of the world—to any world, including the better world of utopia." *Utopia, Limited*, 5.

42. *German Ideology*, MER, 160.

43. Jameson, *Political Unconscious*, 38, 40.

44. Ibid., 40.

45. See, for instance, the sequence in *The Botanic Garden* that culminates in the production of the Wedgwood antislavery cameo: "The bold Cameo speaks" its protest by way of a geological and social production history through which the fossil recrements of primeval animals and plants, the spread of soft clays, England's competition with China in porcelain manufacture, the rediscovery of Etruscan pottery, technological innovation in glazes, and convict labor in Australia's Botany Bay shine in its commodified enamel. Canto 2, 6.277–320.

46. Jameson *Political Unconscious*, 19; Shelley, *Defence of Poetry*, §3, *Shelley's Poetry and Prose*, 512.

47. Jameson, *Political Unconscious*, 19.

48. Chakrabarty, "Climate of History," especially thesis 1: "Anthropogenic Explanations of Climate Change Spell the Collapse of the Age-Old Humanist Distinction between Natural History and Human History," 201–7.

49. Latour, *We Have Never Been Modern*, 6.

50. "Language as it has been spoken, natural creatures as they have been perceived and grouped together, and exchanges as they have been practiced," as Foucault puts the trifold manifestations of an episteme in *The Order of Things*, xxi.

BIBLIOGRAPHY

Abrams, M. H. *Natural Supernaturalism: Tradition and Revolution in Romantic Literature.* New York: Norton, 1971.

———. "Structure and Style in the Greater Romantic Lyric." In *From Sensibility to Romanticism: Essays Presented to Frederick A. Pottle,* edited by Frederick W. Hiles and Harold Bloom, 527–60. New York: Oxford University Press, 1965.

Ackroyd, Peter. *Blake.* London: Sinclair-Stevenson, 1995.

Adorno, Theodor. "On Lyric Poetry and Society." In *Notes to Literature,* vol. 1, edited by Rolf Tiedemann, translated by Shierry Weber Nicholson, 37–54. New York: Columbia University Press, 1991.

Allewaert, Monique. *Ariel's Ecology: Plantations, Personhood, and Colonialism in the American Tropics.* Minneapolis: University of Minnesota Press, 2013.

———. "Toward a Materialist Figuration: A Slight Manifesto." *MLN* 51.2 (Fall/Winter 2013): 61–77.

Althusser, Louis. *For Marx.* Translated by Ben Brewster. London: Verso, 2005.

———. *Philosophy of the Encounter: Later Writings, 1978–1987.* Edited by François Matheron and Oliver Corpet. Translated by G. M. Goshagarian. London: Verso, 2006.

———. *Reading Capital.* London: New Left Books, 1970.

Amrine, Frederick. "The Metamorphosis of the Scientist." In *Goethe's Way of Science: A Phenomenology of Nature,* edited by David Seamon and Arthur Zajonc, 33–54. New York: State University of New York Press, 1998.

Appel, Toby A. *The Cuvier-Geoffroy Debate: French Biology in the Decades before Darwin.* Oxford: Oxford University Press, 1987.

Arac, Jonathan. *Critical Genealogies: Historical Situations for Postmodern Literary Studies.* New York: Columbia University Press, 1987.

Arendt, Hannah. *The Human Condition.* Chicago: University of Chicago Press, 1958.

Armstrong, Charles I. *Romantic Organicism: From Idealist Origins to Ambivalent Afterlife.* New York: Palgrave Macmillan, 2003.

Asmis, Elizabeth. *Epicurus' Scientific Method*. Ithaca: Cornell University Press, 1984.

Auerbach, Erich. *Scenes from the Drama of European Literature: Six Essays*. Gloucester, MA: Peter Smith, 1973.

Azzouni, Safia. *Kunst als praktische Wissenschaft: Goethes Wilhelm Meisters Wanderjahre und die Hefte zur Morphologie*. Cologne: Böhlau Verlag, 2005.

Bach, Thomas. "Für wen das hier gesagte nicht gesagt ist, der wird es nicht für überflüssig halten. Franz Joseph Schelvers Beitrag zur Naturphilosophie um 1800." In *Naturwissenschaften um 1800: Wissenschaftskultur in Jena-Weimar*, edited by Olaf Breidbach and Peter Ziche, 65–82. Weimar: Böhlhaus, 2001.

Barad, Karen. "Posthumanist Performativity: Toward an Understanding of How Matter Comes to Matter." *Signs* 28.3, *Gender and Science: New Issues* (Spring 2003): 801–31.

Batsch, August J. G. K. *Grundzüge der Naturgeschichte des Gewächs-Reichs: Ein Handbuch für Lehrer auf Gymnasien, und für Naturfreunde zum eignen Unterricht*. Weimar: Verlage des Industrie-Comptoirs, 1801.

Beer, Gillian. *Darwin's Plots: Evolutionary Narrative in Darwin, George Eliot and Nineteenth-Century Fiction*. 3rd edition. Cambridge: Cambridge University Press, 2009.

———. "Has Nature a Future?" In *The Third Culture: Literature and Science*, edited by Elinor S. Schaffer, 15–27. Berlin: De Gruyter, 1998.

Behrendt, Stephen C. *Shelley and His Audiences*. Lincoln: University of Nebraska Press, 1989.

Beiser, Frederick C. *The Fate of Reason: German Philosophy from Kant to Fichte*. Cambridge, MA: Harvard University Press, 1993.

Ben-David, Joseph. "The Profession of Science and Its Powers." *Minerva* 10.2 (July 1972): 362–83.

———. *The Scientist's Role in Society: A Comparative Study*. Englewood Cliffs, NJ: Prentice-Hall, 1971.

Benjamin, Walter. *The Arcades Project*. Revised edition. Translated by Howard Eiland and Kevin McLaughlin. Cambridge, MA: Harvard University Press, 2002.

———. "On the Concept of History." *Selected Writings*, vol. 4: *1938–1940*, edited by Howard Eiland and Michael W. Jennings, 389–400. Cambridge, MA: Harvard University Press, 2006.

———. *The Origin of German Tragic Drama*. Translated by John Osborne. Introduced by George Steiner. London: Verso, 1998.

———. *Reflections: Essays, Aphorisms, Autobiographical Writing*. Edited by Peter Demetz. Translated by Edmund Jephcott. New York: Schocken Books, 1986.

———. *Selected Writings*. Edited by Marcus Bullock and Michael W. Jennings. 4 vols. Cambridge, MA: Harvard University Press, 1996–2003.

———. "Über den Begriff der Geschichte." *Gesammelte Schriften*, 1.2: 691–704. Edited by Rolf Tiedemann, Hermann Schweppenhäuser, Theodor W. Adorno, and Gerschom Scholem. 14 vols. Frankfurt am Main: Suhrkamp, 1991.

———. *Ursprung des deutschen Trauerspiels. Gesammelte Schriften*, 1.1: 203–409. Edited by Rolf Tiedemann, Hermann Schweppenhäuser, Theodor W. Adorno, and Gerschom Scholem. 14 vols. Frankfurt am Main: Suhrkamp, 1991.

Bennett, Andrew. "Shelley and Posterity." In *Shelley: Poet and Legislator of the World*, edited by Betty T. Bennett and Stuart Curran. Baltimore, MD: Johns Hopkins University Press, 1996.

Bennett, Betty T., and Stuart Curran. *Shelley: Poet and Legislator of the World*. Baltimore, MD: Johns Hopkins University Press, 1996.

Bennett, Jane. *Vibrant Matter: A Political Ecology of Things*. Durham, NC: Duke University Press, 2010.

Berlant, Lauren. *Cruel Optimism*. Durham, NC: Duke University Press, 2011.

———. *The Female Complaint: The Unfinished Business of Sentimentality in American Culture*. Durham, NC: Duke University Press, 2008.

———. "Thinking about Feeling Historical." *Emotion, Space and Society* 1 (2008): 4–9.

Berman, Morris. *Social Change and Scientific Organization: The Royal Institution, 1799–1844*. Ithaca, NY: Cornell University Press, 1978.

Bewell, Alan. "'Jacobin Plants': Botany as Social Theory in the 1790s." *Wordsworth Circle* 20.3 (1989): 132–39.

———. *Romanticism and Colonial Disease*. Baltimore, MD: Johns Hopkins University Press, 1999.

———. "Erasmus Darwin's Cosmopolitan Nature." *ELH* 76.1 (2009):19–48.

Bichat, Xavier. *Physiological Researches upon Life and Death*. Translated by F. Gold. Notes by F. Magendie. Translated by George Hayward. Boston: Richardson and Lord, 1827.

Bishop, Elizabeth. *Geography III*. New York: Farrar, Straus and Giroux, 1976.

Bishop, Paul, and R. H. Stephenson. "The Cultural Theory of Weimar Classicism in the Light of Coleridge's Doctrine of Aesthetic Knowledge." In *Goethe 2000: Intercultural Readings of His Work*, edited by Paul Bishop and R. H. Stephenson, 150–69. Leeds: Northern Universities Press, 2000.

———. *Friedrich Nietzsche and Weimar Classicism*. Rochester, NY: Camden House, 2005.

Blackmore, Richard. *Creation: A Philosophical Poem. Demonstrating the Existence and Providence of a God. In Seven Books*. 2nd ed. London: S. Buckley, and J. Tonson, 1712.

Blake, William. *Blake's Illuminated Books*. 6 vols. Gen. ed. David Bindman. Princeton: The William Blake Trust and Princeton University Press; London: The William Blake Trust and Tate Gallery Publications, 1991–95.

    "All Religions Are One." In vol. 3: *The Early Illuminated Books*, edited by Morris Eaves and Robert N. Essick, 21–69.

    The Book of Los. In vol. 6: *The Urizen Books*, edited by David Worrall, 195–224.

    Europe: A Prophecy. In vol. 4: *The Continental Prophecies*, edited by D. W. Dörrbecker, 141–283.

    The First Book of Urizen. In vol. 6: *The Urizen Books*, edited by David Worrall, 19–152.

    Jerusalem, The Emanation of the Giant Albion. In vol. 1, edited by Morton D. Paley.

    Milton a Poem. In vol. 5: *Milton a Poem and the Final Illuminated Works*, edited by Robert N. Essick and Joseph Viscomi, 9–217.

    Songs of Innocence and of Experience. In vol. 2: *Songs of Innocence and of Experience*, edited by Andrew Lincoln.

———. *Commercial Book Illustrations: A Catalogue and Study of the Plates Engraved by Blake after Designs by Other Artists*. Edited by Robert N. Essick. Oxford: Oxford University Press, 1991.

———. *The Complete Poetry and Prose, Newly Revised Edition.* Edited by David V. Erdman. Commentary by Harold Bloom. New York: Random House / Anchor Books, 1988.

Blandford, Edward James [E. J. B.]. "A REAL DREAM; *Or, Another Hint for Mr. Bull!*" *Medusa*, April 17, 1819. http://tinyurl.galegroup.com/tinyurl/3eUZV9

Blechschmidt, Stefan. *Goethes Lebendige Archiv: Mensch-Morphologie-Geschichte.* Heidelberg: Universitätsverlag, 2009.

Blumenbach, Johann Friedrich. *A Short System of Comparative Anatomy, Translated from the German of J. F. Blumenbach.* Translated by William Lawrence. 1819; reprint, Salem, 1828.

———. *Über den Bildungstrieb und das Zeugungsgeschäfte.* 2nd ed. Göttingen: Johann C. Dieterich, 1789.

Böhme, Gernot. *Alternativen der Wissenschaft.* Frankfurt am Main: Suhrkamp, 1980.

Bortoft, Henri. *The Wholeness of Nature: Goethe's Way toward a Science of Conscious Participation in Nature.* Hudson, NY: Lindesfarne Press, 1996.

Borushko, Matthew C. "Aesthetics of Nonviolence: Shelley, Adorno, Rancière." *Romantic Circles.* September 2015. https://www.rc.umd.edu/praxis/shelley_politics/praxis.2015 .shelley_politics.borushko.html.

———. "Violence and Nonviolence in Shelley's 'Mask of Anarchy.'" *Keats Shelley Journal* 59 (2010): 96–113.

Bourdieu, Pierre. *Outline of a Theory of Practice.* Cambridge: Cambridge University Press, 1977.

Braunstein, Jean-François. "Le concept de milieu, de Lamarck à Comte et aux positivismes." In *Jean-Baptiste Lamarck, 1744–1829*, edited by Goulven Laurent, 557–71. Paris: Éditions du CTHS, 1997.

Breidbach, Olaf. *Goethes Metamorphosenlehre.* Munich: Wilhelm Fink Verlag, 2006.

Breithaupt, Fritz. *Jenseits der Bilder: Goethes Politik der Wahrnehmung.* Freiburg: Rombach, 2000.

Brown, Alison. *The Return of Lucretius to Renaissance Florence.* Cambridge, MA: Harvard University Press, 2010.

Burgess, Miranda. "On Being Moved: Sympathy, Mobility, and Narrative Form." *Poetics Today* 32 (2011): 289–321.

———. "Transport: Mobility, Anxiety, and the Romantic Poetics of Feeling." *Studies in Romanticism* 49 (2010): 229–60.

———. "Transporting *Frankenstein*: Mary Shelley's Mobile Figures." *European Romantic Review* 25 (2014): 247–65.

Burkhardt, Richard W. "Animal Behavior and Organic Mutability in the Age of Lamarck." In *Lamarck e il Lamarckismo. Atti del Convegno (Napoli, 1–3 dicembre 1998)*, 63–89. Naples: La Città del Sole, 1995.

Burwick, Frederick. *The Damnation of Newton: Goethe's Color Theory and Romantic Perception.* Berlin: De Gruyter, 1986.

———. Introduction. In *Approaches to Organic Form: Permutations in Science and Culture*, edited by Burwick. Dordrecht: D. Reidel, 1987.

Butler, Judith. *Bodies That Matter: On the Discursive Limits of Sex.* New York: Routledge, 1993.

———. "How Can I Deny That These Hands and This Body Are Mine?" *Qui Parle* 11.1 (Fall/Winter 1997): 1–20.

———. *Notes toward a Performative Theory of Assembly*. Cambridge, MA: Harvard University Press, 2015.

———. *Precarious Life: The Powers of Mourning and Violence*. London: Verso, 2004.

———. *The Psychic Life of Power: Theories in Subjection*. Stanford: Stanford University Press, 1997.

Butler, Marilyn. Introduction. In *Frankenstein, or The Modern Prometheus: The 1818 Text*, by Mary Shelley, edited by Butler. Oxford: Oxford University Press, 1993.

———. *Romantics, Rebels, and Reactionaries: English Literature and Its Background, 1760–1830*. New York: Oxford University Press, 1982.

Bywater, Bill. "Goethe: A Science Which Does Not Eat the Other." *Janus Head* 8.1 (2005): 291–310.

Cameron, Kenneth Neill. *The Young Shelley*. New York: Macmillan, 1950.

Canguilhem, Georges. "Du singulier et de la singularité en épistémologie biologique." *Études d'histoire et de philosophie des sciences*, 211–25. Paris: J. Vrin 1994.

———. "Le vivant et son milieu." *La connaissance de la vie*, 165–97. 2nd ed. Paris: Vrin, 2006.

———. "The Living and Its Milieu." Translated by John Savage. *Grey Room* 3 (Spring 2001): 6–31.

Caron, Joseph A. "'Biology' in the Life Sciences: A Historiographical Contribution." *History of Science* 26. 3 (Sept. 1988): 223–68.

Caruth, Cathy. *Empirical Truths and Critical Fictions: Locke, Wordsworth, Kant, Freud*. Baltimore, MD: Johns Hopkins University Press, 1991.

Cavarero, Adriana. "Recritude: Reflections on Postural Ontology." *Journal of Speculative Philosophy* 27.3 (2013): 220–35.

Chakrabarty, Dipesh. "The Climate of History: Four Theses." *Critical Inquiry* 35 (Winter 2009): 197–222.

Chandler, James. *England in 1819: The Politics of Literary Culture and the Case of Romantic Historicism*. Chicago: University of Chicago Press, 1998.

———. "The Pope Controversy: Romantic Poetics and the English Canon." *Critical Inquiry* 10.3 (March 1984): 481–509.

Chaouli, Michael. *The Laboratory of Poetry: Chemistry and Poetics in the Work of Friedrich Schlegel*. Baltimore, MD: Johns Hopkins University Press, 2002.

Chase, Cynthia. *Decomposing Figures: Rhetorical Readings in the Romantic Tradition*. Baltimore, MD: Johns Hopkins University Press, 1986.

Cohen, Tom, ed. *Material Events: Paul de Man and the Afterlife of Theory*. Minneapolis: University of Minnesota Press, 2000.

Coleridge, Samuel Taylor. *Biographia Literaria, or Biographical Sketches of my Literary Life and Opinions*. Edited by James Engell and W. Jackson Bate. Bollingen Series 75. Princeton, NJ: Princeton University Press, 1983.

———. *Coleridge's Criticism of Shakespeare: A Selection*. Edited by R. A. Foakes. London: Athlone Press, 1989.

———. *Coleridge's Poetry and Prose*. Edited by Nicholas Halmi, Paul Magnuson, and Raimonda Modiano. New York: Norton, 2004.

——. *Collected Letters of Samuel Taylor Coleridge*. Edited by Earl Leslie Griggs. 6 vols. Oxford: Oxford University Press, 1956–71.

——. *Marginalia*. Edited by H. J. Jackson and George Whalley. 6 vols. *The Collected Works of Samuel Taylor Coleridge*. [London]: Routledge and K. Paul, 1969.

——. *Philosophical Lectures of S. T. Coleridge*. Edited by K. Coburn. London: Pilot, 1949.

——. *The Statesman's Manual; or, The Bible the Best Guide to Political Skill and Foresight: A Lay Sermon Addressed to the Higher Classes of Society. Collected Works of Samuel Taylor Coleridge*. Bollingen Series, 75. Edited by R. J. White. Princeton, NJ: Princeton University Press, 1972.

Connolly, Tristanne. *William Blake and the Body*. New York: Palgrave Macmillan, 2002.

Coole, Diana, and Samantha Frost. Introduction to *New Materialisms: Ontology, Agency, and Politics*. Durham, NC: Duke University Press, 2010.

Corsi, Pietro. *The Age of Lamarck: Evolutionary Theories in France, 1790–1830*. Translated by Jonathan Mandelbaum. Berkeley: University of California Press, 1988.

Cowper, William. *The Poetical Works*. Edited by H. S. Milford. London: Oxford University Press, 1963.

Craciun, Adriana, and Luisa Calè. Introduction to special issue on "The Disorder of Things." *Eighteenth-Century Studies* 45 (2011): 1–13.

Crary, Jonathan. *Techniques of the Observer: On Vision and Modernity in the Nineteenth Century*. Cambridge, MA: MIT Press, 1990.

Culler, Jonathan. *The Pursuit of Signs: Semiotics, Literature, Deconstruction*. Ithaca, NY: Cornell University Press, 1981.

Cunningham, Andrew, and Nicholas Jardine. Introduction. In Cunningham and Jardine, *Romanticism and the Sciences*.

——, eds. *Romanticism and the Sciences*. Cambridge: Cambridge University Press, 1990.

Cunningham, Andrew, and Perry Williams. "De-centring the 'Big Picture': 'The Origins of Modern Science' and the Modern Origins of Science." *British Journal for the History of Science* 26.4 (1993): 407–32.

Cvetkovich, Ann. *An Archive of the Feelings: Trauma, Sexuality and Lesbian Public Cultures*. Durham, NC: Duke University Press, 2003.

Dalzell, Alexander. *The Criticism of Didactic Poetry: Essays on Lucretius, Virgil, and Ovid*. Toronto: University of Toronto Press, 1996.

Damon, S. Foster. *A Blake Dictionary: The Ideas and Symbols of William Blake*. Revised edition, with Morris Eaves. Hanover, NH: University Press of New England, 1988.

Damrosch, Leo. *Symbol and Truth in Blake's Myth*. Princeton, NJ: Princeton University Press, 1980.

Danesi, Marcel, and Thomas Sebeok. *The Forms of Meaning: Modeling Systems Theory and Semiotic Analysis*. Berlin: De Gruyter, 2000.

Darwin, Erasmus. *The Botanic Garden, A Poem. In Two Parts. Part I, Containing the Economy of Vegetation. Part II, The Loves of the Plants. With Philosophical Notes*. London: Joseph Johnson, 1799.

——. *The Golden Age and The Temple of Nature; or, The Origin of Society*. Introduction by Donald H. Reiman. New York: Garland, 1978.

———. *Zoonomia; or, The Laws of Organic Life*. Vol. 1. London: J. Johnson, [1794]–96.

Daston, Lorraine, and Galison, Peter. *Objectivity*. New York: Zone Books, 2007.

Davis, William Stephen. "Subjectivity and Exteriority in Goethe's 'Dauer im Wechsel.'" *German Quarterly* 66.4 (Autumn 1993): 451–66.

Dawkins, Richard. *The Selfish Gene*. 1976. Oxford: Oxford University Press: 1989.

Deleuze, Gilles. "The Simulacrum and Ancient Philosophy." *The Logic of Sense*, edited by Constantin V. Boundas, translated by Mark Lester with Charles Stivale, 253–79. New York: Columbia University Press, 1990.

Deleuze, Gilles, and Claire Parnet. *Dialogues II*. Rev. ed. New York: Columbia University Press, 2007.

De Man, Paul. *Aesthetic Ideology*. Edited by Andrzej Warminski. Minneapolis: University of Minnesota Press, 1996.

———. *Allegories of Reading: Figural Language in Rousseau, Nietzsche, Rilke, and Proust*. New Haven: Yale University Press, 1982.

———. *Blindness and Insight: Essays in the Rhetoric of Contemporary Criticism*. 2nd ed. London: Routledge, 1996.

———. *The Resistance to Theory*. Foreword by Wlad Godzich. Minneapolis: University of Minnesota Press, 1986.

———. *The Rhetoric of Romanticism*. New York: Columbia University Press, 1984.

De Quincey, Thomas. *The Works of Thomas De Quincey*. Edited by Barry Symonds and Grevel Lindop. 21 vols. London: Pickering & Chatto, 2000–2003.

Derrida, Jacques. "*My Chances* / Mes chances: A Rendezvous with Some Epicurean Stereophonies." *Psyche: Interventions of the Other*, vol. 1, edited by Peggy Kamuf and Elizabeth Rottenberg, 344–60. Stanford, CA: Stanford University Press, 2007.

———. *Specters of Marx: The State of the Debt, the Work of Mourning, and the New International*. New York: Routledge, 1994.

Desmond, Adrian. *The Politics of Evolution: Morphology, Medicine, and Reform in Radical London*. Chicago: University of Chicago Press, 1989.

Diderot, Denis. *Le rêve de d'Alembert*. Edited by Colas Duflo. Paris: Garnier-Flammarion, 2002.

Dimock, Wai Chee. "Nonbiological Clock: Literary History against Newtonian Mechanics." *SAQ* 102.1 (Winter 2003): 153–77.

Doane, Mary Anne. "Indexicality: Trace and Sign: Introduction." *d i f f e r e n c e s : A Journal of Feminist Cultural Studies* 18.1 (2007): 1–6.

Donohue, Harold. *The Song of the Swan: Lucretius and the Influence of Callimachus*. Lanham, MD: University Press of America, 1993.

[D'Oyley, George.] Review of *Lectures on Physiology, Zoology, and the Natural History of Man Delivered at the Royal College of Surgeons*, by William Lawrence. *Quarterly Review* 22 (1820): 1–34.

Dryden, John. *Lucretius: A Poem against the Fear of Death. With an ode in memory of the accomplish'd young lady Mrs. Ann Killigrew, excellent in the two sister arts of poetry and painting*. London: H. Hills, [1709].

Duchesneau, François. *Genèse de la théorie cellulaire*. Montreal: Bellarmin-Vrin, 1987.

———. *La physiologie des lumières: Empirisme, modèles et theories.* The Hague: M. Nijhoff, 1982.

Duff, David. *Romanticism and the Uses of Genre.* Oxford: Oxford University Press, 2009.

Duman, Daniel. "The Creation and Diffusion of a Professional Ideology in Nineteenth Century England." *Sociological Review* 27.1 (February 1979): 113–38.

Engelstein, Stefani. *Anxious Anatomy: The Conception of the Human Form in Literary and Naturalist Discourse.* Albany: SUNY Press, 2008.

Esposito, Roberto. *Bíos: Biopolitics and Philosophy.* Translated by Timothy Campbell. Minneapolis: University of Minnesota Press, 2008.

Essick, Robert N. *Blake's Commercial Book Illustrations: A Catalogue and Study of the Plates Engraved by Blake after Designs by Other Artists.* Oxford: Oxford University Press, 1991.

Evelyn, John. *An Essay on the First Book of T. Lucretius Carus, De rerum natura. Interpreted and Made English Verse by J. Evelyn, Esq.* London: Gabriel Bedle and Thomas Collins, 1656.

Everest, Kelvin, ed. *Shelley Revalued: Essays from the Gregynog Conference.* Leicester: Leicester University Press, 1983.

Faflak, Joel. "The Difficult Education of Shelley's 'Triumph of Life.'" *Keats-Shelley Journal* 58 (2009): 53–78.

Farley, John. *The Spontaneous Generation Controversy from Descartes to Oparin.* Baltimore: Johns Hopkins University Press, 1977.

Favret, Mary. *War at a Distance: Romanticism and the Making of Modern Wartime.* Princeton, NJ: Princeton University Press, 2010.

Ferguson, Frances. *Solitude and the Sublime: Romanticism and the Aesthetics of Individuation.* New York: Routledge, 1992.

Foot, Paul. *Red Shelley.* London: Sidgwick and Jackson with Michael Dempsey, 1980.

Ford, Philip. "Lucretius in Early Modern France." In Gillespie and Hardie, *Cambridge Companion to Lucretius,* 227–41.

Förster, Eckart. "Die Bedeutung von §§ 76, 77 der 'Kritik der Urteilskraft' für die Entwicklung der nachkantischen Philosophie [Teil I]." *Zeitschrift für philosophische Forschung* 56.2 (April–June 2002): 169–90.

———. *The Twenty-Five Years of Philosophy.* Cambridge, MA: Harvard University Press, 2012.

Forster, Thomas. *Observations on the Casual and Periodical Influence of Particular States of the Atmosphere on Human Health and Diseases.* London: Underwood, 1817.

Foster, John Bellamy. *Marx's Ecology: Materialism and Nature.* New York: Monthly Review Press, 2000.

Foucault, Michel. *The Order of Things: An Archeology of the Human Sciences.* New York: Vintage Books, 1994.

———. "Society Must Be Defended." In *Society Must Be Defended: Lectures at the Collège de France, 1975–1976,* translated by David Macey, edited by Mauro Bertani and Alessandro Fontana, 239–40. New York: Picador, 2003.

Fraistat, Neil. "Shelley Left and Right: The Rhetorics of the Early Textual Editions." In *Shelley: Poet and Legislator of the World,* edited by Betty T. Bennett and Stuart Curran. Baltimore, MD: Johns Hopkins University Press, 1996.

François, Anne-Lise. "'O Happy Living Things!' Frankenfoods and the Bounds of Words-worthian Natural Piety." *diacritics* 33.2 (2003): 42–70.

———. *Open Secrets: The Literature of Uncounted Experience.* Stanford, CA: Stanford University Press, 2008.

———. "Unspeakable Weather, or the Rain Romantic Constatives Know." In *Phantom Sentences: Essays in Linguistics and Literature Presented to Ann Banfield*, edited by Robert S. Kawashima, Gilles Philippe, and Thelma Sowley, 147–61. Bern: Peter Lang, 2008.

Freire, Paulo. *Pedagogy of the Oppressed.* Translated by Myra Bergman Ramos. New York: Seabury Press, 1974.

Fulford, Tim. "Britannia's Heart of Oak: Thomson, Garrick and the Language of Eighteenth-Century Patriotism." In *James Thomson: Essays for the Tercentenary*, edited by Richard Terry. Liverpool: Liverpool University Press, 2000.

———. "Cowper, Wordsworth, Clare: The Politics of Trees." *John Clare Society Journal* 14 (1995): 47–59.

Fulford, Tim, Debbie Lee, and Peter J. Kitson, eds. *Literature, Science and Exploration in the Romantic Era.* New York: Cambridge University Press, 2004.

Gale, Monica, ed. *Lucretius.* Oxford: Oxford University Press, 2007.

———. "Lucretius and Previous Poetic Traditions." In Gillespie and Hardie, *Cambridge Companion to Lucretius*, 59–75.

———. *Myth and Poetry in Lucretius.* Cambridge: Cambridge University Press, 1994.

Garani, Myrto. *Empedocles REDIVIVUS: Poetry and Analogy in Lucretius.* New York: Routledge, 2007.

Gasking, Elizabeth B. *Investigations into Generation, 1651–1828.* Baltimore: Johns Hopkins Press, 1967.

Gay, Peter. *The Enlightenment, an Interpretation.* 2 vols. New York: Knopf, 1966–69.

Geoffroy Saint-Hilaire, Étienne. "Mémoire sur la Théorie Physiologique designée sous le nom de VITALISME." *Gazette médicale de Paris*, no. 11, extract 2, vol. 2. Paris: Dezanche, 1830. Goethe-Schiller-Archiv 26/LV, 10, 9.

Geulen, Eva. *Aus dem Leben der Form: Goethes Morphologie und die Nager.* August Verlag, preprint.

———. "Metamorphosen der Metamorphose: Goethe, Cassirer, Blumenberg." In *Intermedien: Zur kulturellen und artistischen Übertragung*, edited by Alexandra Kleihues, Barbara Naumann, und Edgar Pankow, 203–17. *Medienwandel-Medienwechsel-Medienwissen*, vol. 14. Zurich, 2010.

———. "Serialization in Goethe's Morphology." *Compar(a)ison. An International Journal of Comparative Literature* (2013): 53–70

Gigante, Denise. *Life: Organic Form and Romanticism.* New Haven, CT: Yale University Press, 2009.

Gilbert, Scott F., and David Epel. *Ecological Developmental Biology: Integrating Epigenetics, Medicine and Evolution.* Sunderland, MA: Sinauer Associates, 2009.

Gillespie, Stuart, and Philip Hardie, eds. *The Cambridge Companion to Lucretius.* Cambridge: Cambridge University Press, 2007.

Gilpin, George H. "Blake and the World's Body of Science." *SiR* 43.1 (Spring 2004): 35–56.

Goethe, Johann Wolfgang. *Collected Works*. Vol. 12, *Scientific Studies*, edited and translated by Douglas Miller. New York: Suhrkamp, 1988.

———. *Faust: A Tragedy*. Edited by Cyrus Hamlin. Translated by Walter Arndt. New York: Norton, 1976.

———. *Goethe's Botanical Writings*. Translated by Bertha Mueller. Woodbridge, CT: Ox Bow Press, 1952.

———. *Goethes Gespräche: Eine Sammlung zeitgenössischer Berichte aus seinem Umgang*. Edited by Wolfgang Herwig. Zürich: Artemis Verlag, 1965.

———. "Notes Related to Knebel's Lucretius." MS. Goethe-Schiller Archiv, Weimar, 78/134, Reimar Bestand.

———. *Sämtliche Werke nach Epochen seines Schaffens*. Edited by Karl Richter with Herbert G. Göpfert, Norbert Miller, and Gerhard Sauder. Münchner-Ausgabe. 33 vols. Munich: Hanser Verlag, 1995-98.

———. *Die Schriften zur Naturwissenschaft*. Edited by Dorothea Kuhn, Wolf von Engelhardt, and Irmgard Müller. Leopoldina-Ausgabe. 26 vols. Weimar: Hermann Böhlaus Nachfolger, 1947-2011.

———. *Selected Verse, With Plain Prose Translations of Each Poem*. Edited and translated by David Luke. London: Penguin, 1986.

———. *Werke*. Vol. 4. Weimarer-Ausgabe. Weimar: Hermann Böhlaus Nachfolger, 1889-96.

Goldberg, Jonathan. *The Seeds of Things: Theorizing Sexuality and Materiality in Renaissance Representations*. New York: Fordham University Press, 2009.

Goldsmith, Steven. "Almost Gone: Rembrandt and the Ends of Materialism." *New Literary History* 45, no. 3 (Summer 2014): 411-43.

———. *Blake's Agitation: Criticism and the Work of Emotion*. Baltimore, MD: Johns Hopkins University Press, 2013.

———. *Unbuilding Jerusalem: Apocalypse and Romantic Representation*. Ithaca, NY: Cornell University Press, 1993.

Goldstein, Amanda Jo. "Growing Old Together: Lucretian Materialism in Shelley's 'Poetry of Life.'" *Representations* 128.1 (Fall 2014): 60-92.

———. "Irritable Figures: Herder's Poetic Empiricism." In *The Relevance of Romanticism: Essays on German Romantic Philosophy*, edited by Dalia Nassar, 273-95. Oxford: Oxford University Press, 2014.

———. "Obsolescent Life: Goethe's Journals on Morphology." *European Romantic Review* 22.3 (2011): 405-14.

———. "Reluctant Ecology in Blake and Arendt: A Response to Robert Mitchell and Richard Sha." *Wordsworth Circle* 46.3 (2015): 143.

Golinski, Jan. *British Weather and the Climate of Enlightenment*. Chicago: University of Chicago Press, 2007.

———. *Science as Public Culture: Chemistry and Enlightenment in Britain, 1760-1820*. Cambridge: Cambridge University Press, 1992.

Good, John Mason, trans. Appendix. *The Nature of Things: A Didactic Poem, Accompanied with the Original Text and Illustrated with Notes Philological and Explanatory*. 2 vols. Translated by Good. London: Longman, Hurst, Rees, and Orme, 1805.

Goodman, Kevis. "Conjectures on Beachy Head: Charlotte Smith's Poetics and the Ground of the Present." *English Literary History* 81.3 (Fall 2014): 983–1006.

———. *Georgic Modernity and British Romanticism: Poetry and the Mediation of History.* Cambridge: Cambridge University Press, 2004.

———. "Romantic Poetry and the Science of Nostalgia." In *The Cambridge Companion to British Romantic Poetry*, edited by James Chandler and Maureen McLane. Cambridge: Cambridge University Press, 2008.

———. "'Uncertain Disease': Nostalgia, Pathologies of Motion, Practices of Reading." *Studies in Romanticism* (Summer 2010): 197–227.

Goss, Erin. *Revealing Bodies: Anatomy, Allegory, and the Grounds of Knowledge in the Long Eighteenth Century.* Lewisburg, PA: Bucknell University Press, 2013.

Gottlieb, Gilbert. "A Developmental Psychobiological Systems View: Early Formulation and Current Status." In Oyama, Griffiths, and Gray, *Cycles of Contingency*, 41–54.

Grave, Johannes. "'Beweglich und bildsam': Morphologie als implizite Bildtheorie?" In *Morphologie und Moderne. Goethe's "anschauliches Denken" in den Geistes- und Kulturwissenchaften seit 1800*, edited by Jonas Maatsch, 57–74. Berlin: De Gruyter, 2014.

Greenblatt, Stephen. *The Swerve: How the World Became Modern.* New York: Norton, 2011.

Gregg, Richard B. *The Power of Nonviolence.* 3rd ed. Foreword by Martin Luther King Jr. Canton, ME: Greenleaf Books, 1960.

Greisemer, James. "What Is 'Epi' about Epigenetics?" In Van Speybroeck, Van de Vijver, and de Waele, *From Epigenesis to Epigenetics*, 61–81.

Gurton-Wachter, Lily. *Watchwords: Romanticism and the Poetics of Attention.* Stanford, CA: Stanford University Press, 2016.

Guyer, Sara. "Biopoetics, or Romanticism." *Romanticism and Biopolitics.* December 2012. https://www.rc.umd.edu/praxis/biopolitics/HTML/praxis.2012.guyer

———. *Romanticism after Auschwitz.* Stanford, CA: Stanford University Press, 2007.

Hadot, Pierre. *The Veil of Isis: An Essay on the History of the Idea of Nature.* Cambridge, MA: Harvard University Press, 2006.

Hall, Brian K. *Evolutionary Developmental Biology.* Dordrecht: Kluwer, 1999.

Haller, Albrecht von. *First Lines of Physiology.* 3rd ed. Edited by William Cullen. 1786. New York: Johnson Reprint, 1966.

———. Vorrede [Preface to the German translation of the second volume of Buffon's *Histoire Naturelle* (1752)]. *The Natural Philosophy of Albrecht von Haller.* Edited by Shirley A. Roe. New York: Arno Press, 1981.

Hamilton, Paul. *Metaromanticism: Aesthetics, Literature, Theory.* Chicago: University of Chicago Press, 2003.

———. "The Romantic Life of the Self." *The Meaning of "Life" in Romantic Poetry and Poetics*, 81–102. New York: Routledge, 2009.

Hanssen, Beatrice. *Walter Benjamin's Other History: Of Stones, Animals, Human Beings, and Angels.* Berkeley: University of California Press, 2000.

Haraway, Donna. "Situated Knowledges: The Science Question in Feminism and the Privilege of Partial Perspective." *Feminist Studies* 14.3 (Autumn 1988): 575–99.

Hardie, Philip R. *Lucretian Receptions.* New York: Cambridge University Press, 2009.

Harvey, William. *Anatomical Exercitations concerning the Generation of Living Creatures.* London: James Young, 1653.

Hayley, William. *An Essay on Epic Poetry: In Five Epistles to the Rev. Mr. Mason, with Notes.* London: J. Dodsley, 1782.

———. *The Life, and Posthumous Writings of William Cowper.* Vol. 4, *Supplementary Pages to the Life of Cowper Containing the Additions Made to That Work on Reprinting It in Octavo.* London: Printed by J. Seagrave (Chichester) for J. Johnson, 1806.

Heinroth, Johann Christian August. *Lehrbuch der Anthropologie.* Leipzig: Vogel, 1822.

Hennig, John. "Goethes Begriff 'Zart.'" *Archiv für Begriffsgeschichte* 24 (1980): 77–102.

Henry, Thomas. *Memoirs of Albert de Haller, M.D.* Warrington, UK: W. Eyre for J. Johnson, 1783.

Herder, Johann Gottfried Herder. "Fragments on Recent German Literature." *Selected Writings on Aesthetics.* Princeton, NJ: Princeton University Press, 2006.

———. "On Image, Poetry, and Fable." *Selected Writings on Aesthetics.* Princeton, NJ: Princeton University Press, 2006.

———. "On the Cognition and Sensation of the Human Soul." *Philosophical Writings,* edited and translated by Michael N. Forster. Cambridge: Cambridge University Press, 2002.

———. *Werke in zehn Bänden.* Edited by Martin Bollacher, Gunter Grimm, Ulrich Gaier, et al. Frankfurt am Main: DKV, 1985.

Heringman, Noah. *Romantic Rocks, Aesthetic Geology.* Ithaca, NY: Cornell University Press, 2004; Oxford University Press, 2013.

———, ed. *Romantic Science: The Literary Forms of Natural History.* Albany: SUNY Press, 2003.

———. *Sciences of Antiquity: Romantic Antiquarianism, Natural History, and Knowledge Work.* Oxford: Oxford University Press, 2013.

Hilton, Nelson. *Literal Imagination: Blake's Vision of Words.* Berkeley: University of California Press, 1983.

Hoagwood, Terence Allan. *Skepticism and Ideology: Shelley's Political Prose and Its Philosophical Context from Bacon to Marx.* Iowa City: University of Iowa Press, 1988.

Hobbes, Thomas. *The Leviathan.* Edited by C. B. Macpherson, London: Penguin Classics, 1987.

Hoffmeyer, Jesper. *Biosemiotics: An Examination into the Signs of Life and the Life of Signs.* Edited by Donald Favareau. Translated by Hoffmeyer and Favareau. Scranton, PA: University of Scranton Press, 2008.

Hogle, Jerrold. *Shelley's Process: Radical Transference and the Development of His Major Works.* New York: Oxford University Press, 1988.

Höhne, Horst. "Shelley's 'Socialism' Revisited." In *Shelley: Poet and Legislator of the World,* edited by Betty T. Bennett and Stuart Curran, 201–12. Baltimore, MD: Johns Hopkins University Press, 1996.

Holland, Jocelyn. *German Romanticism and Science: The Procreative Poetics of Goethe, Novalis, and Ritter.* New York: Routledge, 2009.

Holmes, Brooke. "Deleuze, Lucretius, and the Simulacrum of Naturalism." In *Dynamic Reading: Studies in the Reception of Epicureanism,* edited by Holmes and W. H. Shearin, 316–42. New York: Oxford University Press, 2012.

Hopkins, David. "The English Voices of Lucretius." In Gillespie and Hardie, *Cambridge Companion to Lucretius*, 254–73.

Horace. *Epistles, Book II, and Epistle to the Pisones (Ars Poetica)*. Edited by Niall Rudd. Cambridge: Cambridge University Press, 1989.

Hörmann, Raphael. *Writing the Revolution: German and English Radical Literature, 1819–1848/49*. Berlin: LIT, 2011.

Hunter, John. "The Progress and Peculiarities of the Chick." In *Descriptive and Illustrated Catalogue of the Physiological Series of Comparative Anatomy Contained in the Museum of the Royal College of Surgeons in London*, vol. 5: *Products of Generation*, edited by Richard Owen, viii–xvii. London: Printed by R. Taylor, 1840.

Hutchings, Kevin D. *Imagining Nature: Blake's Environmental Poetics*. Montreal: McGill-Queens University Press, 2002.

Hutchinson, Lucy. *Works*, vols. 1–2: *The Translation of Lucretius*. Edited by Reid Barbour and David Norbrook. Latin text by Maria Cristina Zerbino. Oxford: Oxford University Press, 2012.

Iovino, Serenella, and Serpil Oppermann. Introduction to *Material Ecocriticism*. Bloomington: Indiana University Press, 2014.

Jablonka, Eva. "The Systems of Inheritance." In Oyama, Griffiths, and Gray, *Cycles of Contingency*, 99–116.

Jablonka, Eva, and Marion J. Lamb. *Epigenetic Inheritance and Evolution: The Lamarckian Dimension*. Oxford: Oxford University Press, 1998.

———. *Evolution in Four Dimensions: Genetic, Epigenetic, Behavioral, and Symbolic Variation in the History of Life*. Cambridge, MA: MIT Press, 2005.

Jablonka, Eva, Marion J. Lamb, and E. Avital. "'Lamarckian' Mechanisms in Darwinian Evolution. *Trends in Ecology and Evolution* 13.5 (1998): 206–10.

Jackson, Noel. "Rhyme and Reason: Erasmus Darwin's Romanticism." *MLQ* 70.2 (June 2009): 172–94.

———. *Science and Sensation in Romantic Poetry*. Cambridge: Cambridge University Press, 2008.

Jackson, Virginia. *Dickinson's Misery: A Theory of Lyric Reading*. Princeton, NJ: Princeton University Press, 2005.

Jacob, François. *The Logic of Life: A History of Heredity*. Translated by Betty E. Spillman. New York: Random House, 1973.

Jacyna, L. S. "Immanence or Transcendence: Theories of Life and Organization in Britain, 1790–1835." *Isis* 74 (1983): 311–29.

Jakobson, Roman. "Closing Statement: Linguistics and Poetics." In *Style in Language*, edited by Thomas A. Sebeok, 350–77. Cambridge, MA: MIT Press, 1960.

Jameson, Fredric. *The Political Unconscious: Narrative as a Socially Symbolic Act*. Ithaca, NY: Cornell University Press, 1981.

Janowitz, Anne F. *Lyric and Labour in the Romantic Tradition*. Cambridge: Cambridge University Press, 1998.

———. "'A Voice from Across the Sea': Communitarianism at the Limits of Romanticism." In *At the Limits of Romanticism: Essays in Cultural, Feminist, and Materialist Criticism*, edited by Mary A. Favret and Nicola J. Watson, 83–100. Bloomington: Indiana University Press, 1994.

Jardine, N., J. A. Secord, and E. C. Spary, eds. *Cultures of Natural History*. Cambridge: Cambridge University Press, 1996.

Jarvis, Simon. "Blake's Spiritual Body." In *The Meaning of "Life" in Romantic Poetry and Poetics*, edited by Ross Wilson, 13–32. New York: Routledge, 2009.

———. "Thinking in Verse." In *The Cambridge Companion to British Romantic Poetry*, edited by James Chandler and Maureen McLane, 98–116. Cambridge: Cambridge University Press, 2008.

———. *Wordsworth's Philosophic Song*. Cambridge: Cambridge University Press, 2007.

Jégou, Bernard. Preface. In *Epigenetics and Human Reproduction*, edited by Sophie Rousseaux and Saadi Khochbin. Berlin: Springer Verlag, 2011.

Johnson, Barbara. *Persons and Things*. Cambridge, MA: Harvard University Press, 2008.

Johnson, Christopher. *System and Writing in the Philosophy of Jacques Derrida*. Cambridge: Cambridge University Press, 1993.

Jones, Donna V. *The Racial Discourses of Life Philosophy: Négritude, Vitalism, and Modernity*. New York: Columbia University Press, 2010.

Jones, Howard. *The Epicurean Tradition*. New York: Routledge, 1989.

Jones, Steven E. *Shelley's Satire: Violence, Exhortation, and Authority*. De Kalb: Northern Illinois University Press, 1994.

Kant, Immanuel. *Critique of the Power of Judgment*. Edited by Paul Guyer. Translated by Guyer and Eric Matthews. Cambridge: Cambridge University Press, 2000.

Kaufman, Robert. "Aura, Still." *OCTOBER* 99 (Winter 2002): 45–80.

———. "Intervention and Commitment Forever! Shelley in 1819, Shelley in Brecht, Shelley in Adorno, Shelley in Benjamin." *Romantic Circles Praxis Series, Reading Shelley's Interventionist Poetry*, edited by Michael Scrivener. May 2001. http://www.rc.umd.edu/praxis/interventionist/kaufman/kaufman.html. Accessed 30 June 2011.

———. "Legislators of the Post-Everything World: Shelley's *Defence* of Adorno." *ELH* 63.3 (1996): 707–33.

Keach, William. *Arbitrary Power: Romanticism, Language, Politics*. Princeton, NJ: Princeton University Press, 2004.

———. "Rise like Lions? Shelley and the Revolutionary Left." *International Socialism* 2.75 (July 1997): 91–103.

———. *Shelley's Style*. New York: Methuen, 1984.

Keats, John. *Complete Poems and Selected Letters of John Keats*. Introduction by Edward Hirsch. New York: Modern Library, 2001.

Keenleyside, Heather. "Personification for the People: On James Thomson's *The Seasons*." *ELH* 76.2 (2009): 447–72.

Keller, Evelyn Fox. *The Century of the Gene*. Harvard, MA: Harvard University Press, 2000.

———. "Organisms, Machines, and Thunderstorms: A History of Self-Organization, Part One." *Historical Studies in the Natural Sciences* 38.1 (Winter 2008): 44–75.

———. "Organisms, Machines, and Thunderstorms: A History of Self-Organization, Part Two. Complexity, Emergence, and Stable Attractors." *Historical Studies in the Natural Sciences* 39.1 (2009): 1–31.

——. *Refiguring Life: Metaphors of Twentieth-Century Biology*. New York: Columbia University Press, 1995.

Kelley, Theresa M. *Clandestine Marriage: Botany and Romantic Culture*. Baltimore, MD: Johns Hopkins University Press, 2012.

——. *Reinventing Allegory*. Cambridge: Cambridge University Press, 1997.

——. "Restless Romantic Plants: Goethe Meets Hegel." *European Romantic Review* 20.2 (April 2009): 187–95.

——. "Romantic Nature Bites Back: Adorno and Romantic Natural History." *European Romantic Review* 15.2 (June 2004): 193–203.

Kennedy, Duncan F. *Rethinking Reality: Lucretius and the Textualization of Nature*. Ann Arbor: University of Michigan Press, 2002.

Khalip, Jacques. *Anonymous Life: Romanticism and Dispossession*. Stanford, CA: Stanford University Press, 2009.

King-Hele, Desmond. *Erasmus Darwin and the Romantic Poets*. London: Macmillan, 1986.

——. *Shelley: His Thought and Work*. London: Macmillan, 1984.

Klancher, *The Making of English Reading Audiences: 1790–1832*. Madison: University of Wisconsin Press, 1983.

——. *Transfiguring the Arts and Sciences: Knowledge and Institutions in the Romantic Age*. Cambridge: Cambridge University Press, 2013.

Klopfer, Peter H. "Parental Care and Development." In Oyama, Griffiths, and Gray, *Cycles of Contingency*, 167–74.

Knight, David. "Romanticism and the Sciences." In Cunningham and Jardine, *Romanticism and the Sciences*, 13–24.

Koranyi, Stephen. *Autobiographik und Wissenschaft im Denken Goethes*. Bonn: Bouvier Verlag Herbert Grundmann, 1984.

Koselleck, Reinhart. *Futures Past: On the Semantics of Historical Time*. Translated by Keith Tribe. New York: Columbia University Press, 2004.

Kramnick, Jonathan. *Actions and Objects from Hobbes to Richardson*. Stanford, CA: Stanford University Press, 2010.

Krauss, Rosalind. "Notes on the Index: Seventies Art in America, Part 2," *October* 4 (Autumn 1977): 59–64, 80.

Kreiter, Carmen S. "Evolution and William Blake." *Studies in Romanticism* 4 (Winter 1965): 110–18.

Kroll, Richard W. F. *The Material Word: Literate Culture in the Restoration and Early Eighteenth Century*. Baltimore: Johns Hopkins University Press, 1991.

Kuhn, Dorothea. "Natur und Kunst." In *Die Schriften zur Naturwissenschaft*, Leopoldina-Ausgabe, edited by Kuhn, Wolf von Engelhardt, and Irmgard Müller), 2.9b: 474–75. Weimar: Hermann Böhlaus Nachfolger, 1947–2011.

——. *Typus und Metamorphose: Goethe Studien*. Edited by Renate Grumach. Marbach am Neckar: Deutsche Schillergesellschaft, 1988.

Kuiken, Kir. *Imagined Sovereignties: Toward a New Political Romanticism*. New York: Fordham University Press, 2004.

——. "Shelley's 'Mask of Anarchy' and the Problem of Modern Sovereignty." *Literature Compass* 8.2 (2011): 95–106.

Lacoue-Labarthe, Philippe, and Jean-Luc Nancy. *The Literary Absolute: The Theory of Literature in German Romanticism.* Translated by Philip Barnard and Cheryl Lester. Albany: SUNY Press, 1988.

Laland, Kevin N., F. John Odling-Smee, and Marcus W. Feldman. "Niche Construction, Ecological Inheritance, and Cycles of Contingency in Evolution." Oyama, Griffiths, and Gray, *Cycles of Contingency*, 117–26.

Lamarck, Jean-Baptiste. *Philosophie zoologique; ou, Exposition des considérations relatives à l'histoire naturelle des animaux.* 2 vols. Paris: Dentu, 1809.

———. *Recherches sur l'organisation des corps vivans: précédé du Discours d'ouverture du cours de zoologie donné dans la Muséum d'histoire naturelle* [1802]. Revised by Jean-Marc Drouin. Paris: Librairie Arthème Fayard, 1986.

Latour, Bruno. "Agency at the Time of the Anthropocene." *New Literary History* 45.1 (Winter 2014): 1–18.

———. *Pandora's Hope: Essays on the Reality of Science Studies.* Cambridge, MA: Harvard University Press, 1999.

———. *The Pasteurization of France.* Translated by Alan Sheridan and John Law. Cambridge, MA: Harvard University Press, 1988.

———. *Reassembling the Social: An Introduction to Actor-Network Theory.* Oxford: Oxford University Press, 2007.

———. "Waiting for Gaia: Composing the Common World through Arts and Politics." Lecture at the French Institute, London, November 2011.

———. *We Have Never Been Modern.* Translated by Catherine Porter. Cambridge, MA: Harvard University Press, 1991.

Laurent, Goulven, ed. *Jean-Baptiste Lamarck, 1744–1829.* Paris: Éditions du CTHS, 1997.

La Vergata, Antonello. "Il Lamarckismo fra riduzionismo biologico e meliorismo sociale." *Lamarck e il Lamarckismo: Atti del Convegno: Napoli, 1–3 Dicembre 1988*, 182–214. Naples: La città del sole, 1995.

Law, Jules David. *The Rhetoric of Empiricism: Language and Perception from Locke to I. A. Richards.* Ithaca, NY: Cornell University Press, 1993.

Lawrence, William. *Lectures on Physiology, Zoology, and the Natural History of Man Delivered at the Royal College of Surgeons.* Salem: Foote & Browne, 1828. [1st ed., London: J. Callow, 1819.]

———, trans. *A Short System of Comparative Anatomy, translated from the German of J. F. Blumenbach.* London: Longman, Hurt, Rees & Orme, 1816.

Lenoir, Timothy. *The Strategy of Life: Teleology and Mechanics in Nineteenth-Century German Biology.* Chicago: University of Chicago Press, 1982.

Lepenies, Wolf. *Das Ende der Naturgeschichte: Wandel kultureller Selbstverständlichkeiten in den Wissenschaften des 18. und 19. Jahrhunderts.* Munich: Hanser, 1976.

———. *The End of Natural History.* Translated by Jean Steinberg. New York: Urizen, 1978.

Levine, Caroline *Forms: Whole, Rhythm, Hierarchy, Network.* Princeton , New Jersey: Princeton University Press, 2015.

Levinson, Marjorie. "A Motion and a Spirit: Romancing Spinoza." *Studies in Romanticism* 46.4 (Winter 2007): 367–408.

———. "Of Being Numerous: Counting and Matching in Wordsworth's Poetry." *Studies in Romanticism* 49 (Winter 2010): 649.

———. "Pre- and Post-Dialectical Materialisms: Modeling Praxis without Subjects and Objects." *Cultural Critique* (Fall 1995): 111–27.

———. "Romantic Poetry: The State of the Art." *Modern Language Quarterly* 54.2 (June 1993): 183–214.

Lewis, Jayne Elizabeth. *Air's Appearance: Literary Atmosphere in British Fiction, 1660–1794.* Chicago: University of Chicago Press, 2012.

Lewontin, Richard. "Gene, Organism and Environment." In Oyama, Griffiths, and Gray, *Cycles of Contingency*, 55–66.

Lezra, Jacques. *Unspeakable Subjects: The Genealogy of the Event in Early Modern Europe.* Stanford, CA: Stanford University Press, 1997.

Lindstrom, Eric. *Romantic Fiat: Demystification and Enchantment in Lyric Poetry.* New York: Palgrave Macmillan, 2011.

Locke, John. *An Essay concerning Human Understanding.* Edited by Alexander Campbell Fraser. 2 vols. New York: Dover, 1959 [1894].

Lucretius Carus, Titus. *De rerum natura.* Translated by W. H. D. Rouse. Revised by Martin Ferguson Smith. Loeb Classical Library. Cambridge, MA: Harvard University Press, 1992.

———. *An Essay on the First Book of T. Lucretius Carus De rerum natura. Interpreted and Made English Verse.* Translated by John Evelyn. London: Gabriel Bedle and Thomas Collins, 1656.

———. *The Nature of Things.* Translated by A. E. Stallings. London: Penguin Classics, 2007.

———. *The Nature of Things: A Didactic Poem, Accompanied with the Original Text and Illustrated with Notes Philological and Explanatory.* Translated by John Mason Good. 2 vols. London: Longman, Hurst, Rees, and Orme, 1805.

———. *Of the Nature of Things, in Six Books, Translated into English Verse. Explained and Illustrated, with Notes and Animadversions, Being a Compleat System of the Epicurean Philosophy.* Translated by Thomas Creech. 6th ed. 2 vols. London: T. Warner and J. Walthoe, 1722.

———. *On Nature.* Translated by Russel M. Geer. Indianapolis: Bobbs-Merrill, 1965.

———. *Von der Natur der Dinge.* Translated by Karl Ludwig von Knebel. Edited by Otto Güthling. Leipzig: Philip jun. Reclams Universal-Bibliothek, 1901.

Lyon, John, and Philip R. Sloan, ed. and trans. *From Natural History to the History of Nature: Readings from Buffon and His Critics.* Notre Dame, IN: University of Notre Dame Press, 1981.

Maatsch, Jonas, ed. *Morphologie und Moderne. Goethe's "anschauliches Denken" in den Geistes- und Kulturwissenschaften seit 1800.* Berlin: De Gruyter, 2014.

Macpherson, Sandra. *Harm's Way: Tragic Responsibility and the Novel Form.* Baltimore, MD: Johns Hopkins University Press, 2010.

Makdisi, Saree. *William Blake and the Impossible History of the 1790s.* Chicago: University of Chicago Press, 2003.

Mann, Jenny. *Outlaw Rhetoric: Figuring Vernacular Eloquence in Shakespeare's England.* Ithaca, NY: Cornell University Press, 2012.

Marković, Daniel. *The Rhetoric of Explanation in Lucretius'* De rerum natura. Leiden: Brill, 2008.

Marx, Karl. *The Eighteenth Brumaire of Louis Bonaparte*. New York: International Publishers, 1994 [1963].

Marx, Karl, and Friedrich Engels. *Marx-Engels Collected Works*. Vol. 1. Moscow: Progress Publishers; London: Lawrence & Wishart; New York: International Publishers, 1975.

———. *Marx-Engels-Gesamtausgabe*. Vol. 1. Berlin: Dietz Verlag, 1975.

———. *The Marx-Engels Reader*. 2nd ed. Edited by Robert C. Tucker. New York: Norton, 1978.

———. *Marx-Engels Studienausgabe*. Edited by Irving Fletscher. 4 vols. Frankfurt am Main: Fischer Taschenbuch, 1999.

McGrath, Brian. *The Poetics of Unremembered Acts*. Evanston, IL: Northwestern University Press, 2013.

McLane, Maureen. *Romanticism and the Human Sciences: Poetry, Population, and the Discourse of the Species*. Cambridge: Cambridge University Press, 2000.

Mead, Richard. *A Short Discourse concerning Pestilential Contagion*. 5th ed. London: for Sam. Buckley and Ralph Smith, 1720.

Mee, John. *Romanticism, Enthusiasm, and Regulation*. Oxford: Oxford University Press, 2003.

Meeker, Natania. *Voluptuous Philosophy: Literary Materialism in the French Enlightenment*. New York: Fordham University Press, 2006.

Menely, Tobias. "'The Present Obfuscation': Cowper's *Task* and the Time of Climate Change." *PMLA* 127.3 (2012): 477–92.

Mensch, Jennifer. *Kant's Organicism: Epigenesis and the Development of Critical Philosophy*. Chicago: University of Chicago Press, 2013.

Mill, John Stuart. "What Is Poetry?" *Essays on Poetry by John Stuart Mill*, edited by F. Parvin Sharpless, 3–28. Columbia: University of South Carolina Press, 1976.

Milton, John. *The Riverside Milton*. Edited by Roy Flannagan. Boston: Houghton Mifflin, 1998.

Mitchell, Robert. "Cryptogamia." *European Romantic Review* 21.5 (October 2010): 615–35.

———. *Experimental Life: Vitalism in Romantic Science and Literature*. Baltimore, MD: Johns Hopkins University Press, 2013.

———. "Romanticism and the Experience of Experiment." *Wordsworth Circle* 46.3 (Summer 2015): 132–42.

Mitchell, W. J. T. *Blake's Composite Art*. Princeton, NJ: Princeton University Press, 1983.

Montaigne, Michel de. *The Complete Essays*. Translated by Donald M. Frame. Stanford, CA: Stanford University Press, 1965.

———. *Essais*. Vol. 3. Edited by Alexandre Micha. Paris: Garnier-Flammarion, 1979.

Mücke, Dorothea von "Goethe's Metamorphosis: Changing Forms in Nature, the Life Sciences, and Authorship." *Representations* 95.1 (Summer 2006): 27–53.

Müller-Sievers, Helmut. *The Cylinder: Kinematics of the Nineteenth Century*. Berkeley: University of California Press, 2012.

———. *Self-Generation: Biology, Philosophy, and Literature around 1800*. Stanford, CA: Stanford University Press, 1997.

Müller-Wille, Staffan, and Hans Jörg Rheinberger. *A Cultural History of Heredity*. Chicago: University of Chicago Press, 2012.

Nassar, Dalia. "Romantic Empiricism after the 'End of Nature': Contributions to Environmental Philosophy." In *The Relevance of Romanticism*, edited by Nassar. Oxford: Oxford University Press, 2014.

Needham, John Turberville. "A Summary of Some Late Observations upon the Generation, Composition, and Decomposition of Animal and Vegetable Substances; Communicated in a Letter to Martin Folks Esq; President of the Royal Society. Paris, Nov. 23, 1748." *Philosophical Transactions of the Royal Society of London*, 1 January, 1753.

Needham, Joseph. *A History of Embryology*. New York: Abelard-Schuman, 1959.

Nersessian, Anahid. *Utopia, Limited: Romanticism and Adjustment*. Cambridge, MA: Harvard University Press, 2015.

Nietzsche, Friedrich. *Die fröhliche Wissenschaft*. Edited by Alfred Kröner. Stuttgart: Kröner, 1956.

———. *The Gay Science: With a Prelude in Rhymes and an Appendix of Songs*. Translated by Walter Kaufmann. New York: Random House, 1974.

———. "On Truth and Lying in a Non-Moral Sense." *The Birth of Tragedy and Other Writings*, edited by Raymond Geuss, translated by Ronald Speirs. New York: Cambridge University Press, 1999.

Nijhout, H. Frederick. "The Ontogeny of Phenotypes." In Oyama, Griffiths, and Gray, *Cycles of Contingency*, 129–40.

Nisbet, Hugh Barr. "Herder und Lucrez." In *Johann Gottfried Herder, 1744–1803*, edited by G. Sauder. Hamburg: F. Meiner, 1987.

———. "Lucretius in Eighteenth-Century Germany. With a Commentary on Goethe's 'Metamorphose der Tiere.'" *Modern Language Review* 81.1 (January 1986): 97–115.

Nixon, Rob. *Slow Violence and the Environmentalism of the Poor*. Cambridge, MA: Harvard University Press, 2011.

Nussbaum, Martha C. *The Therapy of Desire: Theory and Practice in Hellenistic Ethics*. Princeton, NJ: Princeton University Press, 1994.

Nyhart, Lynn. *Biology Takes Form: Animal Morphology and the German Universities, 1800–1900*. Chicago: University of Chicago Press, 1995.

[Oken, Lorenz]. "Von der Sexualität der Pflanzen, Studien von Dr. August Henschel . . . nebst einem historischen Anhange von Dr. F. J. Schelver." Review of *Von der Sexualität der Pflanze*, by Henschel and Schelver. *Isis* 10 (1820): 662–67. *Digitalen Bibliothek Thüringen*/ journals@UrMEL. http://zs.thulb.uni-jena.de/receive/jportal_jparticle _00137903

Otto, Regine. "Lukrez bleibt immer in seiner Art der Einzige." *Impulse: Aufsätze, Quellen, Berichte zur deutschen Klassik und Romantik* 5 (1982): 229–63.

Oyama, Susan. *Evolutions's Eye: A Systems View of the Biology-Culture Divide*. Durham, NC: Duke University Press, 2000.

———. "Terms in Tension: What Do You Do When All the Good Words Are Taken?" In Oyama, Griffiths, and Gray, *Cycles of Contingency*, 177–93.

Oyama, Susan, Paul E. Griffiths, and Russel D. Gray, eds. *Cycles of Contingency: Developmental Systems and Evolution.* Cambridge, MA: MIT Press, 2001.

———. Introduction. In Oyama, Griffiths, and Gray, *Cycles of Contingency*, 1–11.

Packham, Catherine. *Eighteenth-Century Vitalism: Bodies, Culture, Politics.* Basingstoke, UK: Palgrave-Macmillan, 2012.

Passannante, Gerard. *Lucretian Renaissance: Philology and the Afterlife of Tradition.* Chicago: University of Chicago Press, 2011.

———. "Reading for Pleasure: Disaster and Digression in the First Renaissance Commentary on Lucretius." In *Dynamic Reading: Studies in the Reception of Epicureanism*, edited by Brooke Holmes and W. H. Shearin, 89–112. Oxford: Oxford University Press, 2012.

Peacock, Thomas Love. "The Four Ages of Poetry." *The Halliford Edition of the Works of Thomas Love Peacock*, edited by H. F. B. Brett-Smith and C. E. Jones. Vol. 8, *Essays, Memoirs, Letters & Unfinished Novels*, 3–25. London: Constable / New York: Gabriel Wells, 1934.

Peirce, Charles Saunders. *The Essential Peirce: Selected Philosophical Writings.* Vol. 2 (1893–1913). Edited by The Peirce Edition Project. Bloomington: Indiana University Press, 1998.

Peter-Loo Massacre, containing a faithful narrative of events which preceded, accompanied and followed the fatal sixteenth of August, 1819, edited by an observer. [Periodical]. Manchester: J. Wroe, 1819. http://find.galegroup.com/mome/infomark.do?&source=gale &prodId=MOME&userGroupName=cornell&tabID=T001&docId=U3603542491& type=multipage&contentSet=MOMEArticles&version=1.0&docLevel=FASCIMILE

Pfau, Thomas. "'All Is Leaf': Difference, Metamorphosis, and Goethe's Phenomenology of Knowledge." *Studies in Romanticism* 49.1 (Spring 2010): 3–41.

Picciotto, Joanna. *Labors of Innocence in Early Modern England.* Cambridge, MA: Harvard University Press, 2010.

Pinch, Adela, *Strange Fits of Passion: Epistemologies of Emotion, Hume to Austen.* Stanford, CA: Stanford University Press, 1996.

Pinto-Correia, Clara. *The Ovary of Eve: Egg and Sperm and Preformation.* Chicago: University of Chicago Press, 1997.

Piper, Andrew. *Dreaming in Books: The Making of the Bibliographic Imagination in the Romantic Age.* Chicago: University of Chicago Press, 2009.

Porter, James I. "Lucretius and the Sublime." In Gillespie and Hardie, *Cambridge Companion to Lucretius*, 167–84.

Prentice, Archibald. *Historical Sketches and Personal Recollections of Manchester, Intended to Illustrate the Progress of Public Opinion from 1792 to 1832.* 3rd ed. Introduced by Donald Read. London: Frank Cass & Co., 1970.

Priestley, Joseph. "Observations and Experiments Relating to Equivocal, or Spontaneous Generation, read Nov. 18th, 1803." *Transactions of the American Philosophical Society* 6 (1809): 119–29.

Priestman, Martin. "Lucretius in Romantic and Victorian Britain." In Gillespie and Hardie, *Cambridge Companion to Lucretius*, 289–305.

———. *The Poetry of Erasmus Darwin: Enlightened Spaces, Romantic Times.* Farnham, UK: Ashgate, 2013.

———. *Romantic Atheism: Poetry and Freethought, 1780–1830.* Cambridge: Cambridge University Press, 1999.

Pyle, Forest. "'Frail Spells': Shelley and the Ironies of Exile." In *Irony and Clerisy*, edited by Deborah Elise White. Romantic Circles /Praxis. August 1999. https://www.rc.umd.edu/praxis/irony/pyle/frail.html

———. *The Ideology of Imagination: Subject and Society in the Discourse of Romanticism.* Stanford, CA: Stanford University Press, 1995.

———. "Kindling and Ash: Radical Aestheticism in Keats and Shelley." *Studies in Romanticism* 42.4 (Winter 2003): 427–59.

Quinney, Laura. *William Blake on Self and Soul.* Cambridge, MA: Harvard University Press, 2010.

Rajan, Tilottama. *The Supplement of Reading: Figures of Understanding in Romantic Theory and Practice.* Ithaca, NY: Cornell University Press, 1990.

Redfield, Marc, ed. *Legacies of Paul de Man.* New York: Fordham University Press, 2007.

———. *Phantom Formations: Aesthetic Ideology and the Bildungsroman.* Ithaca, NY: Cornell University Press, 1996.

———. *The Politics of Aesthetics: Nationalism, Gender, Romanticism.* Stanford, CA: Stanford University Press, 2003.

Reill, Peter Hanns. "The Hermetic Imagination in the High and Late Enlightenment." In *The Super-Enlightenment: Daring to Know Too Much*, edited by Daniel Edelstein, 37–51. Oxford: Voltaire Foundation, 2010.

———. *Vitalizing Nature in the Enlightenment.* Berkeley: University of California Press, 2005.

Richards, Robert J. "Kant and Blumenbach on the *Bildungstrieb*: A Historical Misunderstanding." *Studies in History and Philosophy of Biological and Biomedical Sciences* 31.1 (2000): 11–32.

———. *The Romantic Conception of Life.* Chicago: University of Chicago Press, 2002.

Richardson, Alan. *British Romanticism and the Science of the Mind.* Cambridge: Cambridge University Press, 2001.

Rigby, Catherine E. "Art, Nature, and the Poesy of Plants in the Goethezeit: A Biosemiotic Perspective." *Goethe Yearbook* 22 (2015): 23–44.

———. "Earth, World, Text: On the (Im)possibility of Ecopoiesis." *New Literary History* 35.3 (Summer 2004): 427–42.

Riskin, Jessica. *Science in the Age of Sensibility: The Sentimental Empiricists of the French Enlightenment.* Chicago: University of Chicago Press, 2002.

Robert, Jason Scott. *Embryology, Epigenesis, and Evolution: Taking Development Seriously.* Cambridge: Cambridge University Press, 2004.

Roberts, Hugh. *Shelley and the Chaos of History: A New Politics of Poetry.* University Park: Pennsylvania State University Press, 1997.

Roe, Shirley. "The Life Sciences." In *The Cambridge History of Science*, vol. 4: *Eighteenth-Century Science*, edited by Roy Porter. Cambridge: Cambridge University Press, 2003.

———. *Matter, Life, and Generation: Eighteenth-Century Embryology and the Haller-Wolff Debate.* Cambridge: Cambridge University Press, 1981.

Roger, Jacques. *The Life Sciences in Eighteenth-Century French Thought*. Edited by Keith R. Benson. Translated by Robert Ellrich. Stanford, CA: Stanford University Press, 1997.

Rogers, John. *The Matter of Revolution: Science, Poetry and Politics in the Age of Milton*. Ithaca: Cornell University Press, 1996.

Rohrbach, Emily. *Modernity's Mist: British Romanticism and the Poetics of Anticipation*. New York: Fordham University Press, 2016.

Ross, Catherine E. "The Professional Rivalry of William Wordsworth and Humphry Davy." In *Romantic Science the Literary Forms of Natural History*, edited by Noah Heringman. Albany: SUNY Press, 2003.

Ross, S. "Scientist: The Story of a Word." *Nineteenth-Century Attitudes: Men of Science*. Dordrecht, Netherlands: Springer Netherlands, 1991.

Ruston, Sharon. *Shelley and Vitality*. New York: Palgrave Macmillan, 2005.

Sachs, Jonathan. *Romantic Antiquity: Rome in the British Imagination, 1789–1832*. Oxford: Oxford University Press, 2010.

Santner, Eric L. *On Creaturely Life: Rilke, Benjamin, Sebald*. Chicago: University of Chicago Press, 2006.

Schaffer, Simon. "Indiscipline and Interdisciplines: Some Exotic Genealogies of Modern Knowledge." 2010. Lecture. http://www.fif.tu-darmstadt.de/media/fif_forum_inter disziplinaere_forschung/sonstigetexte/externe/simonschaffner.pdf

Schelver, Franz Joseph. *Kritik der Lehre von den Geschlechtern der Pflanze*. Heidelberg: G. Braun, 1812. http://zs.thulb.uni-jena.de/receive/jportal_jparticle_00040703

Schmidt, Alfred. *The Concept of Nature in Marx*. Translated by Ben Fowkes. London: Verso, 2014.

Schuler, Robert, and John G. Fitch. "Theory and Context of the Didactic Poem: Some Classical, Mediaeval, and Later Continuities." *Florilegium* 5 (1983): 1–43.

Schwartz, Peter J. *After Jena: Goethe's "Elective Affinities" and the End of the Old Regime*. Lewisville, PA: Bucknell University Press, 2010.

Scrivener, Michael Henry. *Radical Shelley: The Philosophical Anarchism and Utopian Thought of Percy Bysshe Shelley*. Princeton, NJ: Princeton University Press, 1982.

Seamon, David, and Arthur Zajonc, eds. *Goethe's Way of Science: A Phenomenology of Nature*. Albany: SUNY Press, 1998.

Secord, James A. "The Crisis of Nature." In Jardine, Secord, and Spary, *Cultures of Natural History*, 447–59.

Sedgwick, Eve Kosofsky. *Touching Feeling: Affect, Pedagogy, Performativity*. Durham, NC: Duke University Press, 2003.

Serres, Michel. *The Birth of Physics*. Edited by David Webb. Translated by Jack Hawkes. Manchester, UK: Clinamen, 2000.

Sha, Richard C. *Perverse Romanticism: Aesthetics and Sexuality in Britain, 1750–1832*. Baltimore, MD: Johns Hopkins University Press, 2009.

———. "Romantic Skepticism about Scientific Experiment." *Wordsworth Circle* 46.3 (Summer 2015): 127–31.

Shelley, Percy Bysshe. *The Letters of Percy Bysshe Shelley*. Edited by Frederick L. Jones. 2 vols. Oxford: Clarendon Press, 1964.

———. *A Philosophical View of Reform.* In *Shelley's Revolutionary Year: Shelley's Political Poems and the Essay A Philosophical View of Reform,* edited by Paul Foot. London: Redwords, 1990.

———. *The Poems of Shelley,* vol. 2: *1817–1819.* Edited by Kelvin Everest and Geoffroy Matthews. Harlow, UK: Longman, 2000.

———. *Shelley's Poetry and Prose.* 2nd ed. Edited by Donald H. Reiman and Neil Fraistat. New York: Norton, 2002.

———. *The Works of Percy Bysshe Shelley in Verse and Prose.* Edited by Harry Buxton Forman. Vol. 1. London: Reeves and Turner, 1880.

Solomonescu, Yasmin. *John Thelwall and the Materialist Imagination.* Houndmills, UK: Palgrave Macmillan, 2014.

Stanley, John L. *Mainlining Marx.* New Brunswick, NJ: Transaction Publishers, 2002.

Steele, Edward J., Robyn A. Lindley, and Robert V. Blanden. *Lamarck's Signature: How Retrogenes Are Changing Darwin's Natural Selection Paradigm.* Reading, MA: Perseus (Helix) Books, 1998.

Steigerwald, Joan. "Degeneration: Inversions of Teleology." In *Marking Time: Romanticism and Evolution,* edited by Joel Faflak and Josh Lambier. Toronto: University of Toronto Press (forthcoming).

———. "Kantian Teleology and the Biological Sciences." Edited by Joan Steigerwald. Special issue of *Studies in History and Philosophy of the Biological and Biomedical Sciences* 37: 4 (2006).

———. "Rethinking Organic Vitality in Germany at the Turn of the Nineteenth Century." In *Vitalism and the Scientific Image in Post-Enlightenment Life Science, 1800–2010,* edited by S. Normandin and C. T. Wolfe, 51–75. Dordrecht: Springer, 2013.

———. "Treviranus' Biology: Generation, Degeneration and the Boundaries of Life." In *Gender, Race, and Reproduction: Philosophy and the Early Life Sciences in Context,* edited by S. Lettow, 105–27. New York: SUNY Press, 2014.

Stephenson, R. H. "The Cultural Theory of Weimar Classicism in the Light of Coleridge's Doctrine of Aesthetic Knowledge." In *Goethe 2000: Intercultural Readings of His Work,* edited by Stephenson and Paul Bishop, 150–59. Leeds, UK: Northern Universities Press, 2000.

———. *Goethe's Conception of Knowledge and Science.* Edinburgh: Edinburgh University Press, 1995.

Terada, Rei. "Looking at the Stars Forever." *Studies in Romanticism* 50 (Summer 2011): 275–309.

———. *Looking Away: Phenomenality and Dissatisfaction, Kant to Adorno.* Cambridge, MA: Harvard University Press, 2009.

Teskey, Gordon. *Allegory and Violence.* Ithaca, NY: Cornell University Press, 1996.

Thelwall, John. *An essay towards a definition of animal vitality: read at the theatre, Guy's Hospital, January 26, 1793; in which several of the opinions of the celebrated John Hunter are examined and controverted.* London: printed by T. Rickaby, 1793. http://name.umdl.umich.edu/004788862.0001.000

Thomson, James. *The Seasons*. Edited by James Sambrook. Oxford: Oxford University Press, 1981.

Tiffany, Daniel. *Toy Medium: Materialism and Modern Lyric*. Berkeley: University of California Press, 2000.

Timparano, Sebastiano. *On Materialism*. London/New York: Verso, 1975.

Toohey, Peter. *Epic Lessons: An Introduction to Ancient Didactic Poetry*. New York: Routledge, 1996.

Tresch, John. *The Romantic Machine: Utopian Science and Technology after Napoleon*. Chicago: University of Chicago Press, 2012.

Turner, Paul. "Shelley and Lucretius." *Review of English Studies*, n.s. 10.39 (1959): 269–82.

Unseld, Siegfried. *Goethe and His Publishers*. Translated by Kenneth J. Northcott. Chicago: University of Chicago Press, 1996.

Valenza, Robin. *Literature, Language, and the Rise of the Intellectual Disciplines in Britain, 1680–1820*. Cambridge: Cambridge University Press, 2009,

Van de Vijver, Gertrudis, Linda Van Speybroeck, and Dani de Waele. "Epigenetics: A Challenge for Genetics, Evolution, and Development?" In Van Speybroeck, Van de Vijver, and de Waele, *From Epigenesis to Epigenetics*, 1–6.

Van Speybroeck, Linda, Gertrudis Van de Vijver, and Dani de Waele, eds. *From Epigenesis to Epigenetics: The Genome in Context*. New York: New York Academy of Sciences, 2002.

Vicario, Michael A. *Shelley's Intellectual System and Its Epicurean Background*. New York: Routledge, 2007.

Vila, Anne C. *Enlightenment and Pathology: Sensibility in the Literature and Medicine of Eighteenth-Century France*. Baltimore: Johns Hopkins University Press, 1998.

Vogl, Joseph. "Bemerkung über Goethe's Empiricismus." In *Versuchsanordungen 1800*, edited by Vogl and Sabine Schimma, 113–23. Zürich: Diaphanes, 2009.

———. *Kalkül und Leidenschaft: Poetic des ökononomischen Menschen*. Munich: Sequenzia, 2002.

———. "Wolkenbotschaft." In *Wolken*, edited by Lorenz Engell, Bernhard Siegert, and Vogl, 69–79. Weimar: Verlag der Bauhaus-Universität, 2005.

Von Engelhardt, Wolf. *Goethe im Gespräch mit der Erde: Landschaft, Gesteine, Mineralien und Erdgeschichte in seinem Leben und Werk*. Weimar: Hermann Böhlaus Nachfolger, 2003.

Von Gleichen-Rußworm, Wilhelm Friedrich. *Abhandlung über die Saamen- und Infusions-Thierchen*. Nürnberg, 1778.

Walker, Adam. *Analysis of A Course of Lectures in Natural and Experimental Philosophy*. 10th ed. London: Printed for the author, 1771.

———. *A System of Familiar Philosophy in Twelve Lectures, Being the Course Usually Read by Mr. A. Walker . . .* London: Printed for the author, 1799.

Walmsley, Peter. *Locke's Essay and the Rhetoric of Science*. Lewisburg, PA: Bucknell University Press, 2003.

Wang, Orrin. *Fantastic Modernity: Dialectical Readings in Romanticism and Theory*. Baltimore, MD: Johns Hopkins University Press, 1996.

Webb, Timothy. *The Violet in the Crucible: Shelley and Translation*. Oxford: Clarendon Press, 1976.

Weber, Bruce H., and David J. Depew. "Developmental Systems, Darwinian Evolution, and the Unity of Science." In Oyama, Griffiths, and Gray, *Cycles of Contingency*, 239–53.

Weisman, Karen. "Shelley's Ineffable Quotidian." In *Shelley: Poet and Legislator of the World*, edited by Betty T. Bennett and Stuart Curran, 215–23. Baltimore, MD: Johns Hopkins University Press, 1996.

Wellmon, Chad. "Goethe's Morphology of Knowledge, or the Overgrowth of Nomenclature." *Goethe Yearbook* 17 (2010): 153–177.

Whitman, Walt. "Eidólons." *Leaves of Grass: The 1892 Edition*. New York: Bantam Books, 1983.

Williams, Elizabeth A. *A Cultural History of Medical Vitalism in Enlightenment Montpellier*. Burlington, VT: Ashgate, 2003.

———. *The Physical and the Moral: Anthropology, Physiology, and Philosophical Medicine in France, 1750–1850*. Cambridge: Cambridge University Press, 1994.

Williams, Raymond. *Culture and Society: 1780–1950*. New York: Columbia University Press, 1983 [1958].

Wilson, Catherine. *Epicureanism at the Origins of Modernity*. Oxford: Oxford University Press, 2008.

Wilson, Ross, ed. *The Meaning of "Life" in Romantic Poetry and Poetics*. New York: Routledge, 2009.

———. "Poetry as Reanimation in Shelley." In Ross, *Meaning of "Life" in Romantic Poetry and Poetics*, 125–45.

———. *Shelley and the Apprehension of Life*. New York: Cambridge University Press, 2013.

Wolfe, Charles T. "Do Organisms Have an Ontological Status?" *History and Philosophy of the Life Sciences* 32.2–3 (2010): 195–232.

———. "Why Was There No Controversy over Life in the Scientific Revolution?" In *Controversies in the Scientific Revolution*, edited by V. Boantza and M. Dascal. Amsterdam: John Benjamins, 2011.

Wolfson, Susan J. *Formal Charges: The Shaping of Poetry in British Romanticism*. Stanford, CA: Stanford University Press, 1997.

Wordsworth, William. Prefaces of 1800 and 1802. In *Lyrical Ballads: The Text of the 1798 Edition with the Additional 1800 Poems and the Prefaces*, 2nd Ed., by Wordsworth and S. T. Coleridge. Edited by R. L. Brett and A. R. Jones. London: Methuen, 1991.

Wyder, Margrit. *Goethes Naturmodell: Die Scala Naturae und ihre Transformationen*. Cologne: Böhlau Verlag, 1998.

Wylie, Ian. *Young Coleridge and the Philosophers of Nature*. Oxford: Oxford University Press, 1989.

Zammito, John H. *The Genesis of Kant's Critique of Judgment*. Chicago: University of Chicago Press, 1992.

———. *Kant, Herder, and the Birth of Anthropology*. Chicago: University of Chicago Press, 2002.

———. "The Lenoir Thesis Revisited: Blumenbach and Kant." *Studies in History and Philosophy of Biological and Biomedical Sciences* 43 (2012): 120–32.

———. "Teleology Then and Now: The Question of Kant's Relevance for Contemporary Controversies over Function in Biology." *Studies in History and Philosophy of Biological and Biomedical Sciences* 37 (2006): 748–70.

Zimmer, Carl. "How Microbes Defend and Define Us." *New York Times*, July 12, 2010.

INDEX